TA169 .H38 2005
0134 1109097
Hattangadi, A.

Plant and machi
failure prever
c2005.

MW01518033

2008 07 07

PLANT AND MACHINERY
FAILURE PREVENTION

GEORGIAN COLLEGE LIBRARY 2501

#/04.95

PLANT AND MACHINERY FAILURE PREVENTION

A A Hattangadi

Consulting Engineer, Reliability Engineering

Former General Manager,
Chittaranjan Locomotive Works
Indian Railways

250101

Library Commons
Georgian College
One Georgian Drive
Barrie, ON
L4M 3X9

McGraw-Hill

New York   Chicago   San Francisco   Lisbon   London
Madrid   Mexico City   Milan   New Delhi   San Juan
Seoul   Singapore   Sydney   Toronto

The McGraw-Hill Companies

Cataloging-in-Publication Data is on file with the Library of Congress

Copyright © 2005 by The McGraw-Hill Companies, Inc. All rights reserved. Printed in the United States of America. Except as permitted under the United States Copyright Act of 1976, no part of this publication may be reproduced or distributed in any form or by any means, or stored in a data base or retrieval system, without the prior written permission of the publisher.

1 2 3 4 5 6 7 8 9 0   DOC/DOC   0 1 0 9 8 7 6 5

ISBN 0-07-145791-7

The sponsoring editor for this book was Kenneth P. McCombs and the production supervisor was Sherri Souffrance.
The art director for the cover was Anthony Landi.

Printed and bound by RR Donnelley.

This book was previously published by Tata McGraw-Hill Publishing Company Limited, New Delhi, India, copyright © 2004.

McGraw-Hill books are available at special quantity discounts to use as premiums and sales promotions, or for use in corporate training programs. For more information, please write to the Director of Special Sales, McGraw-Hill Professional, Two Penn Plaza, New York, NY 10121-2298. Or contact your local bookstore.

 This book is printed on recycled, acid-free paper containing a minimum of 50% recycled, de-inked fiber.

Information contained in this work has been obtained by The McGraw-Hill Companies, Inc. ("McGraw-Hill") from sources believed to be reliable. However, neither McGraw-Hill nor its authors guarantee the accuracy or completeness of any information published herein, and neither McGraw-Hill nor its authors shall be responsible for any errors, omissions, or damages arising out of use of this information. This work is published with the understanding that McGraw-Hill and its authors are supplying information but are not attempting to render engineering or other professional services. If such services are required, the assistance of an appropriate professional should be sought.

To
His Holiness Shri Sadyojat Shankarashram Swamiji
of
Chitrapur Math, Shirali
N. Kanara, Karnataka
India

Contents

Foreword

The phenomenal success of A.A. Hattangadi's first book, *Electrical Fires and Failures,* is an indicator not only of its high quality but also of the acceptance of reliability concepts and practices in India and abroad. The fact that the sales of this book were higher in USA than in India, is not a reflection on insufficient appreciation of the need and importance of reliability engineering in our country. It is just that the book's merit has been recognised in USA. Though the book is meant for front-line supervisors and engineers, it is also eminently suitable as a rigorous and excellent textbook on reliability engineering. This emerging branch with unlimited applications will progressively gain acceptance as an integral part of courses on design, manufacturing, operation and maintenance of plant and machinery.

It is with this background that Hattangadi has written this new book on *Plant and Machinery Failure Prevention* for design and maintenance engineers. The book starts with an assurance that Zero-Failure Performance, with regard to electrical and mechanical equipment is an attainable goal. It calls for the determination of failure modes and mechanisms, followed by appropriate corrective measures. The author in his characteristic learning approach writes about the need for an encyclopaedia of component and equipment failures, and provides it within the span of sixty six chapters and discusses the various categories of failures. These chapters reveal the distilled quintessence of his long experience in the electrical department of Indian Railways.

The book further highlights two lessons for design and maintenance engineers. One, there is no such thing as a 'trivial' component or detail of any equipment and machinery because the damage due to its failure may be enormous or even horrendous. Second, there is a need for the training at every level in reliability engineering for learning about failure modes and mechanisms, investigation procedures and degradation processes of mechanical and electrical components.

The author also brings out two facts about reliability engineering:

- It is always possible to determine the true or root causes of failures, and
- Natural phenomena and laws are totally consistent, precise and reliable.

On these two valid assertions is based his faith in the possibility of attaining Zero-Failure Performance if the root causes of failures are correctly determined. The author makes a bold statement that natural laws and phenomena govern failure processes and enumerates those that cover 95 per cent of all failures. He finally makes a very simple but weighty statement that understanding failures is very often a matter of common sense.

A majority of failures are due to design deficiencies and only some due to maintenance lapses. It appears paradoxical but it is also true that both failure mechanisms and normal operations observe the same natural laws.

The greatest value of this book lies in its presentation of numerous case studies and the measures taken to reach Zero-Failure Performance which should ultimately lead to confidence building among engineers in their work.

As an academic, with over fifty years association with technical education, I feel that this book will prove to be an excellent reference material for all students of reliability engineering and serve as a gold mine of rarely recorded case studies.

I am sure Hattangadi who had a scholastic career of exceptional brilliance will be remembered not only as an outstanding engineer but more for his pioneering contribution in providing a new direction to thought processes and concepts in reliability engineering. His two books will help bring about abiding awareness of the need to achieve Zero-Failure Performance amongst designers, manufacturers and maintenance engineers.

Sadanand B. Kumta
Former Director of Technical Education
Gujarat State, Ahmedabad, INDIA

Preface

Nevil Chute, the British Aeronautical Engineer and Novelist has written many novels that combine romance, adventure, mysticism, philosophy and even engineering, in delightful and unforgettable tales. In one of his novels, he defines an engineer as:

An engineer is one who can build for a shilling
what any fool can do for a pound

In early twentieth century, that was a very appropriate and perceptive definition. Engineers have striven continually to reduce the sizes, weights and costs of their creations. This has brought incredibly complex structures, devices and appliances within the reach of the common man. The new technologies are important not only for our comforts and entertainment but also for our survival. However, all this technological progress on the one hand and our growing dependence on it on the other, has thrown up a new challenge for engineers.

It is no longer sufficient to design and build machines and structures at the lowest possible cost; they must also be totally safe and reliable. However, Nevil Chute's definition of an engineer still remains valid and relevant if we take into account the total life-cycle cost including repair costs and costs relating to consequential damages and losses of revenue arising out of failures of the engineer's creations. Often these other costs are far greater than the initial cost of the equipment.

Reliability engineering developed as a distinct discipline in the post-World War II era. The US defence establishment provided the required impetus to its growth. Very wisely they de-classified much of the basic work done for them in this field and placed it in the public domain. Reliability of all kinds of hardware and equipment improved by leaps and bounds thereafter. A discipline that started off with studies in the failures of equipment used by the defence establishment for fighting wars, now finds application in all kinds of plant and machinery, rolling stock, aircraft, home appliances, etc.

It is now possible to aim at Zero-Failure Performance for most of the electrical and mechanical equipment in service of railways, airlines, road transport, power stations and also in industrial/agricultural machines, telephones, home/office appliances, etc. The common man in the developing countries and even some maintenance engineers may find it hard to believe this claim. Chapter 1 specifically tries to remove these doubts.

During my 34 years of service on the Indian Railways and 16 years thereafter as a consulting engineer, I came across a number of curious features of the equipment failure scene. These are as follows:

(a) Many equipment failures including those with serious consequences are caused by visible or measurable defects that are not recognised as such. Some of these are apparently trivial.

(b) There is little published material about defects and failures—trivial or otherwise. Although a great deal has been written about phenomena like metal fatigue, metal creep, partial discharge, etc. it is all at post-graduate level. Front-line engineers and technicians are not exposed to even the main conclusions of such topics.

(c) There is little communication between practising maintenance engineers on the subject of failures of equipment. Similar failures continue to occur repeatedly in different organisations.

(d) Whereas some types of failures are eliminated in some units, they continue to occur in other units. Sometimes they start occurring in the same units after changes in maintenance staff.

(e) While the number of types of equipment now in use runs into tens of thousands, the number of common failure modes and mechanisms is less than ten.

I feel, that an encyclopaedia of all equipment failures which highlights all the failure modes and failure mechanisms, including the apparently trivial ones, and which is updated periodically, would help designers to review their designs before authorising bulk manufacture. It would also help maintenance engineers to reach the correct solutions quickly when confronted with repetitive failures of equipment caused by design or maintenance deficiencies. It would further help them to train the younger entrants into the profession in avoiding all kinds of defects that may otherwise be ignored, by explaining the mechanisms of failures caused by such defects.

This book is the result of that feeling. It condenses in a few hundred pages the lessons learnt over many years of experience in the operation and maintenance of a wide variety of electrical and mechanical equipment. I am sure there are many others who could add valuable material to this compendium from the wealth of their own experience.

Although this book is based mainly on observations and experience in the Indian Railways, the analyses and recommendations in this book are relevant to manufacturing industries and also to other forms of transport. The plant and machinery used in railway workshops, power stations and production units are identical to those in other industries. Moreover, the materials, components and devices used in locomotives, coaches and wagons are very similar to those used in stationary installations. For instance, split pins and cotters, threaded fasteners, shafts/axles, bearings, etc. mentioned in the titles of Chapters 7 to 66 are used not only in railway rolling stock and installations but also in general industries. The only exceptions are Chapters 60 and 65 regarding track overhead equipment and pantographs. But even here it will be seen that the components used therein, and their failure modes/mechanisms are not different from those elsewhere.

I hope this book will serve as a reference book for design and maintenance engineers. It would be of special interest to the young engineers entering these professions. To facilitate search, I have grouped the entries component-wise. Although there are thousands of types, makes and models of machinery which include these components, the modes and mechanisms of failures are not only the same for all but also relatively very small in number.

Chapters 1 to 6 deal with general principles, which are applicable to all failures and all the preventive measures discussed in detail from Chapters 7 to 66.

Chapter 4 is about investigation methods. Often young engineers just out of technical schools or colleges are made incharge of maintenance of electrical and mechanical equipment. While they know the constructory details and the theory of operation of the machines in their charge, they are ill-equipped to investigate failures. Chapter 4 will be useful for young engineers, without any experience, to face the task of undertaking investigations into failures of all kinds of mechanical and electrical equipment and components. The methods described in the chapter and a reference to the appropriate equipment chapters and sections that follow will help him to apply his theoretical knowledge to the problems at hand.

Chapters 5 and 6 are about mechanical and electrical degradation processes that are involved in most failure mechanisms. Insight into these phenomena is essential for all those who are investigating equipment and component failures.

The remaining Chapters 7 to 66 deal with different mechanical and electrical components, equipment and systems. These 60 chapters are largely independent of each other with very few cross-references. It is not necessary for the maintenance engineer and manager to read all the chapters in the book but it is desirable that he reads Chapters 1 to 6 fully before going on to the first and the last sections of each of the remaining 60 chapters. He may then return to the relevant chapter as and when required.

The entries in this book are based on my memory, my records, discussions with my colleagues and staff. If I do not give the source or local particulars it is because no one really wants to discuss what is usually perceived as one's own failure in having allowed an expensive failure to occur in the first place. Manufacturers are even more reticent for obvious commercial reasons. When the cause of a failure is trivial and the effects are horrendous, there is usually a conspiracy of silence in regard to the true cause of the failure.

If any readers disagree with something I have said in this book, or if they do not find in it what they need to know, or if they go on to devise their own solutions to their problems, I would request them to write to me at: hattangadianant@hotmail.com.

I am grateful to the large number of engineers, supervisors and workers in the electrical department of the Indian Railways, who knowing my interest in this subject kept me informed of their problems and their successes in the field of Reliability Engineering. In particular, I am indebted to my friend S. Natarajan, Former General Manager of South-Eastern Railway, who went through the manuscript and made many useful suggestions.

And above all, I am grateful to my wife Kumud, who encouraged me to write this book and to my son Arun, who went over the manuscript and made many useful suggestions.

A A HATTANGADI

PLANT AND MACHINERY
FAILURE PREVENTION

Chapter 1

INTRODUCTION

In this chapter we learn

- *Zero Failure performance of electrical and mechanical equipment is a realistic and attainable target.*
- *Through the correct determination of the failure mechanisms and the root causes of failures followed by appropriate corrective measures, that target can be achieved.*
- *The laws of nature apply to mechanical and electrical equipment with absolute consistency and precision during normal operation.*
- *The same laws of nature are equally applicable to failure mechanisms (physical or chemical processes that result in failures).*
- *The cost of the efforts needed for attaining Zero Failure performance is negligible in comparison with the resulting savings in repair costs and consequential damages.*   ❑ ❑

1.1 Zero Failure Performance

𝑀ost engineers, when asked whether it would be possible to attain a state of technology in which equipments never fail, would reply that it is impossible to attain Zero-Failure performance. Their view is based on the following reasoning:

(a) The properties of many materials are variable and the rates at which they degrade in service are even less predictable.

(b) All materials, components and equipment are manufactured, assembled, tested, operated and maintained by humans beings, and they will make mistakes that will eventually result in equipment failures.

(c) It may be possible theoretically, to design and manufacture equipment that will never fail in service but then it would be so expensive as to remain beyond reach of the common man.

Contrary to this general belief, it is proposed to show in this book that, despite the validity of arguments (a) & (b) above,

- It is possible to design, manufacture, operate and maintain mechanical and electrical equipment that would attain Zero Failure performance.
- And that it is possible to do so at reduced overall cost.

The next few sections in this chapter and the remaining 65 chapters of this book will explain the basis of these assertions and I can only ask the sceptical reader to accept for the time being only the following ideas:

- Zero Failure performance is desirable.
- Zero Failure performance should be our goal.

1.2   It is Certainly Possible to Attain Zero Failure Performance

The common man has got inured to the effects of equipment failures. Defective telephones, engine failures, load shedding, vehicle breakdowns, etc. have become so common that they are no longer news in the dailies. Fortunately, train derailments, air-crashes and building collapses still merit headlines but even these are forgotten within a few days. They are still accepted in the same manner as we accept storms, floods or earthquakes. Failures of plant, machinery and systems are considered to be inevitable and the criticism if any is restricted to delays in and paucity of repair or rescue efforts. Very few ask whether these accidents could not have been avoided altogether. That Zero Failure performance is at all possible is inconceivable to the common man.

The public concern, is only about rescue and relief and not about the root cause. Mechanical failures or the so-called Material failures are usually considered to be as inevitable as death. When equipment fails it is either scrapped or repaired.

When there is an electrical fire, the usual explanation is that there was a short-circuit. It is assumed that short-circuits cannot be prevented and must lead to fires.

There is an urgent need for a change in attitude. The common man must demand Zero Failure performance which is well within our reach.

Maintenance engineers have daily encounters with fractured components, charred insulation, melted wires and blown fuses. They would agree that Zero Failure performance of equipment would indeed be very welcome. However, they may find it hard to believe that it is within reach. How is it possible, they may ask, totally to prevent flaws or defects in materials and bad workmanship? There is a very long supply line from manufacturers of raw material, sub-assemblies, assemblies, etc. up to operators and maintainers. This line extends thousands of kilometres. The period of utilisation would also extend over forty years or more and the equipment would be operated and maintained by people who keep changing all the time.

The emphasis, as far as maintenance engineers and even management are concerned, is on speedy rescue, repairs and restoration of services. Maintenance engineers have no time to spend on investigation. Management does not feel the need for assignment of resources to investigation and prevention of failures.

Initially, I was sceptical of the Zero Failure concepts when they first appeared in the technical press. There are many developments that have made me believe that despite the problems mentioned, it is certainly possible to reach the target of Zero Failure and that:

- For every one component that did fail in any equipment, there were many other identical ones that did not.
- Certain components in one particular make or design of equipment failed frequently, but the same type of component in other makes of the same equipment working under identical conditions did not fail at all.
- Sometimes it took a long time to conclude the investigation satisfactorily, but it was always possible to determine the true or root cause of the failures.
- Certain components that used to fail frequently changed overnight into Zero Failure types when the root causes of the failures were correctly determined and eliminated.

Natural laws and properties of materials are totally consistent, precise and reliable. If there are any variations or apparent aberrations they are either due to deficiencies in our understanding of the problem or due to errors in measurement or manufacture. Categorically we are certain that:

- If we take a material with a given composition and manufacture is according to a proven process, the resulting product has the same properties at all times and at all locations around the world. There are no exceptions and no unexpected variations in the finished product as long as the input details are controlled within the specified limits.
- If a conductor is moving in a magnetic field the induced voltage is exactly as predicted by the relevant formula. If two atoms of hydrogen combine with one of oxygen the result is always water and the properties of this material are always the same. The same is true about every other natural law or material.
- The actual results of all natural phenomena are absolutely precise and reliable. If we see any variations in practice they are due to errors in measurement or in some process parameter.
- We can be certain if any equipment or materials 'fail' or behave differently that there is a definite physical or chemical cause or factor at work. All that is required to prevent such failure is to determine that cause or factor and to eliminate it.

The actual fact is that equipment failures occur when design engineers fail to recognise certain natural phenomena or certain limitations of materials as well as operating conditions. The majority of equipment failures are due to such design deficiencies. If the performance of the equipment is monitored closely during the initial service period and defects and failures are thoroughly investigated, it is possible to identify such problems and to evolve suitable modifications or improvements in the design of the equipment. The remaining five percent of the failures are usually due to maintenance deficiencies; but these can also be identified during the initial monitoring and investigation. The maintenance staff can be trained suitably to eliminate all such problems.

1.3 Failure Mechanisms are also Governed by Natural Laws

The equipment under consideration may be either a huge papermaking Fourdrinier or a miniature circuit breaker. These are two different pieces of equipment, different in size, cost, application and principles of operation. But the failures in these totally dissimilar machines could be due to identical failure modes, failure mechanisms and root causes of failure. For instance the failure may be due to the fracture of a helical spring. Similarly, the failure in the terminal board of a very large alternator may be for exactly the same kind of defect as one that causes a failure of the connector of a printed circuit board. Again, the cause of the failure of a tapered roller bearing used in a large earthmoving machine could be the same as that of a tiny ball bearing used in a fractional horsepower electric motor.

When components fail, i.e. when they crack, fracture, overheat, burn, melt, bend or get damaged in any other way, they are actually obeying the same Laws of Nature that they obey when they are working satisfactorily. Hooke's Law, Miner's Law, Lenz's Law, Newton's Laws, Faraday's Laws etc. are all infallible, precise and totally consistent.

There are millions of different types, makes and models of electrical and mechanical machines in use all over the world. However, the number of types of machine elements that go into them is *less than a hundred*. The more important ones of these are listed in the Contents page and discussed in the following Chapters 7 to 66. They cover more than 95 per cent of all failures that take place. Even the remaining failures that have not been specifically mentioned here would be very similar in regard to failure mode and failure mechanism with one or more of those discussed here.

While the number of component types, discussed in this book, are around one hundred, the number of physical and chemical laws and phenomena involved in the failures of these components is even smaller. The more important ones of these are listed below.

1.4 Natural Laws that Govern the Failure Processes

There are just a handful of physical or chemical laws of nature and natural phenomena, which take part in the failure mechanisms covering more than ninety five per cent of all failures of electrical and mechanical equipment. Some of these laws and phenomena are listed below:

Thermal expansion	Friction
Vibration	Brittle fracture
Metal fatigue	Shrinkage
Metal creep	Yield deformation
Corrosion	Erosion or wear
Electrical heating	Contact resistance
Partial discharge	Dielectric breakdown
Electrical tracking	Magnetization
Thermal degradation	Electromagnetic induction
Newton's Law	Faraday's Law
Hooke's Law	Ohm's Law
Miner's Law	Lenz's Law
Arrhenius' Law	Stefan-Boltzmann Law

Those who have studied electrical and mechanical engineering are familiar with these topics and will understand the explanations and terminology given in connection with the failure mechanisms. The theory needed for understanding failure mechanisms is generally nothing more than common sense.

Even more relevant is the fact that there are very few failures in practice that call for a thorough understanding of these theories to prevent the failures from occurring again. In the majority of cases, the causes and the remedies are simple to implement if one accepts certain conclusions on faith without insisting on detailed mathematical proofs. Let us consider a simple example. It is well known to mechanical engineers that metal components that are subjected to high fluctuating stresses are prone to fracture at sharp corners. There is indeed a complex theory that proves why, how and to what extent there is an abnormal increase in stress at sharp corners. However, it is not necessary to understand the mathematical proof to appreciate or understand the conclusion that increase in stress at corners or fillets is inversely proportional to the radius at that point. It is easy, also, to remember that the increase could be by a factor that could be as high as 10. It is also a simple matter to refer to the appropriate handbooks to determine these factors.

At this point it is necessary to emphasise that the natural laws that govern failure mechanisms are exactly the same natural laws which apply in the normal operation of the equipment. There is no separate set of laws for failures and for normal operation. When a component fractures or when an insulator breaks-down it is merely obeying a natural law. There is never any inconsistency or lack of precision in the manner in which natural laws control how things work and also how they fail.

This consistency and precision in the operation of natural laws and the properties of material is one of the two main reasons that give us the confidence to conclude that it is indeed possible to attain Zero Failure performance.

The second reason is that while failures are always due to human error at some stage in design, manufacture, installation, operation or maintenance, it is always possible to pinpoint the exact source of error and to eliminate it.

1.5 Measures to be Taken for Attaining Zero Failure Performance

We shall consider what needs to be done by the engineer who is in charge of the maintenance of plant and machinery to attain Zero Failure performance.

The first and foremost requirement is to develop an optimistic attitude or approach to this question. It is hoped that the foregoing paragraphs will help the reader in this regard. The next few steps are:

- to analyse the current performance;
- to investigate the failures carefully;
- to determine the true and root causes of the failures;
- to determine and implement the corrective steps in the appropriate stage (design, manufacture, installation, operation or maintenance);
- and, finally, to monitor the performance after the corrective action.

Further guidance on these steps is given in Chapters 2 to 6 and the appropriate or relevant chapter amongst Chapters 7 to 66.

1.6 Cost of Attaining Zero Failure Performance

Sceptics may say, "All right, let's assume that Zero Failure performance can be achieved. But at what cost? If you over-design all the components, i.e. operate all of them at very low stresses, there would be no failures. But the equipment would then be far too expensive. It would price itself out of the market". This is a reasonable question and must be answered.

The argument in the preceding paragraph assumes that failures are generally due to excessive stresses and that they can be minimised only by making the components larger and more expensive for reducing the stresses. The reality is quite the opposite. The failures are, more often than not, due to trivial defects the elimination of which would cost very little or nothing at all. It is very rarely necessary to make large scale changes in design. This will become abundantly clear from the practical examples described in Chapters 7 to 66. All of these examples have been taken from actual cases.

We may consider here only a few examples to illustrate this point.

  (a) Many shaft failures are due either to inadequate fillet radius at a change in section or a tool mark (deep scratch) in a highly stressed zone. The cost of increasing the fillet radius or preventing tool-marks is negligible while the costs arising out of shaft failures are thousands of times higher.

  (b) Many electrical failures and fires are due to inadequate tightening of fasteners, insufficient clearances, defects in the design of small hardware, etc. All these types of defects can be eliminated at no cost but the savings effected by the elimination of electrical failures and fires could again be thousands of times higher.

  (3) Many failures of electrical devices such as diodes, capacitors, resistors, windings etc. are caused by local overheating, vibration, thermal expansion etc. The costs of preventing such defects are also relatively insignificant.

It is certainly true that some additional costs are involved at the stages of design, quality assurance and failure investigation; but these costs are insignificant when compared with the costs of repairs, and revenue losses that follow the occurrence of failures. Figure 1.1 shows the relative magnitudes as a function of the failure rates. It will be seen that there is indeed a steep rise in the cost of improving reliability beyond a certain point. However this point is lower than a failure rate of about 0.01 per cent per year which corresponds to one failure per year in a population 10,000 machines or devices. This, for all practical purposes is Zero Failure performance. Figure 1.1 shows also the current situation of India and of developed countries in the failure rate scenario.

Zero Failure performance is, thus, not only desirable but also attainable and, in the long run highly profitable.

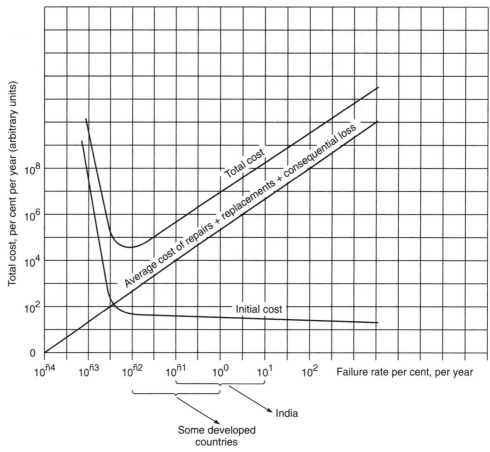

Figure 1.1   Shows relative costs as a function of failure rate. This figure illustrates a general trend. It does not refer to any particular equipment

Chapter 2

HOW TO ATTAIN
ZERO FAILURE PERFORMANCE

In this chapter we learn

- *The importance of determining correctly the failure mechanisms and their root causes.*
- *The difference between proximate causes and root causes.*
- *The root causes of failures are generally trivial deficiencies, even when their effects involve serious consequential damages.*
- *How maintenance engineers have to bear the main responsibility for investigation of failures and for taking corrective action even when the root causes are design or manufacturing deficiencies.*
- *Why maintenance engineers must not justify failures by vague, superficial or generalised explanations.*
- *How effective action can be taken when the root causes are correctly determined and not confused with proximate causes and intermediate stages of the failure mechanisms.*
- *That the above activities have to be in addition to those required under the master maintenance plan and preventive maintenance plan etc. This approach will actually reduce the work load on maintenance engineers.* ❏ ❏

2.1 Proximate Causes and Root Causes of Failures

We have seen in Chapter 1 that Zero Failure performance of electrical and mechanical equipment is a practical and attainable target. We also noted that the only route to it is through the identification of the root causes of failures. We shall examine that route in detail in this chapter.

This chapter as also the rest of this book is mainly concerned with the determination of the true or root causes of equipment failures. It is very important to get to these root causes and not to confuse them with apparent and proximate causes, if we have to take effective measures to eliminate failures altogether and attain Zero Failure performance.

To emphasise this importance and to explain further what we mean by root or true causes of failures as distinct from apparent and proximate causes it is desirable to discuss the latter in some detail. This discussion is essential right at the outset here because, unfortunately, many of these are being put forward by engineers, managers and PROs for public consumption as they help to deflect responsibility and criticism.

It is essential to emphasise what is self-evident because there are many pit-falls on this route. Very often, what may seem to be a root cause turns out, on further investigation, to be only a proximate cause. In such a case, its elimination does not stop the failures from occurring again. The best way to explain this further would be to cite a practical example.

CASE STUDY

In one particular factory there were many 30 h.p. induction motors used for driving certain machines. The failure rate of these motors was too high. It was observed that almost all failures occurred between 7 p.m. and 9 p.m. It was well known that the line voltage was abnormally low during these hours. Induction motors, running at nearly constant speed, drew higher currents at low voltage. It was therefore concluded that these two factors were together the root cause of the poor reliability of the motors. As there was no improvement in the electric supply conditions the management decided to do nothing more than to write periodically to the supply authorities. At this stage there was a change in the maintenance engineer. The new incumbent decided to investigate this problem further. He found that the temperature rise of the motor windings was excessive even when the voltage and frequency were normal. On opening up a few of the motors it was found that the cooling air passages inside the motors were fully blocked by fluff and dust.

The air passages inside these motors were cleaned thoroughly by compressed air jets. The temperature rise dropped by about 30 °C and it was noted that it was now within permissible limits even between 7 p.m. and 9 p.m.

The same treatment was given to all the motors. The failure rate dropped significantly even on the motors that had been in operation with improper cooling for a long time. New and rewound motors now worked without failures. Suitable improvements were made in maintenance practices.

In conclusion it may be stated that overheating of the motors during the low voltage periods was the proximate cause of failures. The root cause was the blockage of the cooling ducts.

It is also necessary to emphasize that investigation into failures should not stop at the technical root cause. Particularly in large organisations, the investigation should continue to determine and to eliminate lacunae in the organisation that allowed the growth of the technical roots of the failure. Again, a practical example will clarify this point.

CASE STUDY

More than twenty years ago, in a certain railway zone, failures of a certain type of relay used in electric locomotives were traced to the inadequacy of a clearance between two parts of the relay. These were corrected on all similar relays working on that railway zone and the failure rate of that relay dropped to zero.

Two years later it was discovered that similar failures continued to take place on another neighbouring railway zone, where similar locomotives were in service. Lack of communication between the railway zones was the obvious lacuna in the organisation. This was effectively removed by the introduction of a system of inter-zonal periodical meetings and circulation of modification sheets by a central service engineering wing.

2.2 Root Causes of Failures are often Apparently Trivial

Many of the causes for break downs seem to be trivial. The effects of failures, even when they result from trivial causes may, however, be far from trivial. The parable of the kingdom that was lost for the sake of a horseshoe nail is only too relevant to many catastrophic failures of electrical and mechanical equipment. After the publication of the first Murphy's Law, maintenance engineers around the world devised similar Laws. One of them states that the cost of a failure is inversely proportional to the cost of preventing it. Here are some examples of this paradox.

CASE STUDY

- Gradual reduction in the tensile forces in the threaded fasteners of a terminal board, due to a defect in its design, led to a major fire in an electric locomotive. The additional cost of rectifying the terminals would have been a few tens of rupees. The cost of repairing the damaged locomotive was several hundred thousand rupees.
- The absence of properly rated fuses in a marriage pandal resulted in a fire involving loss of life. The cost of providing correct fuses would have been nominal. The cost of the fire, and the loss of human lives, was inexpressible.
- Armature shaft failures in a particular design of traction motors were traced to a manufacturing defect. There was an error in the fillet radius at a change in section of the shaft. The cost of providing the correct fillet radius would have been zero. The cost of replacing the shaft and repairing the consequential damages was several hundred thousand rupees.

On account of the triviality of the root causes I have not indicated the local details of the incidents on which the entries in this book are based. They are from memory, records, discussions with my staff and my colleagues. We do not reveal the source or local particulars because no one really wants to discuss and take responsibility for one's own error allowing an expensive failure to occur on account of a trivial and avoidable defect. Manufacturers are even more reticent for obvious commercial reasons. When the cause of a failure is trivial and the effects horrendous, there is usually a conspiracy of silence.

The correct solutions are found after months of following false leads. When the true cause is inconsequential, it is a kind of an anti-climax.

Wherever the failures of equipment led to fires or accidents, the costs of repairs and consequential damages were exorbitant. In some cases there was loss of human life too. At other times identical defects and similar failures caused no problems. Effects of failures depend largely on chance and the local conditions involved. Usually, the effects have nothing to do with the causes. In this book we are concerned only with the root causes and the action taken to prevent the failures from recurring.

A detailed account of the local conditions and the effects of the errors would only add to the volume of the book. Relevance to the reader in search of solutions to his own problems would be nominal.

2.3 The Role of the Maintenance Engineer

This book is written for maintenance engineers. They are the ones who are left behind to deal with all the problems after the sales engineers, designers, manufacturers, and installation/commissioning engineers have all disappeared from the scene. Once the equipment develops defects and fails in service, the maintenance engineers are answerable to the management and the public, not only for restoring normal operation but also for explaining why the failures occurred although, in more than 95 per cent of the cases the root causes lie in design, manufacture or installation.

The manufacturers and designers of the equipment do not generally keep track of the manner in which their equipment performs in service. Some manufacturers routinely return user feedback about problems in service with thinly veiled allegations of bad maintenance or mal-operation. The maintenance engineer is on his own when it comes to investigating failures and finding effective methods of preventing them. This book will be helpful to maintenance engineers who have accepted the challenge of their jobs and decided to attain 'Zero Failure' performance within their domains, inspite of the indifferent attitude of the manufacturers.

Reports or letters to the manufacturers or head-office do not get much response. If the maintenance engineers do not take responsibility for investigation and corrective action, inevitably things go from bad to worse and eventually there is a crisis of plant availability, services or production. Maintenance engineers are in the best position to carry out investigations, to determine the corrective action needed, to implement the required measures. They are the first to reach the site after the occurrence of the failure, and to check the records of previous problems and maintenance. They have the opportunity to examine the condition of associated components while opening the equipment for repairs. They can monitor the performance after making any modifications. Finally and importantly, they are the immediate beneficiaries of Zero Failure performance.

My suggestion to maintenance engineers is, to assume the responsibility for attaining Zero Failure performance of the equipment in their charge and to achieve success with their collective efforts. Most of the actions taken may not be within their direct control. It is necessary for the maintenance engineer to persuade the concerned department or authority to do the needful.

The maintenance engineer's primary responsibility is, to carry out the scheduled maintenance in accordance with the master maintenance plan, which has to be based, on the recommendations of the manufacturer. The basic maintenance philosophy is that all components that degrade during service must be replaced in time by new ones with the same properties as the components fitted on the new machine or equipment.

The manufacturer's recommendations are based on his experience and assumptions regarding the service conditions, the intensity of utilisation and the estimated rates of degradation of different components. The manufacturer can go wrong in these matters, so it is the responsibility of the maintenance engineer to examine the actual degradation rates and the suitability of the recommended maintenance schedule. This becomes vital when there are failures due to component degradation. The schedules and the degradation limits have to be changed, and new materials considered. Service and environmental conditions have to be altered. The maintenance engineer is most suited for this.

A very important point that needs to be emphasised is that, as far as possible, worn or degraded components should be replaced by those obtained from the Original Equipment Manufacturer (OEM). It is theoretically possible to obtain replacement parts from other sources, but then it is necessary to ensure that they are manufactured to the correct design, from the correct materials and with the required quality standards. Hence, unless one has the required technical back-up in the fields of detailed design, manufacture, and quality control, it is wiser and in the long run more economical to get the components from the OEM.

As a rule, the defective or worn out components should be discarded and replaced. No effort should be made to repair, reclaim or re-build the worn or degraded components. Reclamation may often seem to be cheaper but reclaimed parts rarely give adequate service and failure rates are likely to go up. When the costs of failures including the consequential damages and losses of revenue or service are taken into account, it is usually seen that reclamation leads to greater and not lower costs of maintenance and operation. If due to non-availability of new parts, or due to accidental damages to large non-deteriorating parts the old ones are reclaimed under expert supervision, and reused they should be kept under special observation.

In countries where Product Liability Legislation is in force, conditions are better. Manufacturers do take a great deal of interest in the feed-back from the users and often go to great lengths to carry out changes in design in order to prevent failures in service. They have gone even further. Their approach is now to design and manufacture equipment, which will require no maintenance so that they may not fail in service even due to bad maintenance. Safety devices or features are incorporated in the designs to prevent mal-operation.

However, now the problem gets focussed on the designers of the equipment. Due to the good performance of the equipment, which has been perfected over the years, there is little feedback about failures in service from the users. The young design engineer must learn all possible ways in which equipment can fail, so that he can review his new designs to make sure that he is not inadvertently allowing history to repeat itself. Two aphorisms are very relevant here.

If anything can fail, it will fail sooner or later:

Murphy's Law

Those who do not learn from history are condemned to repeat it

While this book was written with the maintenance engineer in view, it is likely to help also the young design engineer who wants to get it right the first time. He can learn from the experience of others and avoid any pit-falls. The maintenance engineer as also the design engineer will be able to investigate failures and determine the root causes and then take the appropriate corrective action. Eventually, they will attain Zero Failure performance from their equipment.

Young engineers in charge of maintenance of electrical and mechanical equipment will find this book useful when confronted with equipment failures for the first time. Under normal conditions, their world consists of control panels with red and green lights in air conditioned rooms, sweetly humming motors or machines, quiescent switch-gear and transmission lines and totally silent mazes of wires and cables. However this benign and apparently harmless world comes crashing when things go wrong. The engineers may then be faced with burnt or charred cables, fractured and twisted components and damaged machines. Services and production come to a stand still. These may lead to injuries and fatalities amongst the staff. Top management will now take over. At this stage the priority is obviously for quick repairs and restoration of services and production.

Once things are under control, it is important for the maintenance engineers to determine the root cause of the problem. Action has to be taken to ensure that a similar problem does not occur again. If the failure was due to some lacuna either in the design or in the maintenance practice, it is bound to occur again until the root cause is eliminated. There are several reasons for failures to occur and the mode of failure is unlikely to be the same. It is unpredictable when a failure mode repeats. The only way to solve the overall problem is to tackle each different failure mode individually. Records must be maintained meticulously so that repetitive cases are taken up for investigation in accordance with their priority.

At this stage, the young engineer faces a practical problem. Chapter 3 and a reference to the appropriate chapters and sections, will help him to apply his theoretical knowledge to the problems at hand. In doing so the trivial and obvious causes of failures should not be overlooked. He must investigate them, determine the root causes and the failure mechanisms.

Non-availability of spares, forces the maintenance engineer to make compromises and to take risks. In such cases he has additional responsibility of taking on the purchase departments to ensure the return to normal maintenance practices.

Initially, assuming responsibility for investigating failures and determining measures for corrective action increases the work-load of the maintenance engineer. He may need a little assistance for collecting data and documents.

In the long run, reduction in failure rates will actually reduce the work load on the maintenance engineer and his staff. In most units, 75 per cent of the existing work-load is for dealing with the effects of failures and carrying out repairs on the failed equipment. Less than 25 per cent of the work-load is scheduled for preventive maintenance. In any case, this type of work usually goes on smoothly without the need for any personal intervention by him. As repetitive failure frequency reduces as a result of failure investigations and corrective action, the overall work-load gets reduced.

2.4 What the Maintenance Engineer must **NOT** do

In the preceding section, we have examined the role of the maintenance engineer. He investigates all failures, determines their root causes, implements corrective action and continues this process until Zero Failure performance is achieved.

Maintenance engineers have not been given adequate training and where top management levels do not appreciate the true value of reliability, failure prevention efforts are given very low priority.

Maintenance engineers must not make excuses for equipment failures through vague, superficial or generalised explanations such as overloading, heavy rain, staff shortage, etc.

CASE STUDY

In the switchyard of a large thermal power station, the insulators supporting the main bus-bars flashed over on the first day of the monsoon. The Grid was overloaded at that time with no reserve on line. All other power stations tripped one after another. Thirty electric trains stalled and major industries shut down in three states for seven hours till the power supply was restored.

The root cause of the failure was that the bus-bar insulators had never been cleaned since installation despite a highly polluted environment.

A high official of the Electricity Board explained this incident away as the result of a thunder storm.

There are many other similar excuses for failures that are often overlooked by the hierarchy. They even reach the summit without challenge. It is very important for those in charge, particularly at higher management levels, to question them and to try to reach the root cause of the failures. However, a few of the more common reasons are enumerated below with brief comments, so that they may be recognised and summarily rejected.

OFFERED EXPLANATION, AS CAUSE OF FAILURE	BRIEF COMMENTS
Equipment has been manufactured by company XYZ	This type of explanation is given when the company XYZ has a poor record. Every failure on equipment supplied by that company is then attributed to that company's reputation; It is necessary to corroborate such conclusions with the details of the manufacturing or design defect and the failure mechanism.
Equipment has been over-hauled by another department PQR of the same organisation	This is an example of an attempt to deflect responsibility to some other department or person; It *may* be a valid and correct reason but then it is necessary to mention clearly the exact nature of the defect and the failure mechanism.
The equipment is of indigenous manufacture	This is a popular excuse in developing countries; see remarks about company XYZ.
The maintenance staff is not technically qualified	While literacy and technical qualifications are certainly useful, very few failures can be said to be due to their absence; routine maintenance requires manual skill and practical knowledge but very rarely does it require engineering knowledge; With appropriate job related training, and provision of proper tools, illiterates and even the handicapped can do good maintenance.

— Contd. —

OFFERED EXPLANATION, AS CAUSE OF FAILURE	BRIEF COMMENTS
The supply voltage or supply frequency was too low (or too high)	Abnormal voltage or frequency can lead to failures of electrical equipment but not every failure subjected to them can with certainty be said to be their result. At the most they may have precipitated failures which would have occurred anyway.
Inadequate strength of maintenance staff, or maintenance shutdowns not allowed by the operating or production department.	Lack of proper maintenance can indeed lead to failures, but such explanations should be accepted only when the failure mode, failure mechanism and details of scheduled maintenance, which was missed, are furnished.
Lack of adequate rest due to continuous operation	Most of the machines do not require rest and they do not get fatigued. There is a phenomenon called Metal Fatigue but this term is misused to explain away equipment failures; Fatigue failures can be totally prevented by appropriate measures during design, manufacture and maintenance.

2.5 Conclusion

It is possible to reduce failure rates to negligible levels and to attain Zero Failure performance only if maintenance engineers assume the responsibility. They must determine the failure mechanisms and their root causes, and then take action to eliminate these and not merely the proximate ones. Above all, they must neither offer nor accept vague or superficial explanations.

Chapter 3

GENERAL FEATURES OF FAILURES

In this chapter we learn

- *Explanation, and examples of the terms 'failure mode' and 'failure mechanism'.*
- *Distinction between the term 'failure', 'defect', 'accident'.*
- *Introduction to a new concept of 'seed-defects'.*
- *Reasons for variability of failure rates.*
- *Effects of variability of 'stress' and 'strength'.*
- *General approach to failure prevention.*
- *Reporting and classification of failures.*
- *Importance of documentation and communication systems in large organisations.* ❏ ❏

3.1 Failure Modes and Failure Mechanisms

The term failure mode refers to the visible, tangible or measurable changes in the appearance or properties of a component, such as the following:

Excessive wear	Cracking	Fracture
Over-heating	Burning	Melting
Charring	Bending	Twisting

In electrical equipment, the following additional failure modes are observed.

Short-circuit	Tracking	Flash-over
Carbonisation	Softening	Open-circuit

The term failure mechanism refers to the physical or chemical process through which the failure mode is reached. Examples of failure mechanisms are given below:

(a) Due to excessive alternating tensile and compressive stresses at a point in a steel component, fatigue cracks develop and grow until the component fractures.

(b) Due to inadequate contact force between two conductors there is excessive contact resistance, over-heating and rapid oxidation of the surfaces in contact. This starts a vicious cycle, which culminates in melting of the conductors and a fire.

(c) Due to chemical action between lubricating oil and a rubber O-ring the latter expands and causes the air-motor to get jammed.

Different failure mechanisms can lead to the same failure mode. For instance:

- Tracking between adjacent terminals can lead to short-circuit and charring of the insulation. The same or very similar failure mode may be observed due to a totally different failure mechanism in which inadequate contact force may cause overheating of the terminal followed by charring of the insulation.

- A seized ball bearing may be the failure mode observed when a machine stops. There are a large number of different root causes and failure mechanisms that will culminate in this failure mode.

- Flashover of the commutator of a direct current motor may be the failure mode. In this case too there are a large number of possible root causes and failure mechanisms.

In the preceding chapter we have discussed the difference between proximate causes and root causes of failures. These root causes are usually in the nature of defects, which are present in the equipment or in its environment.

3.2 Definitions of the Terms: Defect, Failure and Accident

A *defect* is any condition in the component or equipment, which is neither intended nor desirable. It is a condition that can cause a failure of the component or the equipment. Examples of defects are lack of lubrication in a bearing; crack in a highly stressed component; voltage far in excess of the electrical strength of the material; etc. In the chapters that follow we will be listing many more common defects.

A *failure* is an incident or occurrence where a component or equipment fails to perform satisfactorily its intended function. Examples of failures are: seizure of a bearing and stoppage of the rotating machine; fracture of a shaft or axle and stoppage of the rotating machine; short-circuit, burning of insulation, melting of copper wires and interruption in electric supply. In the chapters that follow, we shall discuss many other types of failures. A *defect is a cause, either proximate or root, and a failure is its effect.*

A *seed-defect* is a condition, which has the potential of growing into a *defect* as defined above. It is, as the name suggests, like a seed that may or may not germinate and grow into a plant, depending upon the conditions, or environment in which it is placed. Similarly a seed-defect may or may not grow into a defect and later into a failure. If there is no seed there will be no plant. Similarly, if there are no seed-defects there will be no defects and no failures.

The essential difference between a seed-defect and a defect is this: a seed-defect may or may not grow into a defect but a defect will certainly grow into a failure sooner or later. The identification (and elimination) of seed-defects and defects is, therefore, the key to the prevention of failures.

An *accident* is a failure, which results in severe damage not only to the component or equipment that fails but also to other associated equipment, which is generally but not necessarily in the vicinity. It may involve very high costs in regard to loss of property and even loss of human lives. Examples of accidents are derailments of railway trains; aeroplane crashes; electrical fires in buildings, locomotives or coaches; collapsing of buildings or bridges. All these are the results of failures of components or equipment. Such *accidents* are not accidental. They are the results of avoidable defects and hardware failures.

In the ultimate analysis, accidents are the results of certain types of seed-defects. Whether a particular seed-defect will germinate and cause just a failure, or whether it will develop into an accident is all a matter of chance. The identification of certain types of seed-defects and their systematic elimination is thus the key to the prevention of accidents.

The word accident has several meanings but the common and implied meaning is that it is an event which was unforeseen, unintended and beyond human control. Most accidents of the type mentioned above are unforeseen and unintended but they are certainly not beyond human control. They are the result of some human error at some stage in the design, manufacture, installation, operation, or maintenance. They are caused by human error and they can be prevented only by human effort in the right direction. They cannot and should not be treated as *accidents*.

Truly accidental 'accidents' are very rare. Examples of such true accidents would be those caused by earthquakes, unprecedented floods, direct lightning strokes, and similar convulsions of nature. All the rest are avoidable, preventable and susceptible to analysis.

3.3 General Plan of This Book

The discussions in this book are based on actual case histories. I have tried to be brief in giving sufficient details to help the readers find solutions to their own problems. I shall refer to the factors, which influence the relevant failure mechanisms, and to the appropriate theoretical background. References to some of the published literature which is generally easily available are also given for those who would like to study the theory in greater detail.

While the chapters that follow and the prescriptions for reliability given therein are all based on actual case histories, no attempt has been made to give details such as place, date and time of failure; make, model and name of machine; name and address of the organisation; names of the people involved; consequences and costs of failures. The emphasis is on describing the failure modes, the failure mechanisms, the root causes and the corrective action.

The reader may apply his common sense and engineering judgement to the suggestions given in the book and to look for the possible causes in the actual case under investigation. It is meant to be a reference for those who wish to find lasting solutions to the problems of repetitive failures of electrical and mechanical machines.

3.4 Variability of Failure Rates

While the specific causes of failures in different types of components and equipment will be dealt within the following chapters, we must first examine some frequently asked questions relating to failures.

(a) Why do some components or equipment fail quite some time after commissioning while some others fail very soon?

(b) Why do some components fail while identical components in identical equipment working under identical conditions continue to work satisfactorily?

(c) Why do some components or equipments continue to work satisfactorily for months or even years and then fail suddenly for no apparent reason?

(d) Why do some components fail even though adequate factors of safety are allowed in the design of the components?

There are three main reasons for the apparently erratic behaviour of materials and equipment. These are:

(a) Variance in the properties of materials due to variations in composition and in process parameters.

(b) Variance in the environmental and operating conditions.

(c) Variance in the rates of degradation in the properties of materials and in the quality of maintenance.

These variations can be minimised to any desired level but they cannot be totally eliminated.

Despite the fact that some components can degrade and eventually fail, it is still possible to attain Zero Failure performance of the machine, equipment or system by timely replacement of the components that degrade beyond safe limits. It is not necessary to bring the variations down to zero for attaining Zero Failure performance. The safe limits have to be determined. All that needs to be done is to keep them in view during design of the equipment and to control the limits within which they are allowed to vary.

3.5 Fracture

As seen in Section 3.1 previously, there are a number of failure modes but the most common one is fracture. We come across fracture very early in life even as a child. When we write with a lead pencil, if we press it too hard on paper the lead breaks. If we try to make too fine a point on the lead it breaks. Sometimes, the lead breaks even when the size of the point and the applied force are both normal. We grasp instinctively certain basic principles of Applied Mechanics and Strength of Materials.

Fractures of machine components are certainly a more complex phenomena. The materials used are much stronger than pencil leads but the basic principles are the same and easy to understand. Components fail when the force applied exceeds the strength of the component. The uncertainties raised by the three questions posed in Section 3.4 are due to the following facts.

Relating to Strength

- There are small variations in the properties of different batches of the same raw material.
- Different components of the same material may have different properties due to manufacturing variations.
- The strength of a component may deteriorate with the passage of time and the effects of usage and service strains.

Relating to Stress

- The stress in a component may vary in magnitude, due to changes in loads, speeds and environment.
- The stress in a component may increase with the passage of time and the effects of usage and degradation of components.
- The stress in a component may increase due to changes in operating conditions.

When the stress in a component is higher than its strength the component will fail. Therefore, it follows that failure will not take place if the minimum strength is maintained at all times at a level higher than the maximum likely stress. However the same result can be achieved if the stress is not allowed to exceed the strength of the component. These may seem to be rather obvious and self-evident truths. However they are the starting points for analysing the long-term reliability of a system comprising a large population of similar components.

The two curves shown in Figure 3.1 represent this overall picture. The left-hand curve shows the probability of occurrence of different stresses acting on the components at different times and in different machines. The right hand curve represents the probability of occurrence of different strengths in different components of different machines.

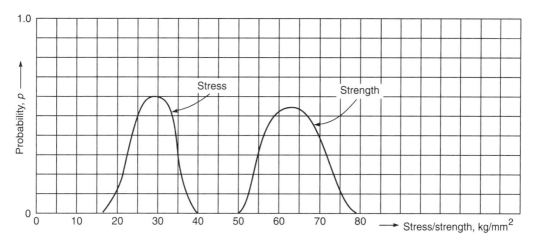

Figure 3.1 Probabilities of strengths of and stresses in a typical component

If the two curves are well apart as shown in Figure 3.1 and do not intersect, there will be no failure ever. On the other hand, if they do intersect as shown in Figure 3.2, there is a chance that at some time, for some component, the strength will be less than the stress and failure will then occur at once. The greater the overlap of these two curves , the greater the probability of failure.

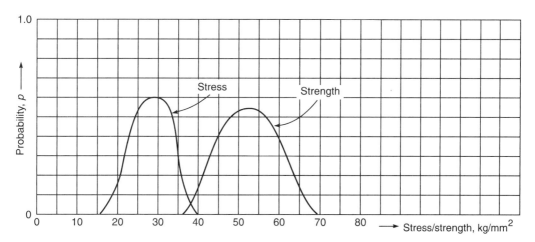

Figure 3.2 Probabilities of strengths of and stresses in the same component as in Figure 3.1 after degradation and shift of strength curve to the left

There is always an attempt, usually successful, by the designer to ensure that the two curves do not overlap anywhere. The equipment then works without failure, at least initially. However, due to either manufacturing errors or material degradation that may take place in service, the strength curve may move to the left and intersect the stress curve as shown in

Figure 3.2. Similarly, due to changes in operating conditions or defects in other associated components, the stress curve may move to the right and intersect the strength curve as shown in Figure 3.3. When, either way, such an intersection occurs, the probability of failure increases from zero to some value between zero and one. Failure will now occur sooner or later.

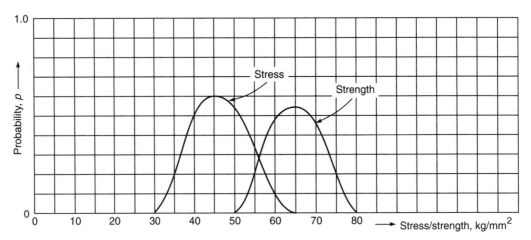

Figure 3.3 Probabilities of strengths of and stresses in the same component as in Figure 3.1 after changes in operating conditions causing shift of the stress curve to the right

Since an increase in the distance between the two curves entails increase in the size or improvement in the quality of the material, any such effort involves additional cost. Therefore the designers' effort is to keep the distance between the curves just sufficient to ensure no intersection. It is now the responsibility of the maintenance staff to ensure that this initial distance between the two curves is not reduced as a result of either degradation of materials or worsening of environmental or operating conditions beyond the limits assumed in design.

3.6 Stresses/Strengths can be of Different Types

We have discussed so far, a comparison of mechanical stress versus mechanical strength. Failure in this context meant fracture or deformation of the component. The same kind of comparison can be made in the context of electrical breakdown or short circuits, between electrical strength and electrical stress. Other failure mechanisms will, naturally, involve comparison of other parameters.

Fractures and short circuits are 'catastrophic' failures of the mechanical and electrical kind. There can be other types of 'degradation' failures in which there is no fracture or short circuit, but where some parameter (such as clearance, resistance, etc.) of a component falls-outside certain limits which is critical for the satisfactory performance of the equipment. There is a similar but somewhat different comparison between probabilities of the actual and the limiting conditions.

Books on reliability engineering explain the mathematical analysis of the above relationships. It is not necessary for the practising maintenance engineer to delve into that theory; but it will certainly help him to correctly assess the evidence before him while investigating failures.

3.7 Prevention of Failures

There are several possible ways of preventing failures. Some ways may be more economical than others. Some ways may be more feasible than others in the prevailing circumstances. In any case the engineer in charge has to ensure that the strength of the component is higher at all times than the stress it has to bear. Either the stress has to be reduced or the strength has to be increased. If the strength is deteriorating, such deterioration must be prevented or slowed down. If degradation of the material cannot be prevented, maintenance schedules have to be established to replace the degraded component before the degradation reaches the level at which failure is probable. The engineer has to use his best judgement and after considering the different ways available, must select the most suitable one. A simple example will make this clear.

CASE STUDY

Some bolts in a certain type of machine were prone to frequent failures. Investigation showed that these failures were due to machining error in the shape of the bolt. The obvious and long term solution was to replace all such bolts by new ones with correct machining. However, this would call for shutting down each machine for several days.

It was found possible to reduce the operating speed of the machines by about 10 per cent without affecting the overall production rate; but the bolt stresses were expected to be reduced by about 19 per cent and the failure rate by a considerably higher percentage. The entire population of about 120 machines were dealt with, within about eight months by reducing their operating speed by 10 per cent and taking out one machine at a time for overhaul and replacing all the defective bolts.

On reducing the operating speed by 10 per cent, the failure rate dropped immediately by about 60 per cent and further to zero after completing the replacement of all the defective bolts. The normal operating speed was then restored. There was no further incidence of bolt failures.

Another method of containing failure rates is by Condition Monitoring. There are various possible methods of detecting defects in their incipient or formative stages. Ultrasonic testing, radiography, dissolved gas analysis, partial discharge measurements, dye-penetrant testing, infra-red photography, etc. are some of the sophisticated methods now available.

Simple visual examination and inspection of vulnerable components periodically, a method which requires no equipment, is sometimes very effective. The method to be adopted depends on the type of damages or defects that are responsible for the failures. Investigation and determination of the failure mode, the failure mechanism and the root cause of the failure is, the starting point for all efforts to attain Zero Failure performance. Here are a few examples of the application of Condition Monitoring as method for minimising failures.

- Fatigue fractures in rails and railway axles are due to a variety of reasons such as over-age, ballast maintenance deficiencies, machining errors, over-loading etc. All these defects cause invisible fatigue damage which accumulates over the years. It is then impossible to prevent fractures in service except by replacement of the damaged rails and axles. Ultrasonic testing devices are then used for condition monitoring to detect hair line cracks in their early formative stages. The defective rails and axles are then replaced before they can fracture in service. This type of corrective action should be considered to be supplementary action. The effective preventive action is to determine the root causes of fatigue damage and eliminate such root causes.

- Electrical transformer failures are often due to defects or deficiencies in design details such as connectors, fasteners, leads, cleats etc. Fortunately, long before catastrophic failure can occur, certain gases are produced at the point where the defect is situated and these gases get dissolved in the transformer oil. It is possible to analyse these gases by gas chromatographic methods and determine the possible causes. This helps the maintenance staff to eliminate the defect. In this manner major failures of transformers are averted. In this case too, it is more important to determine the root causes and to eliminate them. The Condition Monitoring should be considered a supplementary effort or a second line of defence.

- Tapping of railway wheels and feeling by hand for hot axle bearings is a good example of un-sophisticated, but highly effective condition monitoring methods used by railway staff to minimise failures of axle boxes and tyres.

These condition monitoring methods can be said to be a second line of defence, the first being to determine the root causes and to eliminate them at their source.

3.8 Reporting and Classification of Failures

In large organisations, where failure statistics are often maintained and analysed, the failures are sometimes classified under the heads 'accidental', 'material failure' and 'human failure'. It is necessary to discuss this type of instant analysis in detail because most of it is based on false assumptions. A few common examples of the confusion about the causes of failures may be mentioned here:

CASE STUDY

- The cause of a certain fire, which led to loss of life and property, was declared as a 'short-circuit'. The way in which this was stated implied that no one is responsible and that it could not have in any case been prevented from occurring. Everybody including the media accepted this explanation. The true cause of the fire was a defect in the electrical protection system. 'Short-circuit' was only a proximate cause.
- An outdoor distribution transformer exploded during the rainy season. The electricity supply authority ascribed the failure to heavy rain. The press duly reported this. No one pointed out that outdoor distribution transformers are designed to operate in the open and in heavy rain. The real cause of the failure was a defect in the manufacture and installation of a gasket.

While serious problems such as the two cases mentioned above are reported and classified, many defects and failures that occur are neither reported nor investigated. Repairs are carried out, the equipment restored to normal service and the occurrence forgotten.

Accidental failures can be so classified only if they are totally beyond human control as in the case of those caused by direct lightning strokes, earthquakes or unprecedented floods. These are indeed very rare. More than 99 per cent of all failures of electrical and mechanical equipment are due to avoidable causes. All failures are due to some deficiency in design, manufacture, installation, operation or maintenance. Material failure is a misnomer. If the material is correctly selected, manufactured, utilised and maintained, it will never fail.

If, failure does occur it is always due to some deficiency in one of the four stages mentioned earlier. All failures, other than the rare accidents are due to human errors. Appropriate training of the concerned staff can prevent even these. If in a particular group of machines and equipment, failures do occur repeatedly, it is possible to stop these altogether by identifying their root causes and eliminating them.

In every organisation a change of attitude at every level is required. What is needed in particular is an acceptance that equipment failures are avoidable; they can be prevented;

and that the responsibility for this rests with the engineers in charge. It may take time to achieve such a change but a beginning must be made at once and the initiative for this must be taken by the higher authorities. The first step in this direction is reporting and classification of failures.

Performance must be monitored closely. All failures and defects including those that may not grow into failures must be reported meticulously. Failures and defects must be classified component-wise, equipment-wise, manufacturer-wise. Root causes of each type of failure should be determined by thorough investigation. Corrective measures must be devised and implemented systematically. Effectiveness of these measures should be monitored and evaluated periodically.

Chapter 4 covers how this can be done. The remaining chapters, deal with failure modes, failure mechanisms, causes of failures, and remedial measures in respect of commonly used components and equipment.

3.9 Systematic Documentation and Communication

During my service in the Indian Railways, I discovered that certain types of equipment failures had been totally eliminated in some maintenance units. This was achieved through design or maintenance modifications evolved by the local engineers after carrying out thorough investigations and determining the root causes. However, similar problems continued to plague other units. This problem of communication within the Indian Railways was largely overcome as a result of the circulation of modification notices by a centralised service engineering unit, and discussions in periodical meetings of maintenance study groups.

From 1985 onwards, as a consulting engineer, I had a similar experience in a larger field. Certain problems that had been eliminated in the Indian Railways continued to occur in other organisations. I will mention here only one example of hundreds of such cases that came to my notice.

In Chapter 39 about electrical connections through threaded fasteners, I have described a type of failure caused by an apparently trivial discrepancy in the order of assembly of certain spring washers. See Figure 39.7.

This is based on a fire accident that occurred in an electric locomotive at Kalyan, India in 1955. I read a report recently, about a fire in an aircraft engine in USA, caused by an identical, trivial defect. The thorough investigation, which followed this accident, exposed discrepancies in the maintenance manual and the spare parts lists relating to precisely the same type of defect in the assembly of a spring washer. I do not know how many other such accidents have occurred elsewhere during the last forty five years all over the world, but I have never seen any mention in print of this trivial defect, or of the resulting failure mechanism.

I have observed that problems that had been successfully overcome earlier surface again—usually somewhere else but sometimes even in the same place. There would have been in this period a complete change of staff, supervisors and engineers through retirements, transfers and promotions. Even when some of the old-timers were still in service, they were usually in a different section. The documentation about the earlier modifications would be buried in the archives. The manufacturers or suppliers of the equipment would also be different and totally unaware of the earlier experiences.

The result of the above situation is that design defects and deficiencies continue to repeat. It would take quite some time before maintenance engineers recognise a series of failures as a design problem and devise and implement the same solution that had been reached by their predecessors decades ago. Indian Railways use equipment manufactured by reputed companies all over the world. Their design engineers seem to be repeating the mistakes made by their predecessors and their competitors. It is ironic that by oversight trivial errors are repeated. There is not enough documentation that could be made accessible to the new entrants to the field. Obviously, no manufacturer or maintenance engineer is keen to publicise his blunders.

There are two lessons to be learnt by design and maintenance engineers. The first is there is no such thing as a 'trivial' component or 'trivial' detail. The detail may by itself be worth only one cent but it has the potential of causing damage worth millions of dollars. The second lesson, of particular relevance to the training of maintenance engineers, supervisors, and technicians, is that such training must include systematic study of failure modes, failure mechanisms, and case studies.

Therefore, there is an urgent need for a sort of encyclopedia of equipment failures which highlights the common failure modes, failure mechanisms and their root causes, including the apparently trivial ones. It would help designers to review their designs before authorising bulk manufacture. It would also help maintenance engineers to reach the correct solutions quickly when confronted with repetitive failures of equipment caused by design or maintenance deficiencies. It would also help them to train the younger professionals in avoiding all kinds of defects by explaining the mechanisms of failures caused by such defects.

I have grouped the entries component-wise. There are thousands of types, makes and models of machinery which include these components, but the modes and mechanisms of failures are not only the same but also relatively small in number.

This book does not claim to be able to give precise calculations and instant solutions to all hardware problems. Maintenance engineers will, however, be able to get useful leads and information when confronted with equipment failures.

The emphasis here is necessarily on the key ideas and not on any detailed technical analysis or on the values of the design parameters. It is neither practical nor necessary to consider all kinds of design calculations and material properties applicable to the vast variety of equipment. The reader will have to check them out with reference to the particular circumstances of the problem under investigation. In many cases the full solution may become apparent, but in others the details given there will help the investigator either to confirm or to eliminate various ideas under consideration.

References

O'Connor, Patrick, *Practical Reliability Engineering*, John Wiley, 1992.

Hattangadi, A.A., *Electrical Failures and Fires*, Tata McGraw-Hill, 1999.

Failure Analysis and Prevention. Volume 11 of Metals Handbook, Published by the American Society for Metals, 1991.

Himmelfarb, David, *Product Failures and Accidents*, Technomic Publishing Co., 1985.

Collins, Jack A., *Failures of Materials in Mechanical Design*, John Wiley, 1993.

Fuqua, Norman, *Reliability Engineering for Electronic Design,* Marcel Dekker 1987.

Nishida, Shinichi, *Failure Analysis in Engineering Applications*, Butterworth Heinemann, 1996.

Tanner, Peter and Penma, James, *Condition Monitoring of Electrical Machines*, Research Study Press Ltd., 1986.

Muir, Lawrence, *Material and Component Failure Analysis*, Marcel Dekker, 1987.

INVESTIGATION OF FAILURES

> **In this chapter we learn**
>
> - *Advantages of systematic investigation.*
> - *Systematic methods for the investigation of failures of electrical and mechanical components and equipment.*
> - *Fixing priorities for undertaking investigations when many failure modes are present.*
> - *Step-by-step approach to investigation of failures.*
> - *Drawing up, implementing and monitoring of action plans for preventing failures, fires and 'accidents'.*   ❏ ❏

4.1 In the Ultimate Analysis, All Failures are Caused by Human Errors

Failures and accidents are often explained away as due to either material failure or human error. These are proximate causes but the true or root causes also need to be analysed and determined if we need to prevent recurrence. This should be the objective of all investigations.

Material failure is caused due to one or more of the following reasons:

(a) The materials have not been selected correctly by the designer.
(b) They are stressed beyond capacity due to a defect in the design.
(c) The manufacturing processing and assembling is incorrect.
(d) The shape or size of the component or equipment needs to be modified.
(e) The equipment is not operated correctly within the designed limits.
(f) It is not maintained correctly as envisaged by the designer.
(g) The equipment is used longer than the designed life.

In every case mentioned above, the original failure is a human failure. All failures and accidents that are caused by human failure (usually driver/pilot error or operator error) or those that are said to be due to material failure, in the ultimate analysis, are caused by human error. They are not beyond human control. It is certainly possible to prevent all these mishaps by appropriate selection and training of the staff.

All investigations into accidents and breakdowns must focus on training of the persons in charge. When they occur in other organisations such as the manufacturers of equipment, the users' control on them is indirect and through specifications, stage inspection and final inspection. In large organisations, the conclusions of investigations have to be institutionalised through training material, specifications, inspection formats, working drawings, test programs, etc.

4.2 The Need for Systematic Investigations

Very often decisions relating to corrective action are taken on the basis of pre-conceived notions or convenience of implementation. When this happens, the failures may continue to take place. The ineffectiveness of the modifications becomes apparent only after a long time and perhaps much expenditure. Other methods are then tried out, probably by a different group of engineers.

Systematic investigation according to a well thought out plan helps in two ways. Firstly, the time taken for investigation is much less, and secondly, the probability of an incorrect conclusion is greatly reduced. The second reason is more important because modifications made or action taken on the basis of incorrect conclusions can be ineffective if not counter-productive, and consequently, this could be an expensive exercise.

Failures can occur due to a variety of reasons. It is important to reach the correct conclusion with regard to the mechanism of failure and the root cause of the failure.

In this chapter, we shall discuss certain well-tested methods for carrying out investigations into failures of electrical and mechanical equipment.

In the case of some types of electrical failures, fires are possible. If the system design is correct an electrical failure caused by insulation defects should not lead to a fire. The protective system should prevent fire. However, if there is overheating caused by bad contact in joints, there is no second line of defence and the probability of a fire starting is very high. Special care must therefore be exercised to prevent such defects in electrical machines and installations. In case of electrical fires the investigation must be two-pronged—first to determine why and how the electrical failure occurred and the second to determine why and how there was a fire. Remedial measures must also address both these issues.

Any fire usually involves electrical installations due to their ubiquitous nature. It is a common media and public reaction to assume electrical short-circuit whenever there is a fire. Often even management also falls in line with this presumption due to their fallacious belief that short-circuits are inevitable and that they must in every case cause fires. The investigator must avoid this pitfall. More detailed explanations about fires caused by electrical defects are given in my book, *"Electrical Fires and Failures."*

The best method of arresting failures is by preventing the onset of the degradation processes through the implementation of judicious changes in the design of the equipment. To be able to do so, it is first necessary to decide which of the various degradation processes are actually at work and then to examine how they can be prevented from starting or slowed down sufficiently to prevent failures from taking place in service. Once these questions are answered correctly, the remedial measures usually become obvious at once.

The degradation processes that usually affect or afflict the mechanical and electrical components have already been listed in Section 1.4, Chapter 1. The more important ones have been discussed briefly in Chapters 5 and 6, and in greater detail in the book "*Electrical Fires and Failures.*" Specific measures in regard to various components and equipment are discussed in the following chapters in this book.

4.3 Fixing Priorities for Investigations

In any large organisation, with many types and a large number of equipments in service, there may be a large number of type-failures and type-defects that need to be investigated. (A type-failure is one in which there are three or more failures with the same failure mode and the same failure mechanism.) It is not desirable to undertake the investigation of all such defects and failures, simultaneously. If the available technical resources are limited, as is usually the case, it is desirable not to spread them too thinly over too many problems because the results would then be very poor in quality. It is better to concentrate attention on a few type defects or type failures per engineer or senior supervisor.

Normally, it should be possible to find effective solutions to the problems taken up for study within a month or two. In the initial stages, when the engineer or supervisor is undertaking an investigation for the first time, he may be asked to take up only one problem at a time. As he gains experience and completes successful investigations, he can take up two or three problems at the same time; but three problems at a time should be the limit.

It is, therefore, desirable to fix priorities for the investigation of failures. This can be based on the following considerations:

- Type-failures that have more serious consequences should be given priority. Thus, defects and failures that have the potential of causing serious damages should be given the highest priority.
- Type-failures that have very serious effects on production or services should get the next priority.
- When the above two types have been covered effectively, the other failures and defects may be taken up.

Within each of the three categories given above, type failures and then type defects that have higher repetition rates should get higher priority.

Since in any case all the failures and the defects have to be eliminated totally, it is not difficult to decide the priorities. Often it is possible for the engineers in charge of maintenance to draw up a list of priorities based on their experience and a brief look at the failure statistics. As soon as the type-failures to be investigated are assigned to different engineers and supervisors, it is possible for them to focus their attention to the assigned projects. The remaining sections in this chapter are addressed to these engineers and supervisors who take up systematic investigation of one of the type-failures and type-defects. It is possible to follow exactly the same procedure for two or more problems at the same time.

4.4   Process of Investigation

Having gained experience in the investigation of failures, it is often possible to skip some of the steps discussed in the next section. The inexperienced investigator should follow the suggested method. A very experienced reliability engineer may be able to reach the right conclusion after looking at the failed components and the surroundings. However, for the newcomers in the field, the step-by-step procedure suggested in the next section is recommended, because it is infallible.

It is desirable but not essential to follow the sequence of the steps given below. Depending upon the availability of equipment, documentation, concerned engineers and staff, the sequence may be changed as convenient, but it is desirable to take all the suggested steps before coming to any final conclusion about the cause and the remedial measures.

All equipment operates according to certain immutable laws of nature and all failures also occur strictly according to these natural laws. There are no exceptions and there is never any stage in the failure mechanism that does not follow some natural law. It is only necessary to discover the true failure mechanism and once that is done, it is always possible to devise suitable changes in design, manufacture, operation or maintenance. Those investigating failures of electrical or mechanical equipment need to follow logical steps to reach the correct solutions.

4.5   Failure Investigation: Step-by-Step Approach

It is important to follow a systematic approach for failure investigations. The following steps can serve as guidelines for a step-by-step approach:

(a) The first step in failure investigation is open a separate file or folder for each type defect or type failure being taken up for study. All data, documentation collected from time to time should be placed in the folder. This simple step is more important than what it may seem to be. It helps the investigator to put things in correct perspective every time he returns to it from other routine and unavoidable activities. When maintenance engineers have to take on these investigations in addition to their routine maintenance duties, they are usually not able to devote much time every day to the investigation. These folders or files help to maintain continuity of thinking and action under these conditions. Moreover, it helps to take the investigation forward even when there are changes in the investigating team.

(b) Study the following documents very carefully with special reference to the components that are likely to have failed first:

   (i) the operating and maintenance manual;
   (ii) the manufacturing drawings;
   (iii) failure reports and photographs if any of previous incidents.

(c) Collect and tabulate full details of each failure of the type selected during the preceding year. The details to be tabulated should include the following:

   ■ names and makes of components which failed first;
   ■ date, time, place and load when the failure occurred;
   ■ dates of installation, commissioning, overhauls;
   ■ dates of previous maintenance schedules and repairs;
   ■ name of operator at the time of failure;
   ■ and every other such detail that can be collected with reference to the incident and the component/equipment involved.

Often, such tabulation will throw up common factors that highlight the correct solution regarding the mechanism and cause of failure. In some cases, this is the only method which works in a short time.

(d) Examine the failed components and failed equipment, in spite of considerable damage. Reason out which component failed first, differentiate between cause and effect. If there are several possible components which could possibly be considered for the first place, keep them all under consideration for alternative theories until some new fact helps to eliminate or confirm some of them.

> Although, this item appears as the fourth in this list of steps to be taken, it is probably the most important one. Very often the failure being investigated is the first to have occurred in a long time. There would be no statistics and no previous experience. Careful examination of the failed component, the failed equipment and the surroundings is very important.

(e) Maintain notes on observations, summaries of data, and possible causes of failures.

(f) Make sketches of failed components; or take photographs. Retain failed components in special plastic bags and refer to them until the correct solution dawns.

(g) Examine similar components and equipment

- In service in situ;
- While under maintenance or overhaul;
- If possible, while under manufacture and assembly.

Look for incipient signs of various possible or probable modes and mechanisms of failures, especially when the equipment is badly damaged.

Even when it is not damaged, incipient signs and possible causes of failures may be revealed.

(h) Make a thorough study of previous investigations before reaching a conclusion but do not accept the conclusions and recommendations as yet; consider them as possible solutions to be either confirmed or rejected on the basis of further investigation. Keep an open mind until all data are collected and then accept only those conclusions that fit in with all the available data. Be prepared to accept two or more failure mechanisms operating side by side due to different defects or causes.

(i) Discuss the failures with others in the field who may have knowledge or experience of the subject under consideration. Surf the Internet, access libraries and look up the relevant books and magazines; collect the available data find a theory which fits all the facts.

(j) Compare the experience of other installations where similar or identical equipment is in service. Determine the differences in design, environment, operation, maintenance practice etc.

(k) Consider the observations made during routine maintenance and inspections. Focus completely on probable causes of failures.

(l) Study the operating conditions during actual service. Look for unusual features or conditions, which do not conform to the manufacturer's operating instructions.

(m) By now you would have come to tentative conclusions regarding the possible mechanisms of failures and the possible causes. Verify these hypotheses on equipment in service, make the required measurements. If the hypotheses are confirmed, decide on the modifications necessary to avert the failures.

(n) Every defect is a potential failure and every seed-defect is a potential defect. Every failure starts as a seed-defect which grows into a defect and then into a failure. It is desirable to monitor the incidence and analyse the statistics of defects that are detected during routine maintenance and inspections of the components under consideration. The following questions are on each type of defect that is related to the failure under consideration.

　(i) At what stage was the defect introduced?—During design, manufacture, or maintenance?

(ii) Which is the earliest stage at which the defect can be detected?

(iii) Is any improvement possible in the method of detecting such a defect?

(iv) Is it necessary to improve the design or process of manufacture to minimise such defects?

(v) Is the required know-how and know-why being imparted to the concerned staff in a systematic training schedule?

(vi) Does the relevant documentation (drawings, specifications, process sheets, training material, quality assurance schedules etc.) cover the important aspects relating to the prevention of such defects? In particular, does it cover the preventive action for the defect under consideration?

4.6 Review of Failure Statistics

It must be ensured, while making any comparisons, that the failure statistics are considered for the same period. Further, if there are any differences in the conditions of service such as loading, supply voltage, environment, etc. the failure rates will not be strictly comparable. There may be differences in failure modes. Despite such limitations, it is desirable to maintain statistics of failure rates of different types and makes of equipment and components. When wide variations are noticed between different groups, it is necessary to determine the reasons for the same. The reasons could be in the specification, design, manufacture, operation, maintenance, environment and service conditions. Such comparisons will help to determine possible ways for reducing failure rates.

If the hours worked per year of different equipment are widely different, failure rates should be calculated on a per-hour or per-thousand-hour basis instead of per-year basis.

4.7 Investigation of Fires and Serious Accidents

Fires and accidents can occur due to a variety of reasons such as arson, sabotage, negligence by users or public, natural phenomena and such other factors not within the control of engineers in charge. All these are not within the scope of this book. But it is also a fact that fires and accidents can occur due to failures of electrical and mechanical equipment. For instance, defects in protection systems or in connections can lead to electrical fires. Defects in rolling stock or in railway tracks can lead to derailments. The methods of investigation in all such cases are exactly the same as for any other failures of equipment. The magnitude of the effects makes it more difficult to determine the true cause but it has no bearing on the cause itself. The magnitude of the effects such as loss of property, loss of lives etc. is a matter of chance depending only on the place and time of the failure resulting in the fire or accident, speed and effectiveness of the rescue arrangements, etc.

Defects and failures, which result in fires, are discussed fully in my book *Electrical Fires and Failures, A Prevention and Troubleshooting Guide*. A few of the case studies in this book also cover very briefly electrical fires and accidents but the majority are concerned with equipment failures which affect only plant availability, industrial production and public services.

4.8 Action Plan

Any investigation has to culminate in an action plan—a plan which, when implemented, completely stops the incidence of failures. Implementation may have to be in any one or more of the following areas:

(a) Specification
(b) Design
(c) Manufacture
(d) Installation/Commissioning
(e) Operation
(f) Maintenance

If action needs to be taken in either operation or maintenance, the action plan can be implemented easily. If, however, action is needed in specification, design, manufacture or installation/commissioning the plan is more difficult to accomplish. It may even be impracticable and irrelevant as far as the equipment already installed is concerned. In such cases, it is necessary to devise a separate action plan for the equipment already installed, to carry out judicious modifications.

It is obvious that investigation and even the determination of the preventive action plan will have no effect on the performance in the field until the action plan is implemented. Therefore the action plans must be practicable, economical and effective. Fortunately in most cases it is possible to achieve this objective. Otherwise, the whole investigation exercise would become a futile effort.

A very important step to be taken by the investigator is to monitor the performance of the equipment after implementing the action plan to verify that the types of failures that were occurring earlier have stopped completely.

The last, but not the least, important step that should be taken is to circulate a brief report on the failure statistics, the failure mode, the failure mechanism and the preventive action taken. The reduction achieved in the failure rate should be indicated. This failure investigation report should be circulated to all other maintenance engineers in the organisation.

4.9 Conclusion

The methods described above have been tested and proven as highly effective over many years. They will enable any individual or organisation to get to grips with the problem of failures, fires and accidents in electrical and mechanical hardware in industry, railways, roadways, airlines, shipping, etc. A beginning can be made any time. It is not necessary to wait for a devastating fire or accident, fall in production or public outcry before making a start on the road to Zero Failure. In fact, such occasions will not arise if pre-emptive or pro-active steps are taken immediately—the right time for which is when everything seems to be apparently under control.

Chapter 5

DEGRADATION PROCESSES (MECHANICAL)

> ### In this chapter we learn
>
> - *While the numbers of types of electrical and mechanical machines run into millions, the numbers of failure mechanisms and degradation processes are very small.*
> - *Ten such processes (listed in Section 5.1) that cause failures in the mechanical mode are described.*
> - *With regard to each of these ten processes, the physical basis and the main features are discussed; followed by a few examples.*
> - *Stress concentration, i.e. stress augmentation as a result of changes in cross-section, nicks, surface roughness, etc.* ❑ ❑

5.1 Introduction

$\mathcal{A}$s mentioned previously in Section 1.2, Chapter 1, there are millions of types of machines and equipment but the number of *types of machine elements* that go into them are less than one hundred. There are just a handful of physical and chemical phenomena that take part in failure mechanisms. The mechanical modes are reproduced again below:

Thermal expansion	Friction
Vibration	Brittle fracture
Metal fatigue	Shrinkage
Metal creep	Yield deformation
Corrosion	Erosion or wear

More than 95 per cent of all mechanical failures that take place in mechanical and electrical equipment are covered by one or more of the degradation processes listed above. They will be referred to in Chapters 7 to 66. While Mechanical phenomena are discussed in this chapter, Electrical phenomena will be dealt with in Chapter 6. The manner in which they lead to equipment failures in general will also be explained briefly.

Sometimes more than one of these phenomena are involved in the failure mechanism. They are dealt with in a little greater detail in my book *Electrical Fires and Failures (Tata McGraw-Hill, 1998)*. Many excellent books cover each of these topics and the readers are advised to refer to them. The busy maintenance engineer needs to study only the details given here to tackle prevention of failures in the equipment.

5.2 Thermal Expansion

When the temperature of a metal component increases, its dimensions increase in proportion to the rise in temperature. The increase is very small and not visible to the naked eye but it is sufficient to cause failures under certain conditions. Here are a few examples taken from actual case histories:

CASE STUDY

- The armature shaft of an electric motor runs at a temperature that may be 30°C to 60°C higher than the temperature of the motor casing. This difference causes the length of the armature shaft to increase in relation to the motor casing. In one case, the axial clearances in the roller bearings at the two ends of the shaft did not allow for this relative expansion, causing one of the two bearings to get overheated and fail.

- The contact leaf of a small electromagnetic relay got heated due to the passage of current through it. Its length increased in relation to that of the plastic enclosure. In some relays, the clearance between the end of the leaf and the plastic enclosure was inadequate. As a result, the end of the leaf touched the casing and the relay got stuck in the closed position thereby causing the system to fail.

- The length of a transmission line increases when it gets heated by the passage of current through it and also by exposure to the sun. This causes the sag of the line to increase. In one case the sag was not correctly adjusted with reference to the line temperature at the time of installation, the line sagged excessively and caused a flashover to an earthed structure under it. In another case, in which the sag was too small at the time of installation, its contraction in the coldest part of the night increased the line tension appreciably to cause the line to part.

Thermal expansion and contraction has its uses in many engineering applications. Very high forces are developed when thermal expansion or contraction is prevented. This phenomenon is used for fitting railway tyres on wheels, bearing races on shafts, pinions on shafts and even bullock cart tyres on wooden wheels. The forces are so high that steel tyres and bearing races can fracture if the fitting is not done correctly. Many other types of equipment failures that occur as a result of these forces are described below.

CASE STUDY

- Railway tracks are usually provided with small gaps to allow for expansion. There have been cases where the track, despite weighing hundreds of tonnes, buckled laterally due to these forces as the gaps provided were not adequate.
- Copper bus-bars, which are provided in dc series motors for interconnections between the field coils, operate at temperatures that are about 100 °C higher than the carcasse of the motor to which the bus-bars are fastened. In some cases these bus-bars fractured as a result of repeated heating and cooling and consequent alternations of compressive and tensile forces. There was also another phenomenon called metal fatigue, at work in these cases. This is dealt with in another section of this chapter.
- Thermal expansion of squirrel cage rings in induction motors causes severe bending stresses in the bars. Mainly due to design errors, these forces exceed the fatigue endurance limits of copper. Bar fractures are a chronic problem of badly designed induction motors.

5.3 Vibration

Vibration of equipment and components is often unavoidable in certain applications such as railway locomotives, coaches and wagons. Certain stationary machines, e.g. power looms, stone crushers, power hammers etc. are subject to continuous and heavy vibration. In all such machines, failures of components take place if there are lacunae in design or in maintenance.

Vibration of a component as a whole is usually harmless. It is when the component is not well secured and one part of the component is secured differently from another, that flexing and bending stresses within the component are produced and they lead to failures through metal fatigue. Another mechanism of failure involves relative movement and rubbing between two or more components.

Sometimes, the amplitude of vibration increases greatly due to resonance between the natural frequency and the forcing frequency. This hastens and increases the intensity of the degradation processes. A few typical examples are given below.

(a) A very common type of failure in electric rolling stock on the railways is due to the flexing of cables that are not secured evenly along their entire length. Repetitive bending leads to fractures of copper wires, bus-bars and terminal fittings like sockets.

(b) The field coils of nose suspended traction motors are normally supported very firmly over their cores with the help of heavy springs or adhesive compounds. Many coil insulation failures are due to design or manufacturing deficiencies in the coil support system, which allow relative movement between the coils and the cores, rubbing and erosion of the insulation and eventual electrical failure.

(c) Relative movement of coils, damage to insulation and electrical short-circuits often cause transformer failures. The movement is due to the effect of vibration and defects in the coil support system.

5.4 Metal Fatigue

Metal fatigue is probably the most common of all failure mechanisms in both electrical and mechanical equipment. The final result is a fracture of the metal component. If the component that fractures is an electrical conductor, there is arcing and melting of the conductors and probably also a fire. The more important features or characteristics of metal fatigue and fractures are enumerated as follows.

(a) Fatigue occurs only when the stresses in the metal component are either alternating or fluctuating. There is no fatigue if the stress is steady and constant in regard to its direction and magnitude. See Figure 18.2.

(b) In the case of ferrous materials, fatigue damage occurs only when the peak alternating stress exceeds a certain limit known as the endurance limit of the metal. If the peak stress is kept below this limit, there is no fatigue. The endurance limit is generally about 40 per cent of the ultimate tensile strength of the material. See Figure 18.2.

(c) In the case of non-ferrous metals, there is no such endurance limit. However, the number of permissible alternations without causing fatigue cracks, usually increases by a factor of 10 for every 10 per cent reduction in the peak stress. The endurance limit can be taken as the stress permissible for 10^9 alternations.

(d) The stress to be considered in fatigue calculations is the peak stress multiplied by the Stress Concentration Factor (SCF). The SCF depends on the shape of the component, presence of surface cracks and flaws. The SCF can be in the range 1.5 to 20. Apparently minor changes in the shape can lead to an abnormal increase in the SCF. Many fatigue failures in practice are due to this little known characteristic of fatigue. See Section 5.12.

(e) Fatigue damage is cumulative and also irreversible. If a component has suffered fatigue damage no process of annealing or heat treatment can reclaim it. The only way is to replace the component. In very large components such as in bridge girders or locomotive bogies the fatigued zone, with a little safety margin on either side, can be cut out altogether and a patch of new metal can be welded in. Obviously the best way is to prevent fatigue damage from occurring in the first place.

(f) Fatigue fractures have a distinctive appearance. Part of the fractured surface shows a smooth aspect with tide marks. This represents the propagation of the fatigue crack starting from a very small area. The rest of the fractured surface shows a fibrous or rough appearance that depicts the sudden fracture under normal operating forces acting on a weakened component.

(g) Metal fatigue is a failure mechanism. It is not a cause of failure. Metal fatigue is quite unlike human fatigue. Rest cannot erase the effects of metal fatigue. Metal fatigue is not inevitable. It can be totally eliminated by keeping the stress below the endurance limit. Proper design, manufacture and maintenance can prevent fatigue fractures.

Three cases of failures in which metal fatigue played a prominent part are briefly described below:

CASE STUDY

- The shafts of one particular type of traction motor were subjected to a high rate of fractures in service. Investigation revealed three lacunae in the design: (a) stress concentration at a keyway, (b) inadequate size of the shaft and (c) inadequate endurance limit of the material of the shaft. Improvements were effected in these three features and all the shafts were replaced. During the interregnum ultra-sonic tests were carried out to detect shafts with incipient fatigue cracks and to replace them on priority. Shaft failures in service were thereafter very rare.

- Please refer to the examples (e) and (f) in Section 5.2. While the stresses were produced by thermal expansion, the failure mechanism was due to metal fatigue.

- Heavy metal bolts and nuts are used to fix the drive brackets to the wheel centres of one particular class of heavy freight locomotives. These bolts are subject to very high fluctuating stresses while transmitting the tractive efforts to the wheels. They were prone to fatigue fracture under the bolt head. Investigation showed excessive stress concentration at the points of fracture. Increase in the fillet radius and improvement in the surface finish at the critical locations led to the elimination of this type of failure.

- Certain bus-bars in transformers were bent sharply at right angles. They developed hairline cracks at the bends, which grew into fractures and transformer failures. The failure mechanism was fatigue but the root cause was the sharp bend.

5.5 Metal Creep

Metal components under tension elongate and those under compression contract. This change in linear dimensions is not visible to the naked eye, as it is usually less than 0.1 per cent. However this phenomenon is absolutely reliable and precise. The exact magnitude can be calculated using Hooke's Law and Young's Modulus as long as the stress is less than a certain limit, known as the elastic limit. The Young's Modulus as also the elastic limit differ from metal to metal, but their values for any given composition of the metal, pure or alloyed, remain constant.

The phenomenon referred to in the preceding paragraph is reversible and the dimensions return to their original values as soon as the stress is removed. This holds good even after long periods of time. This property is called elasticity. There is however a proviso to this rule. It is valid only if the temperature of the component is less than a certain temperature called the Creep Temperature Limit (CTL). This CTL is also a property of each metal. It is approximately one-third of the melting point of the metal when the temperatures are stated in the absolute scale.

The behaviour of the metal at temperatures above the CTL is different. Hooke's Law still applies but there is an additional factor. There is an additional elongation or contraction, which is dependent on the magnitude of the stress, the time for which the stress is applied and the temperature of the component. This additional change in dimension is not reversible. When the stress is removed, the part of the change in accordance with Hooke's Law disappears but the additional change remains as a permanent deformation. This phenomenon is known as Metal Creep.

The CTLs for a few commonly used metals are given below.

Iron and steel	270 °C
Copper	135 °C
Aluminium	7 °C
Lead	–70 °C

From the CTLs given above it will be clear that metal creep is not relevant for ordinary applications in the case of iron and steel, but they are very important when used for high temperature equipment such as power station boilers, receivers, pipes etc. Similarly, copper is vulnerable at temperatures above 135 °C—a temperature that is often attained in modern electrical machines. Aluminium and particularly Lead can and do create problems even at room temperatures, due to the effects of metal creep.

The phenomenon of metal creep leads to equipment failures mainly through the loss of tension in threaded fasteners. Here are a few examples:

CASE STUDY

- Aluminium wires that were held with threaded fasteners developed bad contacts, overheating and burning due to relaxation of contact force as a result of metal creep.
- Soldered joints under mechanical stress developed cracks and fractures due to metal creep in the solder alloy. Overheating, burning, open circuits, etc. followed.

- High pressure steam receivers and boilers fitted with removable covers were held in place with threaded fasteners and gaskets. As the fasteners were not re-tightened periodically to compensate for metal creep, gasket failures and leakage of steam led to accidents.

5.6 Corrosion

Corrosion is one of the few visible types of degradation. It can reduce the cross-section of the component, increase mechanical stress and lead, eventually, to mechanical failure. This is obvious and does not call for any further explanation. There are, however, some additional effects that are not so obvious.

Corrosion of the surface of a highly stressed component subject to alternating stresses is further affected due to stress concentration at the corroded surface. This effect is also more deadly than the reduction in cross-section due to corrosion. This is one of the reasons why highly stressed components are often given very highly polished finishes to begin with and also during maintenance.

Corrosion acts in other ways too. Oxide scales pass through compressed air into precision equipment and cause malfunctions. The increase in volume during rusting develops extremely high forces, which are sufficient to crack open massive concrete girders. A few examples of equipment failures due to corrosion are given below:

CASE STUDY

- Steel pins of ceramic insulators corroded and fractured under normal operating tension, cause serious dislocations in electric supply.
- Bearing races and rollers corroded due to ingress of water and eventually the bearings seized.
- Defective galvanising of steel wire ropes led to corrosion and failure of the ropes in service under normal tension.

5.7 Friction

Friction increases the resistance to motion and also the heat developed at the point of contact. It can be reduced by improving the smoothness of the surfaces and by providing lubrication at the points of contact. Friction is also useful in certain places e.g. at the brake shoes. A few examples are as follows.

CASE STUDY

- A great deal of heat is developed at the brake shoes of electric locomotives and motor coaches. The surface of the tyre is in any case subjected to very high stresses due to its shrink fit on the wheel and also due to point or line contact at the rail. The alternate heating and cooling of the tyre surface due to passage under the brake shoes created additional thermal stresses. The result of all this was, under certain local conditions, a thermal crack that grew into a tyre fracture.

- Due to wear and tear and defective manufacture, surfaces that normally had a clearance between them actually touched in certain equipment with moving parts. The resulting friction caused malfunction and failures.

- In most cases of bearing failures due to lack of lubrication, excessive friction leads to overheating and eventual failure of the machine.

5.8 Brittle Fracture

We have already discussed the phenomenon of metal fatigue, the end result of which is usually a fracture. There are some cases of fracture that do not involve metal fatigue and these will be discussed in this section.

Ceramics and hardened/tempered metal components sometimes develop brittle fracture. The two-tone appearance of fatigue fractures is absent. The entire fractured surface shows an even and bright aspect. This is a sudden fracture that can occur even on new components as a result of shock or impact loading. In such cases metallurgical examination of the fracture can indicate the presence of flaws or internal cracks. Such cases may call for the assistance of experienced metallurgists and experts in fracture mechanics. The average maintenance engineer should work along these lines:

> To have the material of the component thoroughly tested and the result compared with the specification
> To have the fracture aspect examined by a metallurgist for flaws and inclusions
> To look for possible increases in shock loading

Fractures of this type are rare. A few examples are given below:

CASE STUDY

- Solid core ceramic insulators failed suddenly. The fracture aspect usually showed inclusions of small chips. This was a manufacturing defect.

- Steel rails in sections where temperatures were very low, fractured suddenly partly due to very high thermal stresses caused by contraction, (See Section 5.2) and partly due to reduced impact strength of steel at low temperatures.

5.9 Shrinkage

When the molten metal in the mould for a casting solidifies it shrinks. To allow for this shrinkage the mould is designed to ensure that the solidification takes place gradually from the outer surfaces towards the centre, so that the space created by the shrinkage is continuously fed from reservoirs of molten metal in the feeders and risers. If there are any lacunae or deficiencies in this regard, blowholes and cavities develop in the finished casting. These defects may then develop into fractures or cracks during service.

Even after complete solidification of the casting it continues to contract as the temperature of the casting drops to room temperature (see Section 5.2). If this contraction is not allowed to take place freely due to the presence of cores of excessive strength, the casting may develop what are known as hot-tears. These defects may then cause cracks and fractures in service.

Insulating materials are usually in the form of laminates of paper, glass cloth, or some other fabric bonded together by plastics. They are all subject to slow shrinkage when under mechanical pressure. This phenomenon can lead to equipment failures as in the following examples:

CASE STUDY

- In certain power transformers this shrinkage, in the absence of suitable devices and design features for compensation, led to displacement of spacers and coils and to eventual failure of the transformer.
- In some terminal boards, laminated insulating boards were interposed in the contact force circuits. Their shrinkage led to bad contact, overheating, burning and fires.

5.10 Yield Deformation

If the operating stress in a metal component exceeds the yield point of the metal, the component will deform, and then all kinds of defects can occur depending upon the design of the equipment. Clearances can get reduced; unintended contact between components can take place; the equipment can fail.

Calculation of the stresses and the strengths of the components show the direction in which further steps have to be taken. There can be no general formula in this regard. In some cases the strength may have to be increased, and in some, the stresses may have to be reduced. It may be possible to achieve the desired result by providing additional supports. A few examples may be given here:

CASE STUDY

- In one case short circuits on electrical connections were eliminated by providing additional stays to minimise lateral deformation.
- In another case, push rods that kept the contact wire in place on curved track buckled due to excessive stress. The modulus of section and thickness of the material was increased to improve the stiffness of the push rods.

5.11   Erosion or Wear

Many components in electrical and mechanical equipment are subject to wear during normal operation. Examples of this are bearing liners, slide surfaces, arcing surfaces in switch-gear, plugs and sockets, gears, threads, etc. Normal design usually allows for a certain amount of wear and as long as it does not go beyond specified limits there is no mal-function or failure.

Maintenance engineers should consider abnormal rate of wear as a defect to be investigated, and eliminated. For this purpose it is necessary first to determine all the wearing points, to stipulate the normal intervals for replacements of worn components and then to monitor the replacements vis-à-vis the schedules.

5.12   Stress Concentration

The term 'Stress Concentration' will appear repeatedly in the chapters that follow. It is desirable to understand its significance fully. The theoretical basis is difficult to understand but its application and conclusions are quite simple. A simplified explanation of the phenomenon is given below for front line supervisors and engineers.

Consider a stepped shaft as shown in Figure 5.1 subjected to an alternating tensile/compressive force $+/-F$ expressed in kgf. The cross-section of the larger portion of the shaft is 'A' mm^2, and that of the smaller portion is 'a' mm^2.

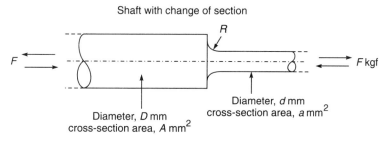

Figure 5.1

The peak stress (tensile or compressive) in the shaft would be

F/A kgf/mm^2: in the larger section, and;

F/a kgf/mm^2: in the smaller section.

Obviously, the stress in the smaller section would be higher. Thus far, the calculation is straightforward and simple. The question that arises now is—what would be the stress at the point of transition, i.e. where the section changes from A to a? The answer is surprising and unexpected. The stress at the change of section will be very much higher than even the higher stress in the smaller section. Advanced mathematical analysis shows that the stress distribution, as a function of the position along the shaft, would be generally as shown in Figure. 5.2.

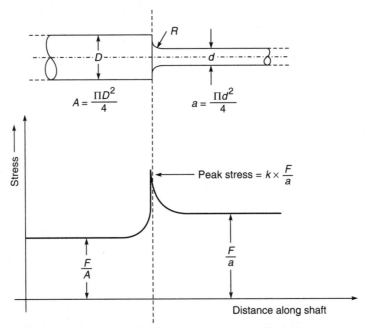

k is the stress concentration factor. It depends on d/D and R.

Figure 5.2 Stress concentration factor in a shaft with charge of section

The peak stress is $K \times (F/a)$, where K could be anywhere between 1 and 10 or even more. **This factor K is known as the stress concentration factor.**

Modern machine shafts have many changes in section, keyways, grooves, etc. all of which produce stress concentration at concave corners such as those shown in Figure 9.1, Chapter 9. At such places the actual stress is many times higher than the stress calculated by simple bending or torsion theory. The factor by which this increase takes place is called the stress concentration factor.

The actual values of K have been determined by experts for all kinds of configurations and put into the form of graphs or tables published in Mechanical Engineering Handbooks.

It is *not* necessary for front line engineers to understand the full theoretical basis for these graphs. It is enough to appreciate the conditions in which these stress concentration factors must be taken into account and to be able to use the published graphs. The calculation of these factors is very complex and difficult. Practising engineers use reference books where these factors are given in the form of graphs and tables. For our purpose here it is necessary only to understand the manner in which these factors are influenced by the shape of the shaft.

The stress concentration factor at a change in section or diameter of the shaft depends on the ratio of the two diameters, d/D and on the radius of curvature R of the fillet between the two diameters. See Figure 5.2.

A typical graph, which shows the relationship between these parameters, is given in Figure 5.3.

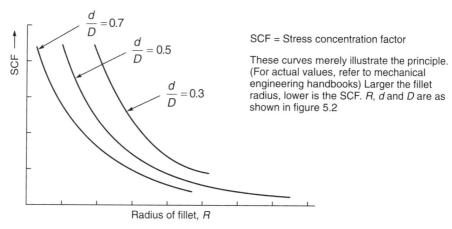

Figure 5.3 Graph showing stress concentration factors at changes in section

The more critical parameter is, obviously, the fillet radius R. It is easy for a machinist to make a mistake in this regard. Such a mistake is unlikely to get detected during assembly, unless the supervisor or quality assurance inspector is specifically looking for it. However, the effects of such a mistake could be very serious. For instance, if the designer stipulated a fillet radius of 3 mm and the machinist made it 2 mm, the stress would increase by more than 100 per cent, which is sufficient to cause a failure sooner or later.

Similar problems can arise in keyways, and grooves as shown in Figure 9.1. The radii of curvature at the corners are all-important. If they are reduced during machining from the values indicated by the designer, the actual stress will increase and this may lead to failure.

Nicks, tool marks and even very shallow grooves also increase the stress very significantly. See Figure 5.4. There is no simple calculation to measure or describe such imperfections, but it can be asserted that their stress concentration factors can be very high, like 2, 3, 5 or even more. Such blemishes must never be allowed to appear on highly stressed shafts either during manufacture, assembly, storage, handling or service.

Welding, brazing, arcing or even weld splatter produces stress concentration of the same type. Fractures have been known to originate from such points.

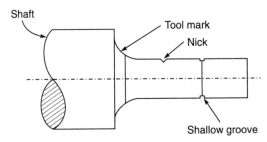

Figure 5.4 Nicks, tool marks and shallow grooves

Surface finish is also important in very highly stressed components. If the designer has assumed surface finish of grade 1.6 microns in a highly stressed area, machining the surface with a rougher finish of say 15-micron grade would increase the probability of fracture through fatigue.

It follows from the foregoing explanations that:

- The designer must, as far as possible avoid sharp changes in section, keyways, grooves etc.
- Where such changes in section, keyways, grooves etc. are unavoidable, the designer must take them into account while calculating the peak stresses and ensure that these increased stresses are well within the strength of the material.
- The manufacturer must ensure that fillet radii, surface finishes and other dimensions of highly stressed components are machined accurately as stipulated in the drawings given by the design office.
- The manufacturer and also the maintenance engineers must ensure that tool marks, nicks, etc. are not allowed to appear on the highly stressed metal surfaces.
- When fractures do occur, the investigator should look for features or defects that could have led to high stress concentration.

5.13 Failures Due to Increase in Stress Concentration Factors

The phenomenon of stress concentration is always present in any component, which is not of a uniform section. Components of uniform section are extremely rare. A transmission line is probably the only common example of a component, which has a uniform section, but even this item has some stress concentration at terminations and supports.

This term, i.e. stress concentration will appear again and again during discussions of failures through fracture of highly stressed components. There are several ways in which this phenomenon will make its presence felt.

- If the designer does not allow for stress concentration or if he makes inadequate provision for it;
- If the manufacturer does not follow critical features of the design such as fillet radius at change of section or surface finish in highly stressed zones or;
- If cracks, nicks, poor surface finish are allowed to appear on highly stressed components due to operational or maintenance errors.

If there is a design error, or if there is a systematic manufacturing defect, failure rates are likely to be high because all the components in service would be vulnerable. If there are stray defects in manufacture or maintenance, the failure rates will not be too high.

Increased stress concentration is not, strictly speaking, a degradation process in the strength of the material. However, if the operating or maintenance conditions result in the appearance of factors that increase the stress concentration factors, there would be effectively, an increase in the stresses and, consequently, a degradation in the margin of safety in comparison with the strength of the components.

Chapter 6

DEGRADATION PROCESSES (ELECTRICAL)

In this chapter we learn

- *While the numbers of types of electrical and mechanical machines run into millions, the number of failure mechanisms and degradation processes are very small.*
- *Eight such processes (listed in Section 6.1) that cause failures in the electrical mode are described here.*
- *In regard to each of these eight processes the physical basis and the main features will be discussed first. This will be followed by a few examples.* ❏ ❏

6.1 Introduction

*I*n Section 1.4, Chapter 1, 18 degradation processes are listed of which ten are mechanical and eight electrical. The ten mechanical processes have been discussed in Chapter 5. The remaining eight electrical degradation processes will be discussed in this chapter. For ready reference, they are listed again here.

Electrical heating	Contact resistance
Partial discharge	Dielectric breakdown
Electrical tracking	Magnetization
Thermal degradation	Electromagnetic induction

One or more of the eight degradation processes listed above cover 95 per cent of all failures that take place in electrical modes. While they will be referred to at the appropriate places in Chapters 7 to 65 some of the more important properties or features of these phenomena will be explained in this chapter. The manner in which they lead to equipment failures in general will also be explained briefly. They are dealt with in a little greater detail in my book *"Electrical Fires and Failures"*.

For the maintenance engineer whose immediate concern is the prevention of failures in the equipment in his charge, the details given here would be adequate for his purpose.

6.2   Electrical Heating

Electrical heating is a process in which electrical energy is converted into heat energy. Electrical heating occurs in three principal ways in electrical machines and equipment. These are:

(a) Passage of electric current through resistance. (Commonly referred to as I square R, or I^2R, losses.)

(b) Passage of electric current through semi-conductors.

(c) Hysteresis and eddy currents in metals placed in fluctuating or alternating magnetic fields.

Electrical heating is not by itself a degradation process. If the heat is carried away and dissipated outside the equipment there is no problem. Natural or forced cooling can limit the temperature rise but a certain rise is inescapable. What causes degradation of materials and consequent failures of equipment is this unavoidable temperature rise of the components. The designers' effort is to limit this rise to safe limits.

- High temperatures adversely affect the insulating properties of most insulating materials. This thermal degradation is discussed further in Section 6.8.
- As already seen in Section 5.5, Chapter 5 metals are subject to creep at temperatures higher than their creep temperature limits.
- Metals oxidize faster at higher temperatures.
- High temperatures affect the mechanical properties of many non-metallic materials like lubricants, plastics and elastomers.
- The life of a primary or secondary cell is reduced by a rise in temperature.

All the five effects mentioned above have an active role in all failure mechanisms in the electrical mode. The rates of deterioration are exponentially dependent on the temperature. For instance, for every 8 °C rise in temperature, the deterioration rate of insulating materials doubles.

CASE STUDY

A few examples of failures of electrical equipment due to the effects of rise in temperature are given below:

- There was a defect in the cooling system for a traction motor as a result of which the motor temperature rose to about 30 °C above the design limit. There was premature failure of insulation.
- The end shield of a motor got over heated due to exposure to radiant heat from a furnace. There was premature failure of the roller bearing in that end shield.
- The base of a high current diode got over heated due to defective thermal contact with its heat sink. The diode failed in service when delivering the full load current due to its overheating.

Any deficiency in the cooling system of an electrical machine reduces the durability of its insulating materials. If the deficiency is excessive, early failure of the machine is certain.

6.3 Contact Resistance

In almost all electrical machines and equipment there are a number of points at which electric current is passed from one conductor to another through simple face to face contact. A few typical examples are as follows:

- contact between cable socket and terminal;
- contact between cable and its socket;
- contact between two bus bars;
- contact between a diode and its heat sink;
- contact between a pin and a socket;
- contact between the contact tips of a relay or a contactor.

The total number of such connections or contacts in any installation is in the wide range of a few tens to hundreds of thousands depending upon the size of the installation. A very large proportion of all electrical failures originates in these contacts. In every such case the failure mechanism is the vicious cycle shown in Figure 6.1, which begins with the loss of contact force and culminates in overheating and fire.

CASE STUDY

There are a very large number of cases of this type amongst the electrical component and equipment failures described in the following chapters. A few examples may be mentioned in this context here:

- The contact force between the socket and the cable was inadequate due to defective crimping. The socket overheated and finally melted. The arcing that followed started a fire.
- The contact force between a pin and its socket in a multi-core coupler was inadequate. This increased the local heating and started the failure mechanism shown in Figure 6.1. The final result was a failure and a fire.
- The force between the contacts of a relay was too small due to a manufacturing defect. Local overheating caused the contacts to get welded together and the relay failed.

6.4 Partial Discharge

When the insulating material between two conductors at a high voltage difference is flawless the voltage gradient inside the insulating material is uniform and at its minimum possible value.

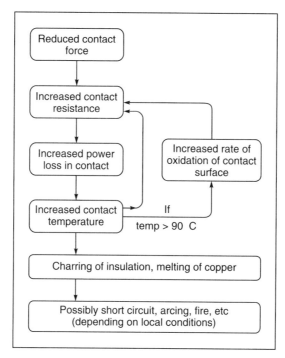

Figure 6.1 Vicious cycle that commences with inadequate contact force and ends in a fire

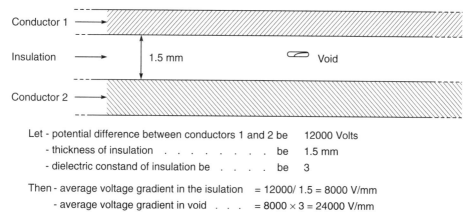

Figure 6.2 Shows how voltage gradient increases and causes partial discharges in voids

However, if there are voids in the insulating material as shown in Figure 6.2, the voltage gradient inside these voids is several times higher than the average voltage gradient. This increased voltage gradient may be higher than the dielectric strength inside the void as a result of which, there would be a momentary short-circuit and a small discharge (a few pico-coulombs i.e. 10^{-12} coulombs) within the void. Since the rest of the insulating material is operating at a voltage gradient that is less than its strength, there is no general breakdown of insulation. There is only a small or partial discharge within the void after which the voltage gradient in the void again rises to its original high value. This intermittent process continues and partial discharges continue to take place within the voids. The material around the void gets damaged progressively, along a line joining the two conductors separated by the insulation. The voltage gradient in the good portion of the insulating material keeps rising as the good length keeps becoming shorter until the entire insulation breaks down. This process is known as degradation and failure of insulation due to partial discharges.

Measurement of partial discharges requires precision instrumentation. These instruments are very expensive but well worth their cost for the prevention of failures in service. With the help of these instruments it is possible to monitor the condition of the insulation and to make pre-emptive or precautionary replacements of vulnerable components. They are useful also in the inspection and testing of new equipment.

CASE STUDY

- Certain instrument transformers rated at 25000 volts on the primary side were prone to failures. Each failure of this type involved consequential losses in addition to the costs of replacements. The high voltage insulation of these transformers consisted of moulded epoxy compounds. Partial Discharge (PD) measurements on all the transformers showed up some in which the PDs were abnormally high in comparison with the rest. They were removed from service and the rest continued in service without failure for many years.

Corona discharge is a form of electrical failure in gases. In some ways it is similar to partial discharges in solid insulating materials. There are many differences too. The similarities are:

- Corona discharge is a precursor of total breakdown.
- Corona discharge takes place in zones where the voltage gradient is higher.
- Corona discharge current is very small in relation to the final breakdown current.

The differences are:

- Corona discharge takes place in the surrounding air and it causes no permanent damage. When the voltage is removed, the air becomes normal again. Partial discharge causes permanent damage.
- Corona discharge is visible and audible. Partial discharge is not.
- The quantity of electricity that flows in corona discharges in gases is very high in comparison with that flowing in partial discharges. The latter are of the order of a few pico-coulombs (10^{-12} coulombs) in solid insulating materials.

By watching for corona and taking immediate corrective action, flashover on high voltage insulators can be averted.

CASE STUDY

- In one traction system, some of the 25000 volt line insulators in a major yard were prone to flashover, particularly in the pre-dawn hours of winter mornings when there were deposits of dew on the insulators and fog around them. This caused repeated tripping of the sub-station circuit breakers and delays to traffic. Inspection of the insulators at midnight showed that some insulators had visible and audible corona around their end caps. These locations were marked and next day, the 'lively' insulators were replaced with new ones that had been thoroughly cleaned. This process was continued for a week or so until all such insulators were dealt with. There were no further cases of insulator flashover and substation tripping in pre-dawn hours.

6.5 Dielectric Breakdown

In the preceding section, we have discussed one form of degradation of solid insulating materials as a result of partial discharges. In this form, the insulation gets progressively damaged, starting from a void or flaw and then going across the full thickness. There is another and equally important degradation mechanism, in which the degradation takes place throughout the insulation due to temperature. We had briefly referred to this phenomenon in Section 6.2. It is important because all electrical machines get heated up during operation. Different insulating materials have different rates of degradation.

In fact, insulating materials are classified and specified according to this property. They are assigned what are known as index temperatures. The index temperature is defined as one at which the insulating material will have a life of 20000 hours when tested under certain standardised test procedures. The maximum permissible temperature rise during full load operation is usually specified by the purchaser, in relation to the index temperature depending upon the operating conditions and expectations of working life from the machines.

It is the normal expectation that insulation breakdown will not occur during the desired life expectation. In actual practice premature breakdown will occur if one of the following things happen:

- There is a defect in the cooling system causing the operating temperature to rise (beyond the designed level).
- There is a local defect causing local overheating of the insulation. It should be remembered that dielectric breakdown will occur if there is degradation in one small zone. It is not necessary for the entire insulation to be damaged.

- There is local mechanical damage to insulation due to relative movement between the insulated conductor and the surrounding metal.
- There is ingress of water or other fluid that can damage the insulation.
- There is an abnormal rise in the applied voltage due to a defect in some other associated equipment.

CASE STUDY

A few examples of dielectric breakdown may be mentioned here:

- The insulation between the main armature winding and the core of a traction motor failed because of over heating caused by reduction in cooling air as a result of torn bellows between the motors and the locomotive body.
- The insulation on the copper conductors in a transformer failed due to relative movement between adjacent conductors caused by failure of the coil support system.
- The enamel insulation on copper wires in mush wound induction motors failed due to switching surge voltages.
- The paper base synthetic resin bonded insulating laminates on bus bars failed due to crushing of the insulating layer. This was due to defective manufacture.

6.6 Electrical Tracking

The dielectric breakdown that we considered in the preceding section was a type of insulation failure where the short-circuit current passed through the body of the insulating material.

There is another type of failure called electrical tracking where the short-circuit current passes over the surface of the insulating material between two adjacent conductors with a high voltage difference. Such locations are found on terminal boards, panels where current carrying conductors are secured mechanically against movement or vibration.

The strength of insulating materials against tracking depends less on the material itself and more on the environment and deposits of dust, moisture, soot, etc. on the surfaces. Almost all failures through electrical tracking are due to such deposits.

In order to prevent tracking failures, it is necessary to provide adequate tracking clearances in the design itself, to minimise such deposits through suitable protection arrangements and to ensure periodical cleaning of the surfaces. Certain insulating materials and certain anti-tracking varnishes are specially designed to resist tracking. They do not allow deposits to accumulate and continuous water films to form.

CASE STUDY

One example of such failures may be mentioned here:

- Certain panels with terminals having high voltage differences were prone to tracking failures. As it was not practicable to increase the clearances between the terminals, insulating barriers were provided between them to increase the tracking distance as shown in Figure 6.3.

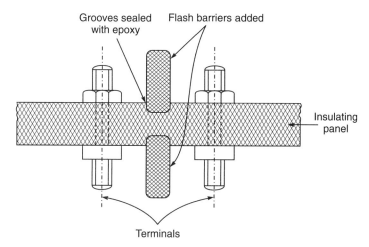

Figure 6.3 Shows how tracking between terminals that were too close, was minimised

- Care was taken to ensure that these barriers were installed in grooves and all crevices were sealed with epoxy adhesives. The panels were originally mounted horizontally. Their mounting was changed to vertical thereby minimising the accumulation of dust on the panels. After carrying out these modifications, there were no further cases of failures through tracking.

6.7 Magnetization

Whenever an electric current passes through a conductor, there is a magnetic field around it. When there are many conductors carrying currents the magnetic fields produced by the different currents add up vectorially. There are precise formulae that connect the magnitudes and directions of the currents and the field strengths. The magnetic field strengths are directly proportional to the currents and inversely proportional to the square of the distance. These are immutable laws.

Any electrical device or installation is, surrounded by magnetic fields. These fields are generally not very strong and their presence does not create any problem. Under certain unusual or rare conditions they can produce unexpected results and one must be aware of such possibilities.

Where electromagnets are used in equipment, the core steel is usually one with very low residual magnetism. When the current is zero, the electromagnet gets demagnetised. Sometimes due to lack of adequate control on the quality of raw materials, electromagnets may NOT get fully demagnetised on switching off the current, and this may lead to failures.

CASE STUDY

A few examples may clarify the above assertions:

- Certain sensitive electromagnetic relays were found to behave erratically. It was then discovered that bus-bars carrying very high direct currents were in the vicinity. The problem was solved fully by shifting the relays.
- Electromagnetic valves were being used for the control of compressed air supply to contactors on some motor coaches. On some occasions they were found to be stuck in the closed position even after being de-energised. Sometimes vibration released them. The problem was solved by the insertion of thin non-ferrous liners on the faces of the electromagnets thereby reducing the effects of residual magnetism.

6.8 Thermal Degradation

This topic was briefly referred to in Section 6.2 about electrical heating. Any kind of heating may increase the temperature of the component. What matters is the temperature attained by the component. The rates of deterioration are exponentially dependent on the temperature. For instance, for every 8 °C rise in temperature, the deterioration rate of insulating materials or the oxidation rate of copper doubles.

Other materials are also vulnerable to degradation of their physical and chemical properties at higher temperatures. For instance thermoplastics get de-plasticised more rapidly at higher temperatures. Local action in secondary batteries is also greatly speeded up at higher temperatures, and their life is reduced. The failure rates of capacitors, semiconductors, resistors and such other devices in electronic equipment increase at higher operating temperatures. These degradations are all different manifestations of the same basic nature of matter, which consists of atoms and molecules that vibrate at rates determined by the temperature.

One of the things that should be done routinely, when investigating failures of electrical equipment is, to check the operating temperature under worst-case conditions. In this context what we need to know is the hot spot temperature and not the general or average temperature. If failures occur at one particular spot, it is desirable to determine the temperature at that spot. A few examples may be considered here:

CASE STUDY

- Some of the components in one particular make of voltage regulators were prone to a high failure rate. As the environment was very dusty, the enclosure was made dust tight in a totally enclosed cabinet. Measurements of the internal air temperature showed that it rose by nearly 40 °C above the outside air temperature due to the heat generated by some of the components themselves. Separation of the high heat generating components into special compartments and provision of externally cooled heat sinks solved the problem.
- The copper bus-bars mounted inside an oven got oxidised more rapidly and there were bad contacts, overheating of bolted joints, melting of copper and failure of equipment.

6.9   Electromagnetic Induction

We referred earlier in Section 6.7 to the fact that any electric current is surrounded by a magnetic field. There is another similar fact. Any change in a magnetic field passing through a substance generates an electromotive force (emf). If the substance is conducting and the circuit is complete, this emf will cause a current to flow. This phenomenon, known as electromagnetic induction, is the basis of electrical machines like dynamos, alternators, transformers, etc. While electromagnetic induction is thus a useful property, it can also create problems when it manifests itself in unintended locations. It then results in energy losses, overheating and unwanted forces. Electrical design engineers endeavour constantly to use this phenomenon for the desired objectives and try to avoid its harmful effects. Sometimes, however, there is a slip up and equipment failures inevitably follow. The following example will clarify this point.

CASE STUDY

- Certain insulated cables carrying heavy alternating currents were installed close to a thick steel plate. As a result of electromagnetic induction eddy currents were produced in the steel plate. The plates got overheated. The insulation on the cables was damaged and eventually the cables developed short-circuits. Re-arranging the cables and increasing the distance between the cables and the steel plate solved the problem.

Chapter 7

SPLIT PIN AND COTTER FAILURES

In this chapter we learn

- *Some of the commonly used designs of split pins and cotters.*
- *Failure modes and failure mechanisms.*
- *Preventive measures and the need for special training of staff.*
- *Case studies.*

7.1 Introduction

$\mathcal{S}$plit pins and cotters are perhaps the lowliest of components used in machinery and plant. Yet, if they are not installed correctly they have the potential of causing expensive failures. They tend to be neglected at every stage from design, manufacture, etc. up to maintenance.

CASE STUDY

- More than 50 years ago, I was travelling in a bullock cart from one village to another. Suddenly, one wheel came off its axle soon after it had passed over a particularly bad part of the road. The cart tilted badly and came to a stop. Fortunately no one was injured. The cart driver showed us the broken cotter. We walked the remaining distance to our destination.

  Every bullock cart driver in India knows about the importance of the cotter outside the wheel, as shown in Figure 7.1. However, he has not heard of low-cycle fatigue, the failure mechanism in this case. It is not only in bullock carts that such problems arise. Cotter failures in brake hangers have led to derailments of trains.

Figure 7.2 shows some of the designs of split pins and cotters now in use in modern machinery and plant. Many other designs are in use but the root causes of their failures are the same and very few in number.

The most common failure mode is one in which the split pin or cotter is either not fitted initially or drops off after some service. This results in the working out of the nut, wheel or

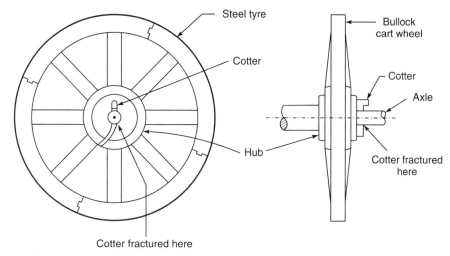

Figure 7.1 Fracture of a cotter at the point where it was bent, probably due to embrittled steel, sharp bend and low cycle fatigue

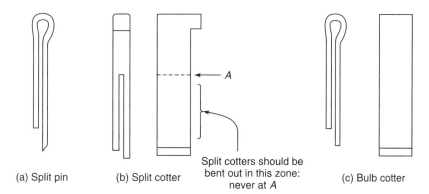

Figure 7.2 Split pin, split cotter and bulb cotter

other component, which the split pin or cotter was supposed to hold in place. The causes of this common type of failure are enumerated below:

(a) failure to install the split pin or cotter;

(b) split pin or cotter not opened out;

(c) cracking and fracture of split pin or cotter.

7.2 Absence of the Split Pin or Cotter

There are millions of castellated nuts now in service, which are obviously meant to be used with split pins. Some of these nuts are in use without these locking devices. Not all of them work out or cause failures. Many such nuts continue to remain tight in service but the fact is that some do work out and the consequences are often serious.

It is the fact that many nuts without split pins work well without failure, which leads to complacence amongst the staff. The designer has thought it fit to provide a castellated nut in a certain location to alert the staff to some feature, which necessitates this additional precaution. Many nuts remain tight even when they are not castellated and without any visible locking device. There are other forms of locking devices such as Nyloc nut, Loctite, bent plate, split nut, spring washers, Belleville washers, etc. All these are as important as split pins or cotters. Whatever be the nut-locking device provided in the original design, it must be meticulously fitted whenever the equipment is re-assembled after repairs or maintenance.

In some applications split pins or cotters are used without any nuts. In such cases, split pin or cotter failures invariably lead to the main pin falling out and causing serious damage.

These aspects must be emphasised during training of all workers who deal with the maintenance of machinery. Most workers instinctively or merely by observation do indeed follow the required disciplines. It is essential to see that all, or 100 per cent, of the staff do so; and this can be ensured only through proper training of all the workers and supervisors with regard to all aspects of their jobs including these apparently trivial items.

7.3 Cracking and Fracture of Split Pin or Cotter

Even when a split pin or cotter is provided as required, failure can occur due to fracture at the point where it is bent. Normally, bending it for the first time is not likely to fracture the split pin or cotter; but due to a phenomenon called low cycle fatigue, it may develop a crack on repeated bending and straightening out at the same point. Bending or straightening the pin or cotter involves stressing beyond the yield point and some plastic flow in the bend zone. Good steels may withstand repeated bending and straightening several times. Pins or cotters made out of recycled steel may develop cracks within just two cycles of bending and straightening.

The situation is particularly dangerous because the pin or cotter may not actually fracture when bent repeatedly; it may merely develop an invisible crack that grows into a fracture during service weeks or even months later.

It is not practicable to keep track of the number of times a pin has been used. It is also not possible to be certain of the quality of the steel. It is safer, to use the split pins and cotters only once and to discard and destroy them when removed for the first time. A highly skilled and conscientious worker can re-use the old pin or cotter if there is no visible crack in it and that this time the bend is in a different place. But as a rule, for the general or average worker, the better option is to replace the pin with a new one every time he has to open it.

It is desirable, also, when accepting supplies of new split pins and cotters, to test the quality of steel by making the bend test on samples drawn from lots in accordance with an approved sampling plan. This is a very simple test that can be made easily. It does not call for any special testing device. The relevant standard *inter alia* says that ...

"The samples selected for test, when cold, shall permit half the length of both legs (of the split pin) being bent back through 180° and closed flat upon themselves without showing any sign of fracture." See figure 7.3

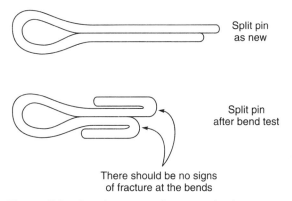

Figure 7.3 Bend test on split pins and split cotters

In the case of cotters there is another point to be checked. The depth of the split in the cotter should be such that its end is inside the shaft or bolt and the bending of the cotter wings should take place in the straight portion and not at the root of the split as shown in Figure 7.4. If the bend is at the root, the maximum stress and the probability of fracture are likely to increase greatly.

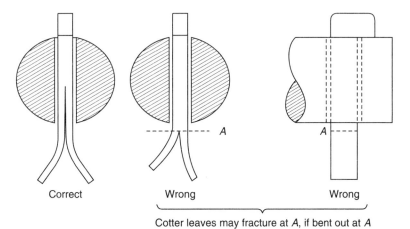

Figure 7.4 Split cotter fracture due to inadequate depth of split

Finally, the cotter must not have the defect shown in Figure 7.5. The machined split must not be eccentric. In this regard bulb cotters of type (c) shown in Figure 7.2 are not prone to this type of defect.

7.4 Corrosion

Split pins and cotters are likely to corrode or rust when exposed to the elements and not always covered by oil. In such cases they are more vulnerable to the problems already mentioned. They should then be replaced without hesitation.

Figure 7.5 Defective cotter. Split not in centre of cotter

7.5 Some General Remarks

The types of defects mentioned above may seem to be not only trivial but also unlikely. Actually, they are all taken from real life case studies one of which was on the most primitive of all vehicles, the bullock-cart. The number of ways in which things can go wrong is very much higher than the few correct paths and it is necessary to exercise care at every step.

Split pins and cotters are, the simplest or lowliest of components in machinery. Very little is written about their failures although they have been in use since the earliest machines were ever used and although many failures of machinery are even today due to failures of split pins and cotters. Training course syllabi and course material generally do not deal with these simple aspects.

7.6 Do's and don'ts for Preventing Split Pin and Cotter Failures

The do's and don'ts for preventing failures of split pins and cotters are:
- Ensure that split-pins and cotters provided in the design are always fitted after maintenance or repairs.
- Use new split-pins or cotters every time. If old ones are being re-used inspect them carefully to see that there are no cracks or corrosion.
- Bend the legs of the split-pin or cotter as shown in the drawing. If there is no indication in the drawing bend each leg through 45°. Avoid making sharp bends.
- Ensure through 180° bend tests on random samples that the metal used in the split-pins or cotters is not defective. Check that the groove in split cotters is central and symmetrical.
- Ensure that the legs of split cotters are not bent at the roots of the grooves.

Chapter 8

THREADED FASTENER FAILURES

In this chapter we learn

- *Different types of threaded fasteners such as screws, bolts/nuts, and studs.*
- *Two basic causes of failures: loss of strength and excessive loading.*
- *Common modes and mechanisms of failures of threaded fasteners.*
- *Measures to prevent such failures.*

8.1 Introduction

$\mathcal{T}$hreaded fasteners are used in almost every kind of electrical or mechanical device, appliance, machine or structure, from the smallest mechanical watch to the largest ship. We will consider in this chapter the failures of bolts, screws and studs. Figure 8.1 shows the basic shapes of these components. There are thousands of other possible shapes, some of which are shown in Figure 8.2. The shapes may be different but their functions, failure modes and failure mechanisms are similar. However, in the detailed discussions that follow, only the basic shapes have been shown.

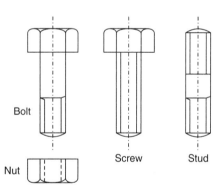

Figure 8.1 Some basic threaded fasteners

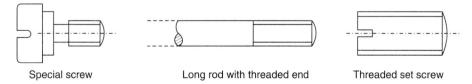

Special screw   Long rod with threaded end   Threaded set screw

Figure 8.2 A few special shapes with threaded ends

Threaded fasteners are more likely to fracture when they are subject to alternating or fluctuating stresses and when the peak stress levels are high in relation to the endurance limits of the material used. Generally, modern machinery and plant are designed with high stress levels in order to reduce the size and the cost of the equipment. It is possible to get reliable service even at high stress levels but it is necessary to pay attention to apparently minor details such as those discussed in the following sections.

High tensile steels are used for fasteners in order to reduce the size and eventually the cost of the equipment. These steels are more sensitive to the effects of stress raisers, surface finishes etc. Therefore, if high tensile steels are used it is necessary to ensure that suitable improvements are made in the factors which can have adverse effects on their endurance limits.

We will be considering first, failures due to loss of strength of the fasteners. The methods by which the strength of threaded fasteners can be maintained or increased will also be mentioned. Increase in stress levels often lead to failures. We will also discuss why they occur and how such failures are to be prevented.

8.2   Fracture Below the Head in Bolts

CASE STUDY

- Figure 8.3 shows the location and the appearance of the fracture of a bolt head. It was a typical fatigue fracture with two distinct zones in the fractured surface: one was smooth, with tide marks and dull, and the other rough and bright. The former represents an area of gradual advancement of a fatigue crack over a considerable period, and the latter shows the sudden fracture of the weakened bolt.

- This bolt had obviously been under a very high alternating or fluctuating stress. By matching together the fractured pieces as in Figure 8.4, the original shape of the bolt and its head is revealed. A few other fasteners in the same location, on identical machines, were examined carefully. A few bolts with visible cracks under the head were detected. On some bolts, hair line cracks were detected on applying dye penetrant. There was a sharp corner at the junction between the head and the shank of the bolt. This resulted in stress concentration by a factor that could be 3,4, or even more. It all depended on the actual radius of curvature and the surface finish at the junction between the head and the shank. These two features varied from bolt to bolt. The remedy was to provide, in the new bolts, a smooth change of section with a fixed and generous radius of curvature as shown in Figure 8.5. The surface was made smooth. The stress concentration factor was thus kept below 1.5 and failures of this type were prevented from occurring again.

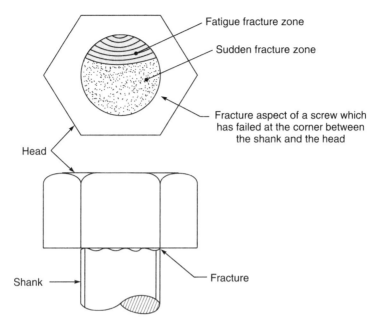

Figure 8.3 Fracture at sharp corner below the head

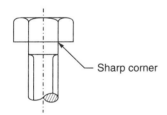

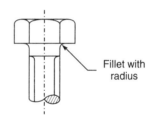

Figure 8.4 Sharp corner between shank and head (WRONG)

Figure 8.5 Fillet with generous radius at corner between shank and head (RIGHT)

It must be ensured, while using bolt heads with generous fillets of the type shown in Figure 8.5 that the mating component has a chamfer as shown in Figure 8.6. Otherwise a sharp corner on the mating component will dig into the bolt fillet and defeat the purpose of the fillet by producing a sharp nick, which may produce an even higher stress concentration.

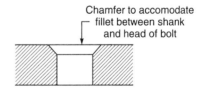

Figure 8.6 Chamfer should be provided on mating component where bolt shank has fillet with the head

Each of the ten sections that follow (8.3 to 8.12) are based on lessons learnt from several case studies.

8.3 Fracture through Roots of Threads

This type of failure can occur in bolts, screws and studs, which have threaded portions. Figure 8.7 shows an enlarged view of a few threads and the location of the fracture. The fracture starts from the root of the thread, which is sharp. As, there is severe stress concentration, fatigue cracks usually start from such points.

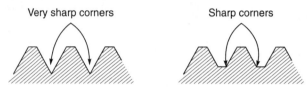

Figure 8.7 Sharp corners at roots of threads can start fatigue fractures in highly stressed bolts or screws

Providing a curved root of thread as shown in Figure 8.8 can prevent failures of this type. This reduces the stress concentration factor. If the threads are produced by rolling instead of machining, the endurance limit of the material gets enhanced and the failure rates reduced.

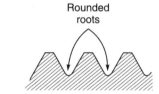

Figure 8.8 Rounded roots of threads are not prone to initiate fatigue cracks

8.4 Fracture through Corners at Changes of Section

Sometimes the shank of a bolt is reduced in diameter to that at the roots of the threads in order to improve its ability to withstand impact loads. In such cases there are two additional changes of section as shown in Figure 8.9. This will inevitably create some stress concentration. The stress concentration factor can be three, four, or even more if there is a sharp corner as shown in Figure 8.9. However it can be made as low as 1.5 by selecting an appropriate radius for the fillet as shown in Figure 8.10, and fractures can be prevented.

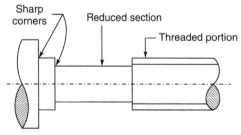

Figure 8.9 Sharp corners at changes of section (WRONG)

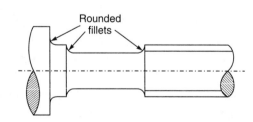

Figure 8.10 Fillets at changes of section (RIGHT)

8.5 Fracture of Threaded Fasteners due to Excessive Stress

We will not consider gross deficiency or error in the basic design of the equipment, though detailed design calculations become necessary if the other more likely causes of failures discussed here have been eliminated.

There may be cases of fracture through one of the three locations mentioned in Sections 8.1 to 8.4, even though the radius of curvature of the fillets is appropriate and the maximum that can be accommodated, since these are the locations of maximum stress. In that event, it would be necessary to consider the use of material of higher endurance limit. Before doing so it is advisable to examine the possibility of the stress having become excessive due to some extraneous factor such as excessive vibration, or inadequate pre-loading of the fastener. These two possibilities may now be considered further.

Whether the vibration of the equipment held in place by the fasteners is excessive, can be determined only by observation or measurement, during actual operation and reference to the specifications. Vibration can become excessive due to a variety of reasons such as:

(a) Excessive clearances in the chain of components involved in supporting the equipment;

(b) Defective or ineffective dampers that may have been provided to minimise vibration or;

(c) Resonance between the natural frequency of the system and the frequency of external impulses.

These possibilities should be examined and appropriate action taken.

Pre-loading of threaded fasteners means tightening the nuts on the fasteners to some specified torque, producing a pre-determined and steady tensile stress in the fastener that is almost equal to the yield point of the material.

It may seem paradoxical that *increasing* the initial or static tension in the fastener to nearly the point where the material starts developing a permanent deformation can actually prevent fracture due to metal fatigue. It can be proved by mathematical analysis as also by actual tests and measurements that this pre-loading results in a *reduction* in the alternating component of the stress in the fastener, thereby minimising the development of fatigue cracks.

Reference to the maintenance manuals or books on machine design will give the appropriate tightening torques for fasteners of different sizes and materials.

8.6 Failures Caused by Relaxation of Tension in Threaded Fasteners

Pre-loading of the fasteners is necessary either to prevent leakage of fluids from pressure vessels or to prevent movement of the components held together as in the case of transformer coils. The force exerted by the fasteners is important for preventing failures of other components. Any phenomenon which can reduce the pre-loading with the passage of time would be the root cause of failures of the equipment though not of the fastener itself. There are in fact two such phenomena. These are metal creep and shrinkage of non-metals.

Both these phenomena viz. metal creep and shrinkage of non-metals can be explained with reference to failures of flanged joints of large pipelines or vessels with high-pressure fluids. See figure 8.11.

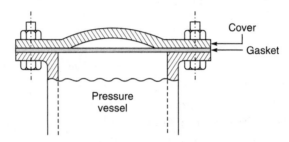

Figure 8.11   Pressure vessel with gasket and bolted cover

The fluid inside the pipe or vessel is prevented from leaking by the flanged joint with a gasket between the flanges, pressed together by the tension in the six or more bolts and nuts. Initially, when the joint is installed full tension, nearly equal to the yield point of the bolt, is applied by tightening the nuts with the appropriate torque. In the process, the bolts elongate and the gasket is compressed. There is a balance of tensile and compressive forces.

The material of the gasket, which may be a laminate of some fibrous material with a natural or synthetic binder, continues to shrink under the influence of the pressure on it and possibly the heat if the operating temperature is high. The rate of shrinkage is very small and even the total shrinkage is also small but it is of the same order of magnitude as the elastic elongation of the bolts and the elastic compression of the flanges. The net result is that these elastic deformations get reduced and consequently the tensile forces exerted by the bolts also get reduced significantly. The joint may start leaking or the gasket may get blown out causing catastrophic leakage of fluid. The bolt may not be damaged. Replacing the gasket and re-tightening the bolts can easily effect repairs. However, the gasket failure may be said to be due to failure of the bolt to provide the required force as a result of the shrinkage.

The remedy, for preventing such failures, is to tighten the nuts periodically in the initial months of service until the gaskets stabilise and the nuts do not turn on application of the specified torque. It is not possible to use fully pre-compressed or incompressible gaskets because the basic need for the gasket arises out of the need to compress the gasket in situ to fill up the imperfections and variations in the surfaces of the flanges.

8.7 Failures of Fasteners due to Relaxation Caused by Metal Creep

If in the case of the flanged joint shown in Figure 8.11, the operating temperature is higher than 300 °C, we have to consider the phenomenon of metal creep, shown in Figure 8.12.

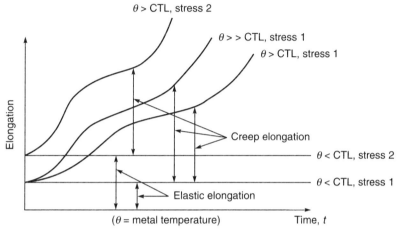

Figure 8.12 Creep characteristics of metals. It may be noted that
– there is no creep if θ < CTL (Creep temperature limit)
-- creep occurs when θ > CTL
– creep rate is higher at higher stress and at higher temperature

Metal creep is a property of metals. Steel bolts elongate continuously when subjected simultaneously to a tensile stress and a temperature higher than their Creep Temperature Limit (CTL). The time rate of elongation depends on (a) the tensile stress and (b) the excess of the operating temperature over the creep temperature limit of the metal. The relationship is not linear as may be seen from Figure 8.12. If allowed to continue in service without periodical re-tightening, eventual failure of such components, is certain. Proper design can help to minimise the creep and timely tightening and replacement of such components can prevent failures in service. It may be reiterated that practical problems arise when the operating temperatures are high in relation to the CTL of the fastener material.

Metal creep can lead to indirect effects in flanged joints. The tension in the bolts will reduce as a result and this will cause leakage and/or failures of gaskets. How frequently the nuts should be re-tightened and the bolts replaced depends on design details, operating temperatures etc. For these details, the advice of the designers and manufactures has to be sought and carried out.

8.8 Locking Arrangements for Nuts in Threaded Fasteners

Spring washers, serrated washers, castle nuts with split pins, nuts with nylon ring inserts, split nuts, etc. and many other devices have been used to prevent nuts from becoming loose. Whenever the nuts have to be tightened sufficiently to produce a stress in the bolt close to the yield point, such locking devices are redundant. The friction between the contact surfaces of the nut is adequate to hold it in position. If there is severe vibration, it is advisable to use locking devices.

8.9 General Design of the Fastener

If the above causes of fastener failures are eliminated from the list of possible causes in any particular application, the basic design of the fastener may be considered to be possibly defective. It may be necessary to use larger fasteners or material of higher mechanical strength may have to be selected. Detailed calculations and analysis will be necessary at this stage. The designers and manufacturers of the equipment will have to be consulted.

8.10 Failures of Screws due to Errors in Lengths

There are some cases of fastener failures due over sight. Failures of screws due to bottoming in blind tapped holes is one of them. See Figure 8.13. The overall length of the screw may be more than the required length, or the depth of the tapped hole may be less than the required depth. It is also possible that some foreign matter such as machining swarf has collected in the blind hole.

The defects could be either in design or manufacturing or in assembly; but whatever the root cause, the effect is that even when the screw S seems to be and actually is tightened fully, the component B which the screw is supposed to hold down remains loose. The effect could be

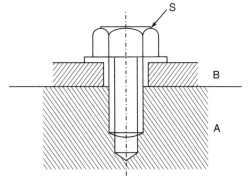

Figure 8.13 Screw appears to be fully tightened but does not serve its purpose of clamping B to A, due to inadequate threaded depth of tapped hole

fracture of either the component B or of the screw S if the assembly is subject to vibration as in rolling stock. If the application involves electrical contact, overheating and burning may be the result.

The same results will follow if the unthreaded length of a bolt is incorrect as illustrated in Figure 8.14.

A gross error may get detected by an alert worker who does the assembly. If he finds that the component, to be held down, is loose even after tightening the fastener fully, he may not overlook it, but if the error is marginal even this chance may not be there. If the application calls for a certain minimum fastening force as in the case of electrical contacts, the problem will not surface until some overheating or burning takes place.

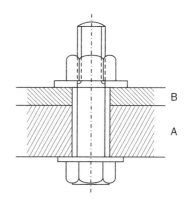

Figure 8.14 Nut appears to be fully tightened but does not serve its purpose of clamping B to A, due to inadequate threaded length on bolt

The use of fasteners of the correct lengths is an obvious technical remedy. On the managerial aspects the remedies are not so distinct. In large organisations to prevent such errors and to meet other requirements modular training systems should be organised for the workers. Refer to Section 20.5, Chapter 20, of my book '*Electrical Failures and Fires*', (Tata McGraw-Hill, 1998) for more details of this system.

8.11 Failures of Threaded Fasteners due to Stripping of Threads

Sometimes, in equipment that has been in service for a long time, threaded fasteners may fail by stripping of threads. The cause is usually corrosion. Either one or both of the internal and external threads may have worn or corroded to such an extent as to reduce the extent of overlap between the two. The remedies include checks on this aspect during periodical overhauls, replacement of fasteners with any visible signs of corrosion, measures to prevent corrosion.

Even new fasteners have failed in this mode due to machining errors, in organisations which did not inculcate basic engineering discipline with regard to dimensions, tolerances, etc.

8.12 Do's and Don'ts for Preventing Fastener Failures

The do's and don'ts for preventing fastener failures are:

- Special care has to be taken when fasteners are subject to vibration, alternating or fluctuating stresses and corrosion.

- If there are changes in section of the fastener, and at corners under the heads, the design should include fillets of ample radius. Routine inspections should include checking for cracks at vulnerable locations.
- In highly stressed fasteners, the roots of the threads should be rounded.
- If fastener design seems satisfactory, failure investigation is directed towards possibilities of excessive stress due to vibration, overloading, unequal load sharing, etc.
- If design calls for control on tightening torque, torque wrenches should be used for tightening the fasteners.
- Look for and eliminate relaxation of fastener tension during service due to effects of metal creep, shrinkage of non-metals, vibration, failures of locking devices, etc.
- Prevent *bottoming* of screws in blind tapped holes and inadequacies in the lengths of threaded portions in bolts.
- Locate signs of corrosion in both internal and external threads. Replace corroded fasteners.
- If none of the above reveal any defect, check the composition and mechanical properties of the material, while investigating fastener failures.

Chapter 9

SHAFT OR AXLE FAILURES

In this chapter we learn

- *Various applications where shafts and axles are used.*
- *Different types of stresses on shafts and axles.*
- *Modes and mechanisms of failures of shafts and axles.*
- *Interpretation of fracture aspects to determine failure mechanisms.*
- *Measures that can be taken for preventing shaft failures.*
- *Condition monitoring of shafts and axles for incipient cracks.*

❑ ❑

9.1 Introduction

*A*ll rotating machines have shafts and all vehicles have axles. (There are in this world, billions of these components in service.) Even the number of types of such applications would run into thousands. The shaft or axle is probably the second machine component to have been invented by man, the wheel being considered his first. In this chapter hereafter, the term shafts will include axles.

Failures of shafts have been reported in every type of application, from the smallest balance wheel shaft in a mechanical watch to the largest propeller shaft in an ocean liner. Somewhere between these extreme examples lie road vehicle axles, engine crankshafts, electric motor shafts and railway axles. The variety of applications, designs and sizes may be large but the failure modes and failure mechanisms of all these are the same and very few in number. The most common one is fracture through the process of metal fatigue.

The effects of shaft fractures are usually catastrophic. The consequential damages to surrounding components are severe. In the case of road and rail vehicles, axle failures lead to serious accidents, the severity of which depends only on the speed and location at the time of the accident.

The vast majority of shafts are made of iron or steel. Some shafts in small instruments are made of non-ferrous materials, but their failure modes and failure mechanisms are the same as those of steel shafts. The simplest shape of a shaft is a plain cylinder as in the case of a bullock-cart or horse-cart axle. Modern machine shafts may have many steps, threads, keyways, grooves, tapers and flats as shown in Figure 9.1.

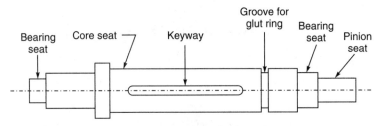

Figure 9.1 Typical motor shaft

Shafts are subjected to tensile, compressive, and shear forces. In addition, if the shafts transmit power, there are torques acting on the shafts. All these forces and torques produce tensile, compressive and shear stresses which can be combined into equivalent principal stresses, which vary from point to point along the length of the shaft. The forces and torques are higher near the middle point of the shaft, but the designer tries to equalise the stresses along the length by increasing the diameter of the shaft where necessary.

The most important characteristic of these stresses in shafts is that they are usually alternating or fluctuating due to the rotation of the shaft, variations in power transmitted and vibrations of the vehicles or machines. Finally, the presence of steps, changes in cross-section, keyways, grooves etc. produce stress concentration, (refer Section 5.12 for the meaning of this term) by factors of the order of 2 to 5 or even more. The designer naturally, takes all these factors into account and, shafts (and axles) may be expected to last the lifetime of the equipment or the machine without fracture. However, the fact is that shaft fractures do take place from time to time.

9.2 Shaft Failures due to Increase in Stress Concentration Factors

The meaning of the term Stress Concentration Factor has been explained in Section 5.12, but the main points are recalled here.

- Modern machine shafts have many changes in section, keyways, grooves, etc.
- All such features develop stress concentration at concave corners.
- At certain places the actual stress is many times higher than the stress calculated by simple bending or torsion theory.
- The factor by which this increase takes place is called the stress concentration factor.
- Practising engineers use reference books where these factors are given in the form of graphs and tables.

- The stress concentration factor at a change in section or diameter of the shaft depends on (a) the ratio of the two diameters, d/D and on (b) the radius of curvature R of the fillet between the two diameters (Reference: Figure 5.2 in Chapter 5).
- The more critical parameter is the fillet radius R. If the designer stipulated a fillet radius of 3 mm and the machinist made it 2 mm, the stress would increase by more than 100 per cent, which is sufficient to cause a failure sooner or later.
- Similar problems can arise in keyways, and grooves.
- Nicks, tool marks and very shallow grooves increase the stress very significantly.
- Welding, brazing, arcing or weld splatter produces stress concentration of the same type. Fractures originate from such points.
- Surface finish is also important in very highly stressed components.

Shaft fractures due to the features mentioned are very common, mainly because of the fact that the increase in the stress concentration factors due to apparently minor variations is very high … 200 per cent to 500 per cent or even more.

If examining the fractured shaft and other shafts of the same type in service in similar conditions does not indicate the presence of any of these problems the following other possibilities may be considered.

9.3 Shaft Failures due to Overloading and Material Deficiency

Overloading and material deficiency are considered together as the fracture aspects would be similar.

If the radial, axial and torsional loads on a shaft increase significantly beyond those assumed in the design of the shaft, obviously the stresses will also increase in the same proportion. If the new stresses exceed the endurance limit, fatigue damage will occur and eventually the shaft will fracture even though there may be none of the defects mentioned in Section 9.2. Since designers normally provide some margins for this type of misuse, it would be incorrect to assume that any overloading above the nominal limits would necessarily result in failure. Detailed stress calculations must be made to verify whether overloading could be the cause of fatigue fractures.

Fatigue fractures have been discussed; but there is another type of fracture, which may be called sudden or brittle fracture. The difference between these two types of fractures is seen from the appearance of the fracture. In fatigue fractures there are two distinct zones in the fracture surface: one is smooth, dull and with tide marks as shown in Figure 9.2(a) while the other is shiny, coarse and even fibrous sometimes as shown in Figure 9.2(b).

In brittle fractures, the entire fractured surface across the whole section of the shaft has the same appearance of a fresh, coarse and bright nature as shown in Figure 9.2(c). To the

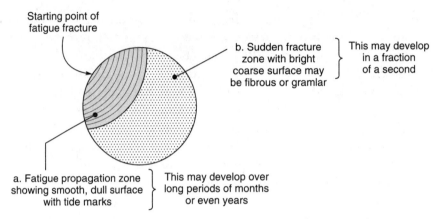

Starting point of
fatigue fracture

b. Sudden fracture
zone with bright
coarse surface may
be fibrous or gramlar

This may develop
in a fraction
of a second

a. Fatigue propagation zone
showing smooth, dull surface
with tide marks

This may develop over
long periods of months
or even years

Figure 9.2 Shows two distinct zones (a, b) in fracture surface

naked eye, the sudden fracture portion of fatigue fractures, as seen in Figure 9.2(b), is similar to the brittle fracture shown in Figure 9.2(c). The only visible difference between Figures 9.2(b) and 9.2(c) would be that the latter covers the entire cross-section whereas the former only a part of it.

Figure 9.2(a) represents the propagation of the fatigue crack over a period of several months or even years and Figure 9.2(b) represents the sudden fracture which takes place when the shaft is weakened so much by the fatigue crack that it fractures suddenly.

When fractures are of the type shown in Figure 9.2(c) or when the fatigue portion is relatively small as shown in Figure 9.3, the possibilities of overloading and material defects must be considered.

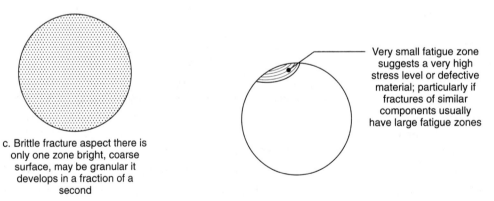

c. Brittle fracture aspect there is
only one zone bright, coarse
surface, may be granular it
develops in a fraction of a
second

Very small fatigue zone
suggests a very high
stress level or defective
material; particularly if
fractures of similar
components usually
have large fatigue zones

Figure 9.2 (c) Brittle fracture aspect

Figure 9.3 Fracture aspect with very small fatigue zone

The extent of overloading is determined only by comparison of the specifications regarding speeds, loads, and power output/input with the actual values during service. Marginal overloading is unlikely to cause failures.

Material defects will be determined by comparison of the specification with actual test results on samples taken from the failed shaft. The properties to be checked are: Ultimate tensile strength, Yield point, Elongation, Brinell or Rockwell hardness number, Izod or Charpy test values. The Izod or Charpy test results should be checked in case of brittle fractures. Microscopic checks to determine grain structure, inclusions etc are also recommended.

Fractures of the type shown in Figure 9.2(c) or Figure 9.3 may also be encountered if impact loads not envisaged in the design are imposed on the shafts. For instance if the spring deflections in a vehicle exceed the designed limits and the springs close solid, there would be excessive impact loading on the axles. Similarly, if the torsional flexibility in couplings is inadequate, sudden torsion loads may be imposed on the shafts.

The probability of brittle fractures increases if the temperatures of the components fall to very low values. Below a certain critical temperature, which depends on the composition and heat treatment of the steel, the impact resistance of the material drops very steeply to a much lower level as shown in Figure 9.4.

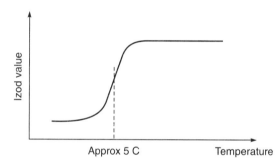

Figure 9.4 Shows the steep drop in impact strengths of some steels at low temperatures

9.4 Preventive Measures

Preventive measures will depend on the root causes of the failure. If there is overloading, it has to be eliminated. If this is not possible then, the design has to be changed by the use of stronger material for the shafts or by increasing the size of the shafts. The shafts will have to be replaced by new ones with the required changes in size or material.

If the failures are due to in-built stress raisers, the shafts have to be modified if possible or else they have to be replaced by new shafts incorporating the required changes in design.

In critical cases such as in aircraft engine shafts or locomotive axles the fleet has to be grounded until replacements are effected. In non-critical cases as in motor shafts, it may be possible to allow them to remain in service. It is usually difficult, sometimes impractical, to replace shafts in a fleet of vehicles or a group of machines in regular productive service.

Sometimes even when a decision is taken to replace all the defective shafts, the process may take time and it then becomes necessary to take interim measures to contain the failure rate.

One possible measure to buy time is to de-rate the operation by reducing speeds or loads as may be convenient. Another more effective method is to undertake ultrasonic testing of the shafts in situ to detect those where the fatigue cracks have developed. Since it may be possible to check the condition of all the shafts within a very short time, it is then possible to identify a few most vulnerable shafts and to replace them on priority. In this way it may be possible to prevent expensive failures while maintaining the service even during the time it takes to replace all the defective shafts.

Ultrasonic testing of shafts is carried out with a special device the size of a TV set. The unit, generates high frequency sound waves, and a probe, which can be placed against the end of the shaft to be tested. Sound waves emitted by the probe travel through and along the length of the shaft. They are reflected by the end of the shaft and also by any fatigue cracks. These reflected sound waves are captured by the probe, fed back to the device which has a measurement unit which can measure the time lag between the emitted and reflected sound pulses. These are displayed on the screen as shown in Figure 9.5. The distance of the crack from either end can be calculated in proportion to the length of the shaft.

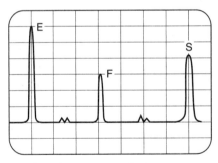

Figure 9.5 Screen of an ultrasonic crack detector
E – emitted pulse
S – reflection from end of shaft
F – reflection from flaw or crack

If there are no reflected pulses like *F* between the emitted pulse *E* and the pulse *S* reflected by the end of the shaft, it confirms that there is no fatigue crack in the shaft. Use of this device requires some skill, but it is possible to train staff in this regard with the help of specially prepared shafts or old defective shafts with known defects.

Shafts that have been withdrawn from the machine are subjected to magnaflux or dye penetrant tests. These tests can detect surface cracks that are not visible to the naked eye. Ultrasonic tests can be made in situ, i.e. without removing the shaft from the machine. More relevant details on these topics can be obtained from books and papers that are available on non-destructive testing.

9.5 Do's and Don'ts for Preventing Shaft and Axle Failures

- Ensure that fillet radius provided in the design of the shaft at each change of section is as large as can possibly be accommodated.
- Ensure that the fillet radius at every change in section of the shaft is machined *exactly* as stipulated in the working drawing. Particular attention should be paid to stress relief grooves and undercuts.
- Ensure that the surface finish of highly stressed zones is exactly as specified in the working drawing. Pay attention to stress concentration zones at changes of section, keyways, slots, holes, etc.
- Ensure that there are no tool marks (grooves, scratches, etc.) in highly stressed zones of the shafts.
- Ensure that shock loads and impacts imposed on the shafts are not increased due to defects in the resilient components of the system, e.g. springs, shock absorbers, elastomeric pads, etc. that may be provided in the design.
- Ensure that the ultimate tensile strength, endurance limit, elongation, Izod value and surface hardness of the shaft are all within the limits specified in the drawing.
- Ensure that highly stressed components are not allowed to corrode or rust.
- When shaft failures occur repeatedly, introduce ultrasonic, magnaflux and dye penetrant tests as may be convenient and practicable, to detect shafts with incipient fatigue cracks.
- Examine the fracture aspects of every failed shaft very carefully. Seek the assistance of metallurgists and shaft designers.

BALL AND ROLLER BEARING FAILURES

In this chapter we learn

- *The special design features of ball and roller bearings relevant to their failures.*
- *Metal fatigue—the most common failure mechanism for ball/roller bearing failures.*
- *Common factors between different causes of ball bearing failures.*
- *Ten different causes of bearing failures.*
- *A few typical case studies.* ☐ ☐

10.1 Introduction

$\mathcal{T}$he term ball bearing will include roller bearings, as their failure modes and mechanisms are the same.

Ball bearings are used in large numbers in all types of machines and their numbers in service would go into billions. Although ball bearings are designed to last the life of the equipment, failures in service are not as rare as they should be. There are several reasons for this problem.

Unlike most machine components, the materials used in ball bearings are always stressed to levels that are close to the yield points and the endurance limits of the materials. This occurs because loads are transmitted through point contacts in ball bearings and line contacts in roller bearings. Theoretically, it might seem that the area of contact is zero and the unit pressures at the contacts should be infinite. See Figure 10.1.

What does happen is that there is deformation of surfaces on the balls (or rollers) and on the races. The contact areas grow until they are large enough to make the stresses nearly equal to the yield point. Metal fatigue is thus endemic to ball bearings. Despite this, modern ball bearings often last the lifespan of the equipment, provided they are manufactured, operated and maintained with great care, leaving no allowance for errors.

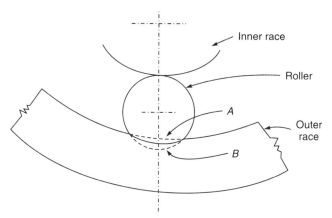

Figure 10.1   Shows what happens at the 'line' of contact between the roller and the outer race. (The deformations are greatly exaggerated for clarity in the drawing.)
- The dotted line *A* represents the original profile of the outer race
- The dotted line *B* represents the original profile of the roller
- The full line between *A* and *B* represents the deformed profile of both roller and race
- The deformation at the upper line of contact has not been shown in enlargement

10.2 The Most Common Mechanism of Failure in Ball Bearings is Fatigue

A great deal of technology goes into the design and manufacture of ball bearings. Good quality materials are used in their manufacture. The steel used has to be free from inclusions, voids and impurities. These are precision components that need to be manufactured, selected, installed, utilised and maintained with great care. However, metal fatigue at the active surfaces of the balls and races of the bearings is present inspite of the bearings being designed, manufactured, utilised and maintained in the best possible manner. A small percentage of the bearings will fail in service unless such failure is prevented by condition monitoring and replacement of bearings with signs of impending failure.

All ball bearings have a limited life and a certain percentage of bearing failures within the declared lifetime of the bearings is unavoidable. The bearing manufacturers guarantee that these failures occur in a number of bearings which would not exceed a specified but small percentage (usually 10) of the total number installed, when the bearings are utilised under rated conditions of load, speed and environment. Using bearings of higher rating can reduce this percentage.

Even this small percentage of bearing failures, which can occur despite care being taken in selection, design, manufacture, utilisation and maintenance, can be averted and reliable performance can be achieved by condition monitoring as follows:

(a) The active surfaces of the balls and rollers as also of the inner and outer races should be very carefully examined visually, using a magnifying glass, during every periodical overhaul. If signs such as excessive wear, chipping, spalling, cracks, pock marks, wash board marks, etc. are detected the bearings should be replaced because these signs indicate failure ultimately. Normally, these surfaces will present a clean, smooth, flawless appearance in which case the bearings can be lubricated and reinstalled.

(b) While the bearing is in normal use, shock pulse meters should be used periodically to detect bearings that are likely to fail. These meters measure certain ultrasonic vibrations, normally produced within the bearing components. When there are surface defects of the type described in (a) above and which are the pre-cursors of fatigue failures, the nature of the ultra-sound vibrations changes and the meter detects and indicates such changes. When the indications go beyond certain empirical limits indicated by the manufacturer of the shock pulse meter, the bearings should be replaced.

(c) In critical applications these devices can be permanently installed to monitor the bearings continuously and to sound an alarm when fatigue damage becomes significant, thereby giving adequate notice for arranging replacement conveniently without affecting service or production.

It is thus possible to get reliable performance from ball bearings in the manner described above. In actual practice ball bearing failures occur mainly because of errors somewhere in the selection, installation, utilisation and maintenance of the bearings. In addition, failures may occur due to design and manufacturing errors in the products of certain bearing manufacturers.

10.3 Common Causes of Ball Bearing Failures

As explained above, all ball bearings fail through metal fatigue. With judicious selection and correct maintenance, reliable service can be obtained. Errors cause premature failure and to ensure that there are no failures in service it is necessary to avoid the factors or conditions that tend to reduce the life of the bearings.

There are many different types of errors that lead to premature failure of ball bearings. In most cases, the failure mode is always the same. The bearing becomes noisy and probably warmer than usual. If allowed to continue in service as is the case, particularly in unattended installations, the bearing seizes and the machine stops. There may be further damage to the equipment if the driving power available is large and automatic devices to stop the machine are not provided. All this is not relevant to the root cause of the failure.

The mechanism of failure is in two stages. The first stage depends on the nature of the root cause but the second stage is always the same. The second stage includes the following:

■ Chipping, flaking etc. from the active surfaces (this is the stage when the bearing becomes noisy);

- Growth in the area and depth of the surface defects (there may be vibration and heating at this stage);
- Seizure of the bearing (fracture and melting of the balls, cages, races and perhaps even the housing, is possible at this stage).

It is in the first stage of failure where all the differences lie. Every possible cause of failure has its own first stage failure mechanism, which ends in the formation of surface flaws on the active surfaces of the balls and/or the races and the commencement of the second stage described above. The different possible causes and the corresponding first stages of bearing failures are described in the following section generally in order of probability of occurrence.

The engineer in charge must assess personally, and determine the root causes in every case of ball bearing failure. This is not easy because the failure mode as also the second stage of failure mechanism is the same for a large number of different first stages. If the root cause is not correctly determined and eliminated, more failures are bound to occur with increasing frequency. It is vital to determine the technical cause as also the organisational or systemic cause and to take suitable corrective action towards zero failure.

Even when the bearing failures are due to the causes described in the following sections, the condition monitoring systems described in Section 10.2 above offer a second line of defence in order to prevent failures of bearings in service.

10.4 Lack of Lubrication

This is probably the most common cause, due to simple oversight by the staff concerned, due to lack of communication between staff, supervisors and engineers. (The basic requirement is the circulation of a maintenance plan, with important details such as brand, grade and quantity of lubricant to be used, frequency of lubrication, etc.)

In ball bearings, it appears that there is only rolling action and no sliding action between the balls and the races and the need for lubrication may not be apparent. In radial bearings, due to the difference in the diameter at the active surfaces of the inner race and the outer race, there is a small amount of sliding action between the balls and the inner races and also between the balls and the outer races. Absence of lubricant creates excessive friction, wear and overheating and corrosion as well. Minor flaws appear on the active surfaces.

They develop very easily, in the extreme conditions at the point/line contacts, into the second stage described in Section 10.1 and then into eventual bearing failure. Sometimes lack of lubricant may be due to defects in grease nipples or blockages in the lubricant passages from the grease nipple to the bearing roller space. All these should be checked and cleaned during overhauls and assembly and also during investigations of bearing failures.

In the case of bearings designed to operate with oil lubrication, lack of lubricant may be due to leakage of oil from the tank, defects in oil pumps, fractures of lubricant pipes, failures of oil seals etc. In such cases the bearing failure is consequential damage and it should be classified under tank failure, oil pump failure, oil pipe failure, oil seal failure etc., as the case may be. These failures will be discussed separately.

10.5 Wrong or Defective Lubricant

Ball bearings, which operate at high speed, are sensitive to the quality of grease. Ball bearing grease is a sort of semi-solid emulsion of lubricating oil held in a matrix of soap. Due to the churning action inside a ball bearing, this emulsion gets separated into its components viz. soap and oil if the selection of the lubricant is incorrect or if the manufacturing process is defective. The oil gets churned out of the soap matrix and thrown out. As the soap which gets left behind is not a good lubricant, failure is then inevitable.

The technology which goes into the design and manufacture of ball bearing greases is confidential. Users have no alternative but to use branded products recommended by the manufacturers for each particular application proven by laboratory tests, endurance tests and service experience. As there are many reputed manufacturers of grease, it is imperative to test them and make the correct choice. For each application, two or more different brands/ grades of greases should be approved. This will control the prices and availability of greases.

The alkali base for the soap constituent of the grease, the type of lubricating oil, additives and the emulsification process, determine the properties of the grease. Suitability of the same for any application depends on the size, speed, load and temperature of the bearing. Test methods have been standardised but these are only for selection of brands. They cannot be applied for acceptance of supplies, since the properties of the grease cannot be determined by non-destructive tests on the finished products.

It is essential, to ensure that the grease used in the bearing is of the approved brand and grade, recommended by the bearing and grease manufacturers and proven by endurance tests and long service. Since very few greases last forever, it should also be checked whether the age of the grease since last refilling has exceeded the limit recommended by the manufacturer of the grease.

In case of ball bearing failures one of the first things to check on is the quality of the grease used. If the bearing has seized, the quality of grease in similar bearings may be checked. Bearings with a service due for grease replacement should be checked. The feel or consistency of the grease removed from the bearings should be compared with that of new grease. If the grease feels harder, drier or less slippery, if oil droplets are seen in bearing overflow vents it is an indication that some of the suspended oil has been churned out and the grease is unsuitable for the application. If the grease is of the approved brands the whole issue of approval of the brand should be reopened and the manufacturer of the grease should be notified.

10.6 Inadequate or Excess Lubricant

Ball bearings are filled with fresh grease initially and thereafter, certain specified quantities of fresh grease are pumped in at specified intervals to replace whatever grease is damaged by normal service. If there is any lacuna in this specified periodical replenishment, the service conditions become less than satisfactory for the bearing surfaces and there are increased chances of bearing failures. The life of the bearing is reduced and premature failure may be the result of defects in maintenance.

The maintenance required for ball bearings is (a) periodical replenishment of grease, (b) periodical replacement of grease and (c) use of approved brands and grades of grease. Many bearings in service work well despite occasional deviations from the specified practice making the staff complacent or casual about them. The fact is that there is long-term damage and reduction in life of the bearings and this becomes apparent years later in the form of premature failures.

Normally, excessive grease does not cause any problem because the excess quantity gets thrown out through overflow vents. Some bearing manufacturers recommend pumping in fresh grease until some of it comes out of the vent. However, in the case of some special, high-speed bearings they recommend that the quantity of grease should not exceed the specified limits. They fear that excess grease can cause overheating; so it is best to follow their instructions.

10.7 Contamination of Lubricant

It should be amply clear that the quality of the lubricant is vital for the reliability and durability of ball bearings. Obviously, contamination of the lubricant by foreign matter is totally unacceptable.

There are, a number of ways in which lubricants get contaminated. All concerned must take positive steps to prevent contamination. Some of these ways are described below:

(a) In ball bearings fitted on outdoor equipment rain water may enter the bearings, cause corrosion on the active surfaces and lead to bearing failure sooner or later by the process discussed in Section 10.1. The bearings must be provided with covers, seals etc. to ensure that water does not penetrate inside.

(b) In ball bearings adjacent to or inside gearboxes the gear lubricant may find its way into the bearings. If the gear lubricant is a heavy bituminous compound and it enters the ball bearing, failure is certain. Effective seals must be provided to keep the gear lubricant out of the bearing housing. In such situations, bearings must be opened out periodically and examined carefully to check on the effectiveness of the seals, at least initially. Contamination could be due to seal failure and bearing failure would only be consequential damage.

(c) Another common mode of contamination is in storage and handling. If grease barrels are stored without proper covers and grease is drawn and filled into guns or machines without proper care, inevitably dust, grit and machine swarf find their way into it and eventually into the bearings. It is desirable to set up small 'Clean Rooms' inside workshops for all maintenance work on ball bearings. As soon as bearings are removed from their machines for periodical overhauls, they should be cleaned thoroughly, oiled and sent into the clean rooms for storage. Detailed visual inspection, re-greasing are required for re-assembly to be done by staff who are specially trained for clean assembling.

(d) Greases manufactured by different companies should not be mixed. If a particular bearing is initially charged with a particular brand/grade of grease, the same brand/grade should be maintained for replenishment. (There is no harm in dismantling and cleaning up a bearing fully and then using another approved brand/grade.) It is also possible that mixing of two particular types of grease may not cause failure; yet it is a risk not worth taking.

10.8 Inadequate Interference between Inner Race and Shaft

The inner races of ball bearings are usually fitted with an interference fit on their shafts. The extent of this may vary with size, speed, load, vibration, operating temperature, etc. It is necessary to ensure that the inner race does not slip and rotate on the shaft, or even creep around because this may soon develop into rotation. If such rotation occurs sliding friction will generate heat and inner race temperature will rise, causing it to expand. Slipping may increase and bearing clearance will get reduced. Lubrication of active surfaces will be affected, surface flaws will develop and get worse. Stage 2 of Section 10.1 will follow.

It is desirable to check the diameter of the inner race before and after fitting onto the shaft. If the 'swell' of the race is less than the specified limits, it is an indication of inadequate interference between the inner race and the shaft.

The radial clearance between the rollers and the races can be measured very easily with the help of a dial gauge and a small jack to lift the rotor of the machine. See Figure 10.2.

Excessive radial clearance (as compared to the limits specified by the bearing manufacturer) on a new bearing after installation is an indication of inadequate interference between the inner race and the shaft. It could also be due to worn bearing. Use the device in Figure 10.3 to check the radial clearance in the bearings after cleaning.

The interference between the inner race and the shaft must be correctly ensured in accordance with the recommendations of the bearing manufacturer determined after long

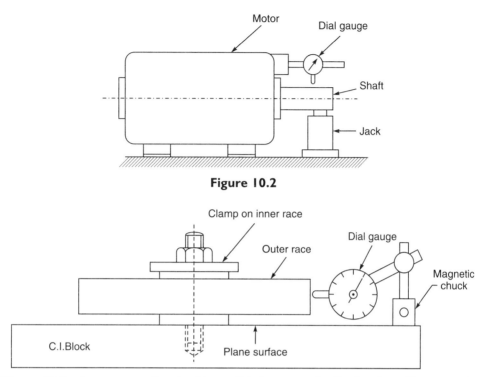

Figure 10.2

Figure 10.3   Arrangement for checking radial clearance of roller bearings. Outer race is moved to and fro to see radial clearance on dial gauge

service trials. The dimensions of the inner race are standardised by them. Errors in these dimensions are very rare. The only control on interference between the inner race and the shaft is through the diameter of the bearing seat on the shaft, the limits for which are specified by the bearing manufacturer. Great care should be taken in ensuring the accuracy of the measuring instruments and in making the measurements. Since surface finish can affect the effective interference it should also be ensured that the surface finish of the bearing seat is as required in the drawing.

10.9   Excessive Interference between Inner Race and Shaft

Excessive interference between the inner race and the shaft (as compared to the limits recommended by the manufacturer of the bearing) will also lead to ball bearing failures. Excessive interference has two effects. Firstly, it increases the hoop stresses in the inner race and it may fracture, in which case bearing failure is certain. Secondly, the swell of the inner race will be excessive and the bearing clearance between the races and rollers will consequently get reduced. When the bearing starts running, the temperature of the rollers and the inner race usually rises faster than the rest of the bearing and housing. As a result the clearance gets further reduced due to unequal thermal expansion and the bearing may seize.

It is desirable, to check the diameters of the bearing seats on the shafts, before installation of the bearing and also the swell of the inner race and the bearing radial clearance after installation. They should be within the appropriate limits specified by the manufacturers.

Inner races are shrunk onto the shaft. They are heated, pushed into position and then allowed to cool. The temperature rise of the inner race during this process over the temperature of the shaft must be monitored and regulated by the use of thermometers and oven thermostats, within the limits specified by the designer of the machine.

10.10 Errors in Axial Positions of Inner Races

Rotor shafts have two bearings, one at each end. The machining of the bearing housings in the body or the carcass of the motor/machine determines the positions of the outer races of the two bearings, whereas the positions of the inner races on the shafts are determined by the machining of the rotor shafts. Shoulders are usually provided on the shafts so the axial positions of the inner races can be located correctly.

Any difference between the relative axial distances of the inner and the outer races, will lead to indeterminate axial forces, rubbing of rollers on the lips of the races and so on. This causes heating, wear and eventual bearing failure. Therefore the relative axial distances between the inner and the outer races as also the axial play of the rotor must be within the limits stipulated by the manufacturer.

The user or maintainer of the equipment has little control on these axial positions and distances. They depend on the accuracy of the machining of the shaft and the body of the machine. The only scope for error is in the positioning of the inner race on the shaft and the outer race in the housing. The former is ensured by making sure that the inner race is pushed down hard on the appropriate shoulder and the outer race is similarly pushed fully into its housing. There must be no foreign material like grit or machine swarf between the critical faces. See Figure 10.4.

After final assembly, the axial play of the rotor should be measured with the help of a dial gauge and a device such as a small jack, to push and pull the rotor axially. The rotor must also rotate easily and noiselessly. The perpendicularity of the bearing relative to the shaft is also important to ensure reliability.

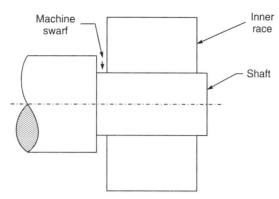

Figure 10.4 Shows how machine swarf or grit between critical surfaces can affect correct assembly of inner races

10.11 Slipping of Outer Race

The outer race of the bearing is usually fitted into its housing with a very light interference fit or a transition fit. Normally, the outer race is stationary, and the rise in temperature of the outer race compared to the housing temperature gives it a tighter fit in service. The possibilities for the rotation of the outer race in its housing are remote. However if it is too loose in its housing, and if there is excessive vibration of the rotor as in rolling stock bearings, the outer race may start creeping or doing a stick-slip rotation. This will cause overheating, wear, eventual bearing failure.

Follow the manufacturer's limits on housing bores. As accurate measurement is a bit tricky use the special devices suggested by them. During re-assembly of outer races during successive overhauls, rotate the race by 90 degrees in order to bring into use fresh unworn surfaces of the outer race.

10.12 Damage to Races during Assembly

Bearing races usually have tight fits. They have to be forced into position. Small bearing races can be pushed into place but larger bearings require shrink fits. The precautions to be taken in these processes though few are very important. They are:

(a) In order to place the races into position, use special screw driven or hydraulic pullers or pushers. The uniform distribution of the applied force around the circumference is important. Refer Figure 10.5 which shows a simple device for removing inner races. Small bearings may be hammered lightly into position; special cup shaped anvils should be used to ensure that the loading around the circumference of the race being fitted is uniform and that no force is applied to the cage or the other race. Refer Figure 10.6.

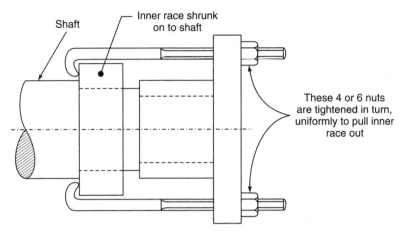

Figure 10.5 Puller for shrunk-on bearing races

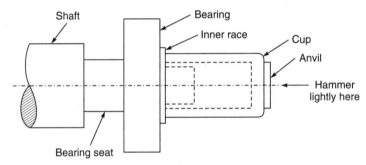

Figure 10.6 Device for fitting inner race of small bearing with light interference fit on shaft

(b) Do not heat the inner races for shrink-fitting by using open flames or fires. Use induction heaters, ovens or oil baths. Heat gradually as specified and control the final temperature accurately as per directions. Overheating can affect the hardness of the active surfaces.

(c) Before attempting to make a force fit, check the dimensions of the 'shaft' and the 'bore' carefully to ensure that the interference does not exceed the limits specified by the bearing manufacturer.

(d) Wherever possible, check active surface diameters of the races after fitting races, to verify that the swell of inner races or shrinkage of outer races is within specified limits.

(e) Ensure that the instruments used for measuring dimensions are accurate by periodical calibration.

(f) Make sure the surface finishes of the fitting faces are within the requirements.

(g) Be sure that machine swarf, grit and such other foreign matter are not allowed to come between fitting surfaces.

10.13 Case Studies

All the possible causes of bearing failures mentioned above in Sections 10.3 to 10.12 are authentic. The most common causes are those in Sections 10.3 to 10.6 relating to lubricants. The causes in Sections 10.7 to 10.12 relating to assembly and installation of bearings are less common but not rare. A few case studies, which cannot be categorised in one or the other of these 10 causes, are described in the following sections. Here the failure mechanism is not very different from those described earlier but the root causes are different.

10.14 Bearing Failure due to Seal Failure-I

CASE STUDY

Most ball and roller bearings run at speeds of the order of a few hundred to a few thousand revolutions per minute. The durability and the reliability of ball and roller bearings are closely connected with the operating speed. Higher the speed, lower is the reliability. In one exceptional case a wheel fitted on needle bearings was subject to bearing failures, although its speed was extremely low—of the order of a few revolutions *per day*.

The case in question is related to a tension regulating equipment fitted on overhead traction lines to maintain a constant tension in them despite the changes in their lengths due to variations in the temperature of the wires.

On examination the bearings showed that the bearing races, the needle rollers, and the cages had all rusted. Due to the outdoor mounting and inadequacy of the seals rainwater was getting into the bearing and washing out the lubricant. Bearing failure under these conditions was inevitable. More frequent overhauling of the needle bearings and use of better seals to prevent entry of water into the bearings resolved the problem.

10.15 Bearing Failure due to Seal Failure-II

CASE STUDY

There were many cases of roller bearing failures on traction motors of motor coaches. These were due to defects in the design of the sealing arrangement provided to prevent entry of gear lubricating compound into the bearing lubricant. The gear lubricant was a thick, almost solid, bituminous compound while the bearing lubricant was a lithium base grease. Mixing of the former into the latter caused the soap in the lubricant to separate from the oil, which then drained away and left behind a hard residue. The bearings got progressively overheated and finally seized.

The original seal design and the improved seal design are shown in Figures 10.7(a) and 10.7(b), respectively. The improved design ensured that the gear compound which entered the seal was thrown back by centrifugal action into the gear case and not into the bearing space.

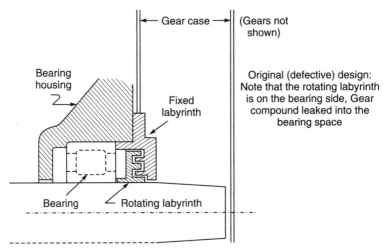

Figure 10.7(a) Original (defective) design of bearing seal

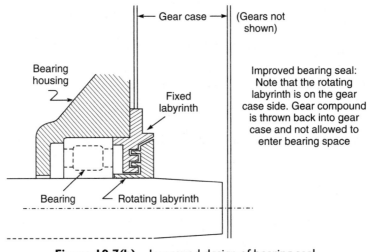

Figure 10.7(b) Improved design of bearing seal

10.16 Do's and Don'ts for Preventing Ball and Roller Bearing Failures

- Select the type of bearing, the lubricant and the schedules for its application carefully in accordance with the recommendations of the manufacturer as applied to the working conditions such as speed, axial and radial loads, temperature, etc.
- During service trials, monitor the condition of the lubricant and the active surfaces. In the event of appearance of any signs of impending failure investigate fully and make modifications until satisfactory performance is achieved.
- Ensure correct fitting after every overhaul.
- Ensure correct lubrication at all times by specifying intervals for replenishment and for replacement of lubricant. These may be based on manufacturers' recommendations, but should be verified for suitability under actual working conditions by periodical examination of lubricant condition.
- Monitor active surfaces and shock-pulse measurements.
- Investigate each case of bearing failure to determine the root causes and then eliminate them.
- Some of the common causes of failures are:

  Incorrect selection of bearing type (10.3)
  Absence of lubricant (10.4)
  Wrong lubricant (10.5)
  Defective lubricant (10.5)
  Inadequate lubricant (10.6)
  Excess lubricant (10.6)
  Contaminated lubricant (10.7)
  Race too loose (10.8)
  Race too tight (10.9)
  Axial and angular position error (10.10)
  Slipping of outer race (10.11)
  Damage during assembly (10.12)
  Seal failure (10.14 and 10.15)

- Provide clean rooms for inspection and assembly of bearings and storage of lubricants.

Chapter 11

SLEEVE BEARING FAILURES

> ### In this chapter we learn
>
> ■ *Differences between failure modes and failure mechanisms in sleeve bearings and those in ball and roller bearings as discussed in Chapter 10.*
> ■ *Common factors between the failures of the two types of bearings mentioned above.*
> ■ *Measures that can be taken to prevent sleeve bearing failures.*
> ■ *Case studies.* ❑ ❑

11.1 Introduction

$\mathcal{B}$earings for rotating shafts can be divided into two main classes: ball/roller/needle and sleeve bearings.

Sleeve bearings can be further classified by the type of material used. They have been listed in order of increasing initial cost:

(a) Bronze bearings
(b) White-metal bearings
(c) Sintered metal bearings

These three types cover more than 90 per cent of sleeve bearings. Only these three will be discussed here.

The selection of the type of bearing depends on many factors like accessibility and availability for periodical inspection and maintenance; speed and type of load on the shaft; environment; cost, etc. Certain failure modes and failure mechanisms are common to all the three types of bearings and these will be discussed first.

11.2 Common Failure Modes

The most common failure mode is overheating. If the bearing is allowed to run despite overheating, the bearing metal may melt, the shaft may get scored, or the shaft may even break. Overheating is the first sign that a bearing failure is imminent.

The normal temperature of the sleeve bearing may be only a few tens of degrees Celsius over the ambient temperature. The normal temperature rise will vary from machine to machine and from design to design. The temperature rise limits for any particular design and application may be ascertained from the manufacturer. It would be ideal if temperatures are measured carefully on 10 to 20 machines. The average temperature rise and the standard deviation should be calculated (See Appendix 11.1). The average plus three times the standard deviation should be taken as the upper temperature rise limit for the case in question.

While measuring temperatures for the purpose of determining the limits, it should be ensured that the temperature measured is always at the same spot, using an identical method of measurement, when the shaft speed and load are at their highest level and allowing enough time for stabilisation of the temperature. For condition monitoring during service, similar procedures should be adopted.

In the case of medium and high speed bearings, the lubricating oil is pulled into the space between the shaft and the bearing and a very thin but significant film of oil separates the shaft and the sleeve bearing. This film helps to minimise the friction, the energy loss and the temperature rise of the bearing.

Failure of this film of oil is the starting point of sleeve bearing failures. There is metal to metal contact, excessive friction, excessive heat and eventual failure of the bearing. All possible causes and circumstances that could lead to failure of the oil film should be investigated. The most common causes generally in order of probability of occurrence are as follows:

(a) Lack of lubricant, though very obvious, is also one of the most common causes of bearing failures. The lack of lubricant could be due to a number of causes:

   (i) Lack of constant or timely replenishment of the lubricant.

   (ii) Leakage of the lubricant due to either a defect in a rubber seal, a fracture or blow hole in the casing of the bearing.

(b) Contamination of the lubricant causing it to thicken; or become very thin; or contain abrasive foreign matter in it.

(c) Excessive or abnormal loading on the bearing.

(d) Excessive or inadequate clearances between shaft and bearing due to initial installation error, or excessive clearance due to wear of the bearing sleeve.

(e) Poor surface finish of the bearing surface and/or the shaft surface.

(f) Heat input into the bearing from an adjacent source of heat, that had not been considered in the original design.

The above list of causes are equally applicable to ball and roller bearings. However, there are certain differences in their failure mechanisms. In the case of sleeve bearings the starting point is the failure of the lubricating oil film, while in the case of ball and roller bearings it is the commencement of fatigue cracks in the active surfaces of the bearing races.

11.3 Failures due to External Heating

CASE STUDY

In an industrial installation, a number of small (Fractional Horse Power) motors were used to drive specialised machines. As there was a special need for noise abatement in the room, the electric motors were fitted with sleeve bearings. Of the 15 motors of this type installed in the room, two motors were prone to frequent bearing failure. Their failure rate was about 0.5 per year per motor whereas the bearing failure rate of all the other motors put together was zero over a period of 5 years.

Comparison of the factors showed that the two motors in question were installed close to electric furnaces whereas the other thirteen were relatively away. In all other respects such as speeds, loads, usage etc. all 15 motors were identical. Measurement of casing temperature on these motors showed that the temperatures on the two motors were at least 30 °C higher than the rest. The problem was solved by the installation of simple vertical baffles between the motors and the furnaces to prevent radiant and convected heat from reaching the motors.

11.4 Failures due to Thefts of Lubricants

CASE STUDY

Traction motors of electric locomotives operating from one shed were subject to higher failure rates on the sleeve bearings through which the traction motors are suspended from the axles driven by them. These bearings are of the white metal type. The lubrication is by oil applied through lubricating pads dipping in oil reservoirs. The oil rises by capillary action in the material of the pads. Oil was replenished in seven different stations within the operating range of the locomotives.

Random checks were made on the oil level in the reservoirs below the bearings. These readings were co-related (with the help of oiling records) with the last station where the oil replenishment was done and the elapsed time between replenishment and the measurement. It was observed that the drop in oil level was proportional to the elapsed time in the case of six oiling stations. However after replenishment was done by one particular station, the oil levels were consistently lower than expected. On making further checks on the history of locomotives that had suffered these bearing failures, it was found that of 16 bearing failures in one year, 11 were traceable to last replenishment by the same oiling station. The oil consumption, per locomotive oiled, was found to be actually higher than the average, in this particular oiling point. This pointed to thefts of large quantities of oil. The workers concerned were transferred to different stations and the matter of suspected thefts was handed over to the department that deals with such problems. The bearing failure rate showed a significant drop within a few months.

11.5 Failures due to Leakage of Oil from Reservoirs

CASE STUDY

In case study 11.4, it may be recalled, five cases of bearing failures were not connected with the suspected oiling point. On comparing the statistics it was observed that three cases, at intervals of a few months, were on the same traction motor as identified by its serial number although working on different locomotives. It was a new traction motor received as spare. The locomotive number changed because every time there was a bearing failure, the motor was replaced, repaired and then installed after several days on another locomotive. Careful examination of this particular traction motor revealed that there was slow leakage of oil from the oil reservoir. (This was not detected earlier as these components are generally oily and dirty being close to the track.) There were small blowholes or porosity in the steel casting, which formed the reservoir. Thorough cleaning of the affected area with solvent, drying, heating with blowtorch and pouring liquid epoxy adhesive helped to eliminate the oil leakage from this casting.

The motor continued to work without bearing failure for at least one year thereafter.

11.6 Failures due to Excessive Stress

CASE STUDY

One class of electric locomotives in use between 1930 and 1980 had what was then called side-rod drive for the transmission of power from the traction motors to the wheels of the locomotive. The locomotive had two three-axle bogies. A pair of large traction motors, which drove a common countershaft through a gear drive, powered each bogie. This countershaft had two crank pins at an angle of 90° between them. Large connecting rods coupled these crank pins to similar crank pins on the outside of one of the three axles of the bogie. The other two axles were connected to the driving axle through similar crank pins and side rods.

The side rods were fitted with white-metal lined sleeve bearings as shown in Figure 11.1.

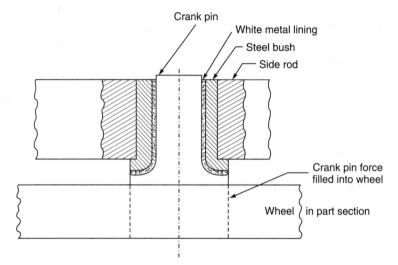

Figure 11.1 Side rod bush with white metal living

These bearings, lubricated with heavy lubricating oil, were known for their short life and high failure rate. There were several different causes of these failures—enumerated in Section 11.2. However, one particular cause, which was identified with difficulty, is discussed here.

Due to distortion of the bogie and inadequate attention paid during periodical maintenance, the centre distances of the different axle-boxes drifted from their original values by several millimeters. The distances between the bearings on the side rods remained correct as per drawings but different from the actual wheel centre distances on the locomotives. This resulted in excessive stresses on the side rod bearings when the crank pins were at the dead centres. The soft white metal gave way, the clearances became excessive and the eventual result was frequent bearing failures. This problem was overcome when the quality of bogie maintenance was improved and the wheel centre distances were maintained within limits (See Chapter 12 for more details).

11.7 Failures of Railway Axle Boxes

CASE STUDY

In the two cases discussed above the bearings consisted of full sleeves going right round the shafts. There is another type of bearing, in which the direction of the load is always the same, the bearing 'sleeve' is only a small part of the full circle. A very common example of this was the railway axle box, which was widely used before the advent of roller bearings.

In this design of bearing only the top half (or even less) of the shaft is in contact with the partial sleeve as shown in Figure 11.2.

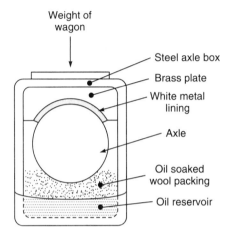

Figure 11.2   Axle box with white metal lined bearing surface

The lower half provides space for a lubricating pad, which was soaked in oil. The principle of operation of this type of bearing is the same as for a full sleeve bearing. An oil film is produced between the white metal lined bearing plate and the shaft.

Lack of lubrication was the most common cause of failure, caused due to one of the following reasons:

(a) Failure to replenish the oil.

(b) Failure to oil-soak and pack the wool pads correctly below the shafts.

(c) Failure to machine and scrape the white metal lining correctly.

Two other common causes, though less frequent, were as follows:

(a) Poor bonding between the bearing steel and the white-metal due to inadequate cleaning, or non-application of flux prior to metalling, or low temperature of white metal. The poor bonding resulted in the flaking off of the white metal and consequent bearing failure.

(b) Axle surface scored, not truly cylindrical or with poor surface finish. These deficiencies interfere with the formation and maintenance of a satisfactory oil film between the white metal and the axle, as a result of which there is excessive wear, higher bearing temperature, and eventual bearing failure.

11.8 Failures due to Excessive Ambient Temperatures

CASE STUDY

Sleeve bearings are used in large numbers in internal combustion engines for the crankshafts, crankpins and piston pins. Sleeve bearings are used in preference to ball or roller bearings mainly because the space available for these bearings is very limited.

The lubricant is a specialised product with a defined viscosity–temperature profile. Lubrication is usually forced oil circulation. Dimensional accuracy and the surface finish of the bearing shells, and the shaft journals are of great importance. If these factors are maintained within the specified limits the bearings give very reliable service. If there is a deficiency in regard to any one of their specifications, bearing failure is certain.

In one particular installation a number of diesel engines were working in a region where the ambient temperature varied between 0 °C in winter and 45 °C in summer. The manufacturer of the engine had recommended only one type of lubricant for all seasons. It was observed that there were a few bearing failures at the peak of summer. Assuming that these may have been due to excessive thinning of the lubricant, two methods were tried out. In a few engines the lubricants were changed to a higher viscosity grade and in the other engines, copper coils with flowing water in them were installed in the crank case to cool the lubricating oil in summer. Both methods were successful in preventing bearing failures.

11.9 Do's and Don'ts for Preventing Sleeve Bearing Failures

- Select the type of bearing, the lubricant and the maintenance schedules for its application in accordance with the recommendations of the manufacturer, as applied to the working conditions such as speed, axial and radial loads, temperature, etc.
- Carry out service trials to confirm the selection made as above. During these service trials, monitor the condition of the lubricant and the active surfaces. In the event of appearance of any sign of impending failure investigate fully and make modifications until satisfactory performance is achieved.
- Ensure correct filling of lubricant after every overhaul. Pay particular attention during overhaul to clearances, cylindricity and surface finishes of the active surfaces.
- Ensure correct lubrication at all times in regard to quality and quantity of the lubricant.
- Monitor active surfaces during overhauls. If irregularities are noticed repair shaft surfaces. If cracks, flaking, roughness etc. are seen in the bearing linings repair or replace the bearing linings.
- Investigate every case of bearing failure in service to determine the root causes and then to eliminate them.

- Some of the common causes of failures are:
  Absence of lubricant.
  Wrong lubricant.
  Defective lubricant.
  Inadequate lubricant.
  Contaminated lubricant.
  Axial and angular position error.
  Damage during assembly.
  Excessive or inadequate clearances.
  Excessive wear on active faces.
  Poor bonding between lining and shell.
  Poor surface finish on active faces.
  External heating of bearing.

Appendix 11.1
Calculation of Standard Deviation

The standard deviation of any set of measurements X_1 to X_n can be calculated in a tabular form as follows.

Measured Values	Deviation from average	Square of deviation
X	$X_{av} - X$	$(X_{av} - X)^2$
X_1		
X_2		
X_3		
–		
–		
X_n		
Sum_1		Sum_2

Step 1—Enter the n measurements in column 1 of the table
Step 2—Calculate the average $X_{av} = Sum_1/n = (X_1 + X_2 + \ldots\ldots X_n)/n$
Step 3—Calculate 'deviations' and 'squares of deviations' in the table
Steo 4—Calculate Sum_2 = Sum of the 'squares of deviations'.

$$\text{Then standard deviation} = \sqrt{(sum_2/(n-1))}$$

(It is easy to do these calculations on a Scientific Calculator. Enter the measurements and the standard deviation appears on the screen at once.)

Chapter 12

FAILURES OF WHITE METAL BEARINGS OF ELECTRIC LOCOMOTIVES

In this chapter we learn

- *A description of the main traction drive on a class of locomotives in service in the first half of the 20th century on the ex GIP Rly.*
- *A description of the types of failures on white metal bearings.*
- *The manner in which the problem was analysed and the conclusions.*
- *The remedial measures implemented with excellent results.* ❏ ❏

12.1 Introduction

White metal bearings were mentioned briefly in Chapter 11, case study 11.6. This particular case is discussed in greater detail in this chapter.

The earliest series of 1500 Volt dc electric locomotives on the Great Indian Peninsula Railway for the operation of freight trains was commissioned around 1930. They were used also as bankers on the Ghat section between Karjat and Lonavala and between Kasara and Igatpuri. In these sections the railway ascended from the western coastal region of India onto the Deccan Plateau at gradients of the order of 3 per cent. For railway operation these are very severe gradients while they may appear ordinary to road users.

In order to get maximum adhesion, the six wheels on the three axles of each bogie were coupled together mechanically through side rods and cranks as shown in Figure 12.1.

To avoid dead centres, the cranks on the two sides of each axle were kept at an angle of 90 degrees. The three wheel sets were driven by connecting rods from a jack-shaft driven through gears by a pair of motors on each bogie. All these arrangements meant that there were on each locomotive, four jack-shaft bearings for the two *jack-shafts*, four big-end bearings and four small end bearings on four connecting rods and finally twelve side rod bearings on each locomotive for the main traction drive only. All these 24 bearings were of the oil lubricated white metal type.

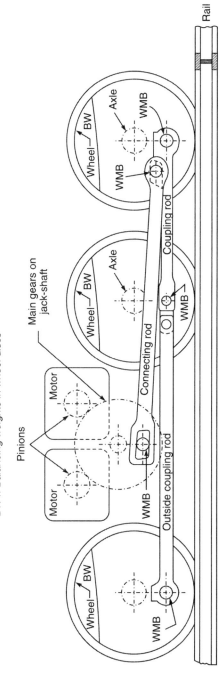

WMB – White metal bearings on connecting rods and coupling rods.
Failures were due to differences in dimensions between axle bearing and WMBs on connecting/coupling rods

BW – Balancing weights in wheel discs

Figure 12.1 Schematic drawing of mechanical transmission of power from motors to wheels through gears, connecting rods and coupling rods, on one side. (Similar arrangement on the other side, with 90° angular shift)

These locomotives could exert starting tractive efforts of the order of 36 tonnes and reach speeds up to 80 km/hour. Despite the presence of heavy reciprocating masses in the side rods and the connecting rods, the balancing weights on the wheels ensured smooth and even riding.

Each locomotive was in three sections:

- One central main frame that housed all the electrical equipment and the driving cabs.
- Two identical bogies each with two traction motors, the jackshaft, and the transmission comprising connecting rods and side rods described above.

12.2 Failure Modes

The following discussion is about the failures of white metal bearings in the traction power transmission system. Initially these bearings functioned well with limited failures. They did require a great deal of maintenance in the form of frequent lubrication and periodical re-metalling of white metal bearings some of which showed excessive wear, cracking or flaking.

However, with the passage of years the failure rate increased gradually and at one stage it was perceived as excessive. The failure modes and the causes of failures were as follows:

Failure Modes	**Perceived Causes**
- Excessive wear of white metal	- Lack of lubricant
- Melting of white metal	- Poor bond between shells and white metal linings
- Excessive wear on crankpins	- Poor quality of lubricant

12.3 Failure Investigation

While there was evidence to show that many failures were due to the perceived causes listed above, there were other contributory or accelerating factors. Statistical analysis of failure data was conducted to detect any co-relation between failure rates and

- (a) Driver.
- (b) Asst Driver.
- (c) Locomotive number.
- (d) Time of the day.
- (e) Date.
- (f) Ambient temperature.
- (g) Lubricant brand.
- (h) Virgin metal and recycled metal.

(i) Position of the bearing on the locomotive.

(j) Names of workers who fitted the bearings.

(k) Age since last schedule of inspection.

(l) Age since last periodical overhaul.

(m) Age since commissioning of the locomotive.

(n) Type of service: passenger, freight and banking.

(o) Section of railway line where the failure occurred.

This analysis did not highlight any particular or significant co-relation although some false leads were followed up. On extending the period of analysis, the co-relation disappeared or even got reversed. There was a significant and sustained reduction in the overall failure rate without any modification in the design or maintenance plan of the locomotives. It was concluded that the staff in charge of lubrication and maintenance had become more careful.

At this stage it was pointed out that the bogies were being extensively interchanged with locomotives and any attempt to co-relate failures with locomotive numbers was meaningless. This being a very valid point, the analysis was extended to bogie numbers instead of locomotive numbers. And immediately a startling picture emerged.

(a) There were some bogies that had been working for months with failure rates that were less than 10 per cent of the average failure rate;

(b) There were some bogies that had been working for months with failure rates in excess of 10 times the average failure rate;

(c) The above pattern was repeated in both parts when the total period of analysis was divided into two equal parts.

It was concluded that there was a definite difference in the group of bogies (a), as compared to the group (b). The first possible difference, which came up for consideration, was in the relevant dimensions. A series of accurate measurements were carried out in regard to the following items:

(i) Distances between axle box guides.

(ii) Clearances between axle-boxes and their guides.

(iii) Distances between centres of the bores of the side rods.

(iv) Distances between the connecting rod bore centres.

(v) Angles between crank-pins on wheels and on the jack-shafts.

(vi) Concentricity of bearing surfaces and shell surfaces.

It was observed that the dimensions in group (a) bogies were by and large much closer to the nominal dimensions than those in group (b). They were in fact outside the permissible limits in the bogies of group (b). It was concluded that the main cause of the increased incidence of white metal bearing failures was the lack of the required precision in the key dimensions mentioned above. Over the years the bogie frames had probably got distorted and the machining of the axle-box guides and liners was not as accurate as specified in the drawings.

12.4   Remedial Measures

A model bogie was then prepared partly by selective assembly and partly by special machining, keeping the dimensions very close to the nominal dimensions and within the limits allowed by the drawings.

This special bogie worked for six months with a bearing failure rate that was less than 5 per cent of the average for the rest of the fleet. A special drive was then undertaken to improve the methods of machining, measurements and selective assembly in order to get the same degree of accuracy on all the bogies as on the trial bogie. There was a dramatic reduction in the overall failure rate of bearings.

12.5   Conclusion

This case proved the use of statistical analysis of failure data as a method for the investigation of failures. This method was referred to in Section 4.5(c), Chapter 4. It is particularly useful when the failure rates are high and when all other direct methods have proved to be ineffective.

For such methods of investigation to be feasible it is necessary to maintain detailed statistics of data relating to routine operation, maintenance and failures. This has to be done right from the day of commissioning of the equipment and maintained throughout its lifetime. The cost and effort involved are negligible in relation to the benefits.

This case also highlights the need for respecting the tolerances indicated in original manufacturing drawings for any equipment whenever major overhauls or repairs are undertaken.

FLAT BELT FAILURES

In this chapter we learn

- *Applications of flat belts for the transmission of power in early twentieth century, some of which are still in service.*
- *Failure modes and failure mechanisms are similar to those in vee-belts discussed in greater detail in Chapter 14.*
- *Measures for prevention of flat belt failures are similar to those for vee-belts.*
- *A few case studies.* ❏ ❏

13.1   Introduction

Flat belts are becoming obsolete these days. Power is transmitted through electric motors, gears, cardan shafts or vee-belts. Flat belts are used only in some isolated pockets in very old installations. Reliability and durability of flat belts may be briefly examined, for those who have to maintain such old installations.

13.2   General Factors that Influence Reliability and Durability

The performance of a flat belt depends on the same factors that influence the performance of a vee-belt discussed in Chapter 14. These may be recalled briefly as follows:

(a) Pulley condition. The active face of the pulley being barrel shaped, makes the belt ride centrally. Sometimes flanges are provided on the pulleys to prevent belts from slipping off.

   (i) The pulley surfaces must be smooth. (Visible signs of defects: excessive wear, scuffing, peeling of active face of the belt.)

   (ii) There must be no sharp edges on the flanges. (Visible signs of defects: frayed edges, belt running in extreme position.)

(b) The belt tension should be kept within the limits specified by the manufacturer.

  (i) If the tension is too low it will lead to slipping of the belt and excessive wear. There would also be energy loss. Any stretching of the belts must be compensated suitably. (Visible signs of defects: excessive wear on active face of belt, overheating of belt.)

  (ii) A very high tension would lead to excessive wear and possibly belt or joint fracture. There may be problems with bearings due to excessive loading. (Visible signs of defects: belt torn near joint, bearings running warm.)

(c) The pulleys must be aligned and parallel. (Visible signs of defects: belts running in extreme positions, rubbing on flanges, worn or frayed edges of belt.)

(d) If belts can be shifted laterally by idler pulleys, the forks or guides used for this purpose should have smooth and rounded faces in contact with edges of the belts. (Visible signs of defects: frayed or worn edges of belts).

13.3 Failures due to Joint Defects

CASE STUDY

Unlike vee-belts, flat belts are usually provided with joints. The majority of flat belt failures are associated with these joints which become the weakest parts of flat belts.

Figure 13.1 shows one particular design of a flat belt joint, which has been in use for many years. Examination of failed belts showed two failure modes: one in which the belt failed in a transverse cut at XX' and the other in which the belt was cut in two longitudinal slits YY'. It was obvious that the sharp corners of the clip components had cut into the belt fabric, which was a composite of woven tapes with a rubber-bonding agent. Modifying the design of the clips to eliminate the sharp edges in contact with the belt, and increasing the distance 'Z' as shown in Figure 13.2, would increase belt life.

There are many designs of flat belt joints. Their selection should be based mainly on the failure rate, which must be examined to determine the exact failure mode, failure mechanism and the measures to prevent such failures. Normally, the joint should be expected to last without failure as long as the belt itself. They should be periodically inspected and replaced in time to prevent failures during operation of the drive.

As the general factors enumerated above were brought under control improvements in belt life and reduction in belt failures followed.

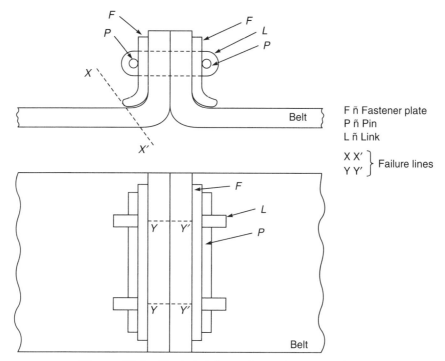

F ñ Fastener plate
P ñ Pin
L ñ Link

X X′ ⎱
Y Y′ ⎰ Failure lines

Figure 13.1 Flat belt failures due to fastener defects

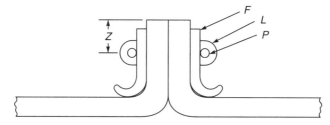

Figure 13.2 Improved design of belt fastener to minimise belt failures

13.4 Failures due to Rubbing on Edges

CASE STUDY

There were many cases in which one or more of the general factors listed above were deficient. These were determined on the basis of visible signs on used or failed belts and then on pulleys.

In one particular case there were signs of fraying of edges on the belt, but the pulleys showed no defect. Checking of the space around the belt throughout its run showed that a part of the belt guard was close to the belt, though not actually touching it. Further examination while the machine was working showed that the belt was touching this part of the belt guard. The belt guard was shifted laterally to provide adequate clearance from the belt, and the problem disappeared.

13.5 Failures due to Belt Stretch

CASE STUDY

Battery charging dynamos on railway coaches were driven through flat belts from axle mounted pulleys. Excessive wear on the active surfaces of the belt was experienced in some divisions of a railway. It was further observed that although the belt tension was satisfactory at the time of installation, within a few days they became slack due to stretching of the belt. This resulted in the belts slipping at low speeds when the transmitted torque was highest, due to the operation of the current regulation system. This problem was not experienced in other sections where the train speeds were higher and the minimum required tensions were lower.

The dynamo suspension was provided with a mechanism for moving the dynamo away from the axle in order to take up the slack. The staff were not aware of this device and they had failed to take care of this aspect. See Figure 13.3.

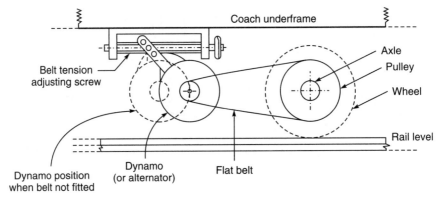

Figure 13.3 Shows dynamo suspension under coach. Belt tension is provided by weight of dynamo

A system was introduced to re-tension the belt two days after installation of belts. This measure resulted in improved reliability of the belt drives and also increased durability of the belts.

13.6 | Do's and Don'ts for Preventing Flat Belt Failures

- Monitor the condition of the belt. Look for signs of abnormal wear on the active surface and at the edges.
- Check the tension in the belt. It should be neither too low nor too high in comparison with the limits specified by the manufacturer.
- Monitor the condition of the pulley. Look for signs of excessive wear, over-heating, cracks and fractures, wobble, looseness on the shaft.
- Monitor the condition of the belt joints. Look for signs of excessive wear, cutting of belt and movement of fasteners.
- Ensure that the belt runs centrally. If there is a defect check pulley alignment, parallelism of shaft.
- Maintain records of belt life and correlate with the make of the belt. Determine the makes that give longer life.

Chapter 14

VEE-BELT FAILURES

In this chapter we learn

- *The advantages of vee-belts over flat belts, which were in use earlier in most industries for the transmission of power.*
- *The special precautions to be taken when vee-belts are used.*
- *The common modes and mechanism of failures of vee-belts.*
- *Measures to be taken to prevent service failures and to prolong the life of vee-belts.*
- *Case studies.* □ □

14.1 Introduction

$\mathcal{T}$ransmission of mechanical power from one shaft to another is often required in all types of industries wherever it is necessary to change the rotary speed or the RPM. In the initial stages of the industrial revolution, flat belt drives and rope drives were widely used for this purpose. These were gradually replaced by gear drives and vee-belt drives on account of their advantages.

Vee-belt drives have the following advantages over the earlier types of drives:

- reduced size and reduced initial cost;
- lower maintenance cost;
- higher reliability and higher durability.

However, the advantages claimed above can be achieved only if great care is taken in the design, selection and maintenance of the drives viz. the belts and the pulleys. The systems are more sensitive to deviations from the correct norms of design or operation. The following features are very important:

- The selection of the belt capacity for power transmission.
- Dimensions and surface finish of the vee grooves.
- The correct maintenance of belt tension.

(With regard to the limits for the items mentioned above, it is best to follow the manufacturer's advice.)

- The matching of the belts in multi-belt drives (See Case Study 14.2).

14.2 Belt Failures due to Mis-matching

CASE STUDY

The most common failure mode for vee-belts is premature fraying, wear and finally, breakage of the belts. Multi-belt drives which have two or more belts operating in parallel are more susceptible to these problems.

In one organisation, where a large number of vee-belt drives were in service for almost 20 hours per day, in some cases the belts lasted more than eight months while some failed within two months. As the belt drives were not easily accessible, the belts could not be inspected often. Careful inspection of the belts on one drive showed that out of six belts in parallel, two belts showed more severe fraying and wear than two others, whereas the remaining two seemed in good condition. This drive was monitored regularly and it was noted that one belt broke and dropped within a month and another a week later. A similar pattern was observed on two other drives.

The manufacturers of the vee-belts advocate use of matched sets of belts, but their distribution networks and the purchase/storage departments were not efficient in making matched sets of belts. This became evident, in this particular case, on checking a few of these sets.

It was, therefore, decided to do the matching at the workshop where the belts were actually fitted. The length of each belt was measured under tension by a simple device as shown in Figure 14.1. The belts were classified into six groups corresponding to each of three standard deviations on either side of the average. The plus or minus one-sigma groups grew rapidly and the plus or minus two-sigma groups grew slowly, but there was no difficulty in getting adequate numbers of matched sets amongst each of these classifications, for use in the installations. The plus or minus three-sigma groups accumulated very slowly but as soon as six belts in one of these two classes became available they were also used up.

The life of the belts in actual service increased to more than one year. The belts were replaced during the annual overhauls and failures in service became very rare.

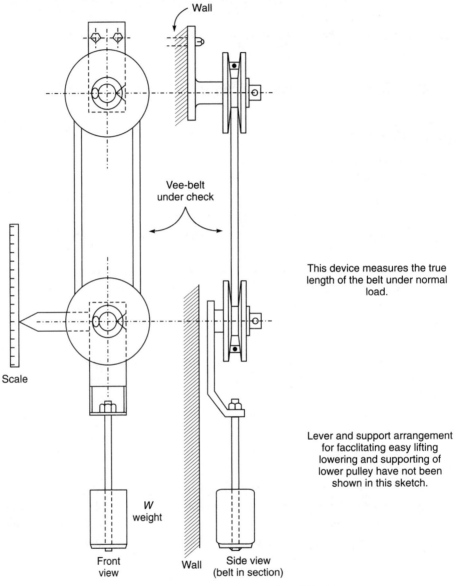

Wall

Vee-belt
under check

Scale

W
weight

This device measures the true
length of the belt under normal
load.

Lever and support arrangement
for facclitating easy lifting
lowering and supporting of
lower pulley have not been
shown in this sketch.

Front
view

Wall

Side view
(belt in section)

Figure 14.1   Device for matching V belts

14.3 Belt Failures due to Pulley Defects

CASE STUDY

In another location, where similar drives were in use, examination of the pulleys showed that the dimensions, surface finish and the angles of the vee-grooves on many of the pulleys were not within the appropriate limits. The alignment of the grooves of the driving and driven pulleys and the parallelism of their shafts were not always satisfactory. As a result, some of the belts were riding high while some were touching the bottoms of the vee-grooves. Even though the belts were well matched, belt tensions were not uniform and the life obtained was much less than what was expected with optimum conditions. Replacement of these defective pulleys and re-alignment of the machines led to an immediate improvement in the belt performance. Belt life quadrupled and belt failures became rare.

14.4 Belt Failures due to Belt Stretch

CASE STUDY

In another installation, an auxiliary machine—a cooling fan—was driven by vee-belts from the main shaft. There were a few cases of failures of the main machine due to failures of these vee-belts, and consequent failure of the cooling fans. Examination of a few of the machines in service showed that the belt tensions were too low and there was excessive slipping of the belts. Further investigation showed that while the initial tensioning of the belts seemed satisfactory, the belts were stretching during the initial operation. As there was no arrangement for automatic regulation of the tension, a practice of re-tensioning the belts, after 10 hours and again after a week by shifting the fan pedestal slightly on its slide-rail, was introduced. Subsequently, belts of another make which stretched less were put into use, but the practice of rechecking and adjusting the tension was continued. These simple measures improved the reliability and the durability of the vee-belt drives.

14.5 Belt Failures due to Poor Belt Quality

CASE STUDY

Vee-belt life depends on a number of factors relating to belt selection, belt quality, pulley design, pulley quality, tension regulation, maintenance practices, speed and torque characteristics of the driving and driven machines, and finally, the environment. Care taken in these regards is amply repaid in terms of increased life and reduced consumption of the belts and improved reliability of the system. However, these belts have a limited life. Since there is a lot of variation in the life of different belts, if service failures must be prevented it is necessary to replace the belts in advance during one of the scheduled maintenance plant outages.

In an industrial unit where a large number of vee-belt drives of different sizes and other characteristics were in regular use, the cost of belt replacements and of belt failures was perceived to be excessive. The following simple steps led to a considerable (more than 50 per cent and 80 per cent, respectively) reduction in these two costs:

(a) All the staff concerned with the maintenance of these belts were given a three-day course on the factors which influence the durability and reliability of the belts. The potential for improvements in these parameters by practices such as matching of belts, control on pulley quality, periodical re-tensioning, etc. were discussed.

(b) Machine-wise records of belt replacements were maintained, with such details as make of the belt, date of replacement and pulley condition. The manufacturer representatives were associated in reviews of these records.

(c) Maintenance decisions regarding belt and pulley replacements, and purchase decisions regarding quantities ordered from three different suppliers were taken on the basis of the patterns emerging from the records.

(d) Decisions regarding offers made by some manufacturers for special quality belts (at higher prices), were taken after comparing the life-cycle costs on the basis of the records mentioned in (a) above.

14.6 Belt Wear due to Gritty Dust

CASE STUDY

In one particular application, even after all the above factors were taken care of and, considerable improvements in belt reliability and durability achieved, the engineers in charge felt that further improvements were possible if certain steps were taken. They felt that the gritty dust in the air from the stone crushers might be causing increased wear of the belt active faces. Fully airtight closed guards replaced the wire mesh belt guards. This resulted in an 80 per cent improvement in belt life. The total increase in life obtained as a result of all the measures taken on the above lines was of the order of 300 per cent and service failures were almost completely eliminated.

14.7 Do's and Don'ts for Preventing Vee-Belt Failures

- Verify the correct selection of the size and number of belts provided, on the basis of the relevant parameters relating to speed, power, environment, etc.
- Examine failed vee-belts and corresponding grooves.
- Examine pulleys for vee-groove angles, pitch diameters and surface condition. In case of multi-groove pulleys verify matching of groove diameters.
- Verify correct matching of multi-belt sets.
- Verify belt tensions after installation and again after a few days of service.
- Ensure the maintenance of registers giving details of names of manufacturers, dates of fitting of belts, dates of failures and replacements.
- Ensure that vee-belt drives are not exposed to water, oil, dust grit, sunlight.

Chapter 15

PULLEY FAILURES

In this chapter we learn

- *Two pulley failure modes, although they are rare.*
- *Pulley defects cause belt failures and reduction in belt life.*
- *The most important requirement for pulley manufacture is accuracy in the dimensions and surface finishes of the active faces and the bores.*
- *Two case studies.*

15.1 Introduction

$\mathcal{P}$ulley failures due to their own internal defects are rare. It is necessary to recall briefly pulley defects that cause excessive wear and finally failures of belts. These are:

(a) Pulleys not aligned correctly.
(b) Pulley axes not parallel.
(c) Active surfaces of pulleys not smooth and truly circular.
(d) Pulley flanges not rounded.
(e) Pulley surfaces not properly crowned.
(f) Vee-belt groove angles as well as groove width incorrect.
(g) Vee-belt groove diameters not equal.

15.2 Defects in Pulley Fabrication

CASE STUDY

Pulleys are generally made of cast iron but large pulleys are often fabricated. One such pulley consisted of a cast iron hub, mild steel spokes and a strip of mild steel bent into a narrow cylinder or rim. The tapped ends of the spokes were screwed into tapped holes in the cast iron hub. The outer ends of the spokes were welded to the mild steel rim (See Figure 15.1).

In one case, the welding between the spokes and the rim cracked. These cracks were discovered early and were restored. The welding process and the design were reviewed before effecting repairs. The rest of the spokes, were also modified. As the active faces of the pulley were not running true, the manufacturing process was reviewed and modified as follows:

The drawing for the pulley was modified to incorporate the improvements in weld design. The sequence of the spokes was indicated. The final or finish boring of the hub and the final machining of the pulley active surface was done last, after the welded pulley was annealed. This ensured a pulley surface that ran true.

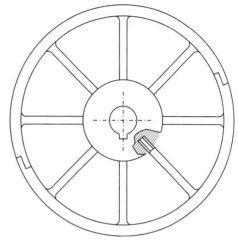

Figure 15.1   Fabricated pulley did not run true. Fabrication process was renewed and machining of bore and active force of pulley done last. This improved belt life.

15.3   Defects in Machining Pulley Bore

CASE STUDY

This was a case of a cast iron pulley with vee-grooves. The pulley was in two halves with heavy bolts/nuts across the hub to hold the pulley on the shaft seat (See Figure 15.2).

The problem experienced was—slipping of the pulley on the shaft. Investigation showed that the dimensions of the pulley bore and the shaft seat diameter did not provide adequate interference and the gap G was not adequate. Some pulleys were scrapped. Some were utilised by selective assembly. The limits for the pulley bore were changed in the drawing to ensure that a small gap G as shown in Figure 15.2 remained between pulley faces inspite of tightening the bolts and nuts.

The machining process was modified. A 2-mm plate was inserted between the two halves of the pulley. The bolts/nuts between the two halves were fully tightened before machining the bore and the vee-grooves. The 2-mm plates were then removed before assembling the pulley halves on the shafts. These changes effectively eliminated the problem of pulleys slipping on the shafts.

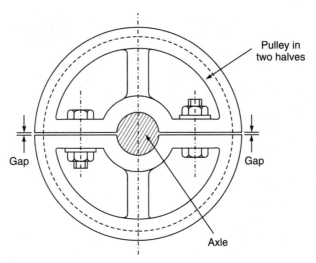

Figure 15.2 The pulley dimensions should be such that a small gap remains after full tightening of bolts and nuts.

15.4 Do's and Don'ts for Preventing Pulley Failures

- Ensure that the defects listed in Section 15.1 are eliminated, to prevent belt failures.
- If necessary examine the fabrication and machining procedures to prevent defects in pulleys themselves.

Chapter 16

GEAR FAILURES

In this chapter we learn

- *Two major failure modes on gears, viz. fracture of teeth and excessive wear.*
- *Mechanisms for each of the above failure modes.*
- *Case studies of gear failures.*
- *Preventive measures.*

❑ ❑

16.1 Introduction

Gear drives are used widely in all types of machines. The smallest most commonly used gears are in the wristwatch. Gears are also used in toys and household appliances. Their failures in these applications are rare, as the power transmitted is small.

Gears used in transportation and industry, as in automobiles, locomotives, ships, aircraft, machine tools, and manufacturing machines transmit power in the range of a few horse-power to tens of thousands horse power. For reasons of economic design, the operating stresses on the gear materials in these applications are closer to their safe limits, than in wristwatches or toys.

Gears are used in large numbers in road vehicles. Over the last fifty years through sustained research, failure investigations and elaborate quality assurance measures, manufacturers have attained high reliability of automobile gearboxes.

Gear technology with regard to tooth profiles, hobbing and grinding machines, and gear inspection has reached a stable state. The few failures that still occur are due to defects in materials, application engineering and the manner in which the machines are used.

The gears of railway traction transmission systems are, like all other components and equipment in railway locomotives and rolling stock, designed with very low margins of safety in order to minimise weight and size. Consequently the design stresses in gear teeth are usually close to the endurance limits of the steel used for gear manufacture. The margins of safety are very small. Any stress raisers on the surface of the gear teeth can become the initiators of fatigue cracks and fractures.

Gear failures are not as rare as they should be in such applications. Zero failure performance can be achieved by paying attention to the critical aspects at the stages of design, manufacture and maintenance. These are highlighted in the following case studies.

16.2 Principal Modes of Failures due to Gear System Defects

There are two principal modes of failures of gears: (a) breakage of gear teeth; (b) excessive wear of gear teeth.

(a) Causes of gear teeth breakage:
- There may be stress raisers due to defects in surface finish or due to grooving at the roots of the gear teeth. In such cases the fracture aspect will show two distinct parts—one which is smooth indicating the fatigue crack propagation period and the other which is rough indicating the sudden failure of the weakened tooth. Improvements in the machining and grinding operations in the manufacture of the gears will eliminate such failures.
- The failure may be due to excess shock loading in the system. This could be due to defects or deficiencies in the resilient elements in the power transmission route. In such cases the fractured surface is likely to show brittle fracture aspect.
- Gear teeth fracture may also occur due to falling of extraneous metallic or other hard substance components between the teeth. If such possibilities exist, the broken components should be carefully examined or screened to detect foreign material. Such cases can be prevented by provision of gear cases.

(b) Excessive wear of gear teeth surfaces could be due to one or more of the following causes, generally in order of probability:
- Gritty or hard contaminating substances getting into the gear lubricant.
- Defects or deficiencies in the lubrication of the gear.
- Incorrect distances between gear centres or gear axes not being parallel.
- Defective quality of the material of the gear or improper heat treatment.

16.3 Failures of Railway Traction Gears

CASE STUDY

The main traction power transmission in electric and diesel electric locomotives with nose-suspended, axle hung traction motors is through simple pairs of pinions and gears as shown in Figure 16.1.

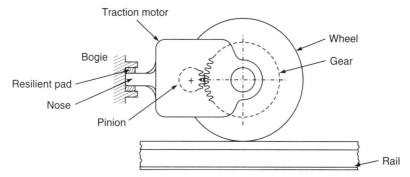

Figure 16.1 Axle mounted traction motor, nose suspension on bogie. The gear and pinion are subject to shock loads from axle vibration

Apart from the torques and forces on account of the main power transmission, these gears are subject to inertia torques that arise due to the epicyclic gear effect of the vertical oscillations of the wheels. These are produced when the locomotive passes over rail joints and other track irregularities. The resulting forces on the gear teeth consist of fluctuating forces with impact components.

Investigations showed that impact stresses due to the epicyclic gear effects of the vertical movements of axles were responsible for some failures. These were eliminated by providing more resilient nose suspension.

16.4 Gear Failures due to Machining Defects

CASE STUDY

This section deals with a case of failures of gear teeth in an application where the gear train transmitted several hundred horse power. There were three cases spread over a period of about four months. The failures led to extensive damage to the gearboxes and in one case to associated equipment. Examination of the failed gears, gear boxes and the associated components showed that the original failure in each of these three cases must have been due to the fracture of one gear tooth, all other damages being consequential. Closer examination of the fractured gear teeth showed the following features:

(a) In each case, the fractures were at the roots of the teeth.
(b) The fracture showed two distinct zones-one with a smooth, striated, stained aspect and the other with a rough and brighter aspect. This led to the conclusion that they were fatigue fractures.

(c) The ratio of the fatigue propagation zone to the sudden fracture zone was small showing that the stresses in the teeth were close to the endurance limit of the material.

(d) Examination of the undamaged teeth of the gears showed sharp grooves on all the gear teeth at the exact location of fractures. These could have become stress raisers and led to the initiation of fatigue cracks.

(e) Dye penetrant tests on the undamaged gear teeth at the suspect zones showed hairline cracks in similar locations on two other teeth.

(f) Examination of the records revealed that all the gears, which failed, had been installed during the previous year and were all from one particular batch of supplies.

Figure 16.2 shows the fracture aspects and the appearance of the stress raisers at the roots of the gear teeth. These grooves had obviously been introduced by some defect in the machining tools used by the manufacturer of the gears.

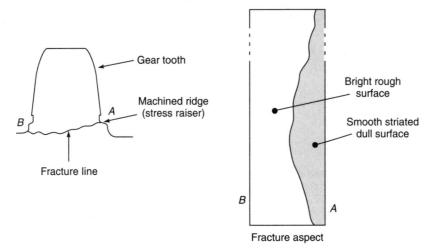

Figure 16.2 Fracture of gear tooth due to fatigue crack

- The dates of commissioning and the locations of all the gears from the same batch of supplies were determined from the records. The equipment having the suspect gears with the longest service were withdrawn from service, a few at a time, and the gear wheel were replaced with new ones having gears of another make that had no history of failures.

- The defective gears were cleaned and subjected to dye penetrant tests at the roots of the teeth around the grooves. Those with signs of hairline cracks were replaced and discarded. The gears that passed the dye-penetrant tests were reclaimed by grinding away some material around the grooves to provide a smooth transition and remove all stress raisers as shown in Figure 16.3.

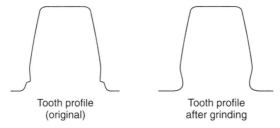

| Tooth profile (original) | Tooth profile after grinding |

Figure 16.3 Grinding of teeth to remove all stress raisers

- The reclaimed gears were put back into service and kept under observation, i.e. annual examination by dye penetrant tests. This practice was discontinued only after ten gears with more than one year's service were found free of cracks.
- The relevant gear drawings were amended with the active faces of the gear teeth free of any ridges, grooves or such other stress raisers. The gear inspection and maintenance staff were advised about these failures and their causes. The gear manufacturer was persuaded to make free replacements of all broken and cracked gears.

16.5 Gear Failures due to Defective Heat Treatment

CASE STUDY

In another case of gear failures due to tooth fracture, there were no stress raisers on the teeth and no visible differences between the teeth. On examining the surface hardness of all the gear teeth on each wheel wide variations were observed, and fractured teeth had excessive hardness. Investigation of the manufacturing process (heating, quenching and tempering) showed defects, which produced differences in the hardness of the teeth.

The manufacturer replaced all such gears. Suitable changes were introduced in the manufacturing process. Gear inspection staff were alerted about such problems and the relevant drawings were amended with the remark that gear teeth hardness should be measured on at least five teeth around the circumference. Reduced tolerances for the differences in hardness on each gear were indicated.

16.6 Gear Failures due to Grit Entry in Gear Boxes

CASE STUDY

In another case, the problem was not fractures but excessive wear on the teeth of some locomotive gears. Examination of the gear compound showed the presence of grit. Inspection of the gear boxes of other locomotives with the same design of gear box revealed that some of the covers of the inspection windows on top of the gear boxes had fallen off, and much of the gear compound had also been thrown out. Enquiries with the maintenance staff revealed that this was a chronic problem due to a defect in the latch mechanism.

The following steps were taken:

(a) The latch mechanism was studied and modified on the lines of other more reliable inspection covers on gearboxes.

(b) Gear compound samples from all other gearboxes were checked and those with grit content were cleaned out and replaced.

(c) Gearbox cover inspection was added to the list of specific checkpoints in the schedule for daily visual inspection of locomotives.

Although the proximate cause of the gear wheel failures was grit entry into the gear lubricant, the root cause was defective latch mechanisms on the gearbox covers.

16.7 Gear Rim Failures due to Machining Defects

CASE STUDY

There were some cases of failures of gears where the problem was not the gear teeth but the gear rim itself. The rim was provided with internal projections, which were used, for the transmission of torque from the gear centre through resilient bushings. The main purpose of providing this arrangement was to introduce flexibility, thereby equalising power transmission between gears in parallel and absorbing shock loads. Figure 16.4 shows a schematic view of the arrangement and an enlarged section of the projections on the gear rim.

Failures of the gear rim were through YY', as shown in Figure 16.4. As in the case of the gear tooth failure described in case (I) above, these were fatigue fractures initiated from stress raisers in the corner at R , where the radius of curvature was too small. Increasing the radius of curvature at the corner from 1 mm to about 5 mm as shown in Figure 16.4(c) solved this problem. Gears already in service were withdrawn, checked for cracks by dye-penetrant tests and grinding of undercuts done on rims without cracks. Those with cracks were discarded.

16.8 Gear Failures due to Other Causes

CASE STUDY

There are other ways in which gears are damaged. These are rare but have been experienced. They are due to the negligence of maintenance staff. Unlike the large scale problems created

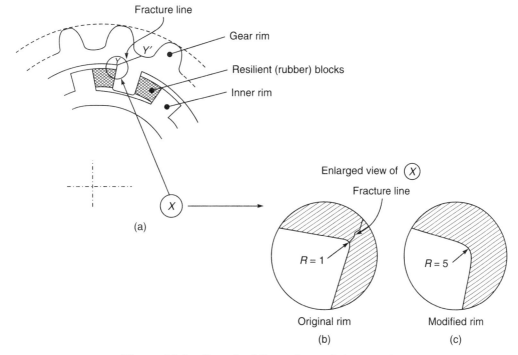

Figure 16.4 Gear rim failures due to fatigue cracks

by design or manufacturing errors, the problems due to maintenance deficiencies are usually in small numbers. They are detected and eliminated very soon. For the benefit of the new members, here is a list of such cases.

■ Gear failures caused by total lack of lubricant in gear box due to:

    (i) Leak in oil lubricated gear box.

    (ii) Sheer negligence of maintenance staff.

    (iii) Contamination of lubricant due to entry of foreign material from the environment.

■ Excessive wear and fractures of gear teeth due to errors in the centre distances between gears.

■ Gear fractures due to small tools or hardware left inside gearboxes by maintenance staff.

To prevent failures of this type, it is necessary to provide comprehensive and continuous training of maintenance staff in regard to not only knowledge of correct maintenance practices, but also attitudes regarding failures and their responsibilities. Maintenance supervisors have a great role to play. Conscientious supervisors can keep the maintenance organisation under their control by ensuring the following:

■ General cleanliness of the machinery and the surroundings.

■ Alertness to unusual sounds and vibrations.

- Investigation of every defect, failure and accident to determine the root causes.
- Monitoring of temperatures, vibrations, and effluents.
- Inspection of worn and discarded components, etc.

16.9 Do's and Don'ts for Preventing Gear Failures

The following list is prepared on the assumption that the material, tooth-profiles, heat treatment, and dimensions of the gears and their boxes are all correct. Such an assumption is generally valid. The following list deals mainly with operation and maintenance.

- Ensure the correct quality and quantity of the gear lubricant.
- Ensure that there is no contamination of the gear lubricant by other lubricants, water, grit, dust, etc.
- In case of gear tooth breakage or excessive wear, investigate carefully. Look for the probable causes discussed above in Sections 16.2 to 16.6.
- If none of the defects in operation or maintenance are detected, consider the possibilities of manufacturing defects. Check the gears against the design details, i.e. the working drawings, in regard to dimensions, tolerances, tooth profiles, material and heat treatment.
- If there are no manufacturing defects then consider the possibilities of defects in the design. This will require reference to experts in gear design.

Chapter 17

STEEL WIRE ROPE FAILURES

In this chapter we learn

- *Different types of stranded wire ropes.*
- *Failure modes of wire ropes.*
- *Failure mechanisms such as corrosion, fatigue, inertia forces.*
- *Failure prevention measures.*
- *Case studies.*

17.1 Introduction

Wire ropes are used in a number of applications such as lifting cranes; automatic tension regulators in traction overhead equipment; terminating wires and span wires which are not required to carry any current; elevator lift ropes; etc.

Wire ropes may be made of soft iron, steel, special steels, stainless steel and many other alloys, depending upon the application, space available and environmental conditions.

The wire rope may have hemp cores or the core may also be of steel. The wire rope may be composed of a large number of strands, which may be 6, 7, 19 etc. wound together in groups and layers. There are many other forms and variables connected with the design of wire ropes but these are not relevant in discussing the failures of wire ropes.

17.2 Failure Modes and Mechanisms

There are just a few failure modes and mechanisms which can be relevant to all the different types and sizes of wire ropes. These common modes and mechanisms are categorized below:

(a) Fracture due to excessive load, continuous or momentary.
(b) Fracture due to degradation of strength due to:
  (i) Corrosion.
  (ii) Abrasion.
  (iii) Fatigue.

Wire rope failures are very rare but if they occur there may be a great deal of consequential damage. Wire ropes are designed with factors of safety in the range 5 to 8 depending upon conditions of service and the consequences of failures.

The factors to be considered on safety measures are:

- rope speed;
- design of rope terminations;
- number and size of ropes in parallel;
- number and size of pulleys, sheaves and drums over which the ropes have to pass;
- length of rope;
- possibilities of corrosion and abrasion;
- variations in tension.

An important consideration for calculating the maximum force on the rope is the acceleration and retardation of heavy supported masses during starting and stopping. The inertia forces can rise to very high levels under certain conditions.

When investigating failures of wire ropes the factors of safety and the maximum forces should be considered. If a rope failure occurs it can be assumed that the cause lies in one of seven factors listed above. A process of elimination can determine the true cause.

It is not often realised that a simple component like a wire rope is in fact almost like a small machine. It has all the features of a machine. It has wearing parts, which rub on each other. They need to be lubricated. The components are highly stressed and subject to corrosion fatigue. It require regular inspection to look for evidence of abrasion, fretting, corrosion, deformation and fatigue cracks. The stresses are due to a combination of:

(a) Simple tensile stresses due to the main load,

(b) Bending stresses due to bending over sheaves, pulleys or ferrules,

(c) Hertzian stresses at the points of contact between ropes and sheaves/pulleys or between the strands.

Some parts of the ropes may normally be inaccessible for inspection while subject to some or all the degradation processes mentioned above. Failures often occur at such inaccessible locations. Another common location of failure is at the terminations where the rope is either bent around a ferrule or crimped/swaged in a socket. In some cases the failure may be of the terminating socket itself and not of the rope. It is very important, therefore, to determine the exact location of the failure and also the precise failure mechanism and the factors that contribute to the degradation process, before deciding on the preventive steps. Mere replacement of the rope is not sufficient. If recurrence of failures is to be prevented, the causes of the failures must also be eliminated by suitable changes as may be required in the material specification, design of the rope and the terminations, or the maintenance systems.

Even after all deficiencies in design, manufacture and installation of wire ropes are eliminated, periodical inspection and periodical replacement of highly stressed wire ropes must be ensured for attaining total reliability.

Steel ropes can be subjected to non-destructive testing through electromagnetic examination in accordance with ASTM 1571. Dye penetrant tests on ferrules and sockets are advisable when there is a history of failures in these components.

The selection of the material of the wire rope is also important. In corrosive environment, stainless steel or galvanised steel may be used. There are controversial views in this regard. Although stainless steel is often preferred, some users have reported that galvanised steel is more reliable and more durable than stainless steel for wire ropes because the latter has a lower fatigue strength.

17.3 Case Studies

The generalities discussed in Section 17.2 can be illustrated by the following case studies.

Case Study Number	Failure Mode	Failure Mechanism and Remedy
17.3.1	Hoist rope snapped during initial tests on a new design of hoist	The hoist rope was wound on a drum driven by a worm gear. The rope snapped while stopping a downward movement of the full load. The worm drive being non-reversible, the load stopped abruptly when the motor was switched off. The inertia force snapped the rope. The provision of an elastic coupling between the worm gear and the drum solved the problem. Measurements with a maximum tension indicator between the hoist hook and the load confirmed that the maximum force was now 25 per cent more than the static load instead of 300 per cent more. The rope size was also increased to give a factor of safety of 8 instead of the earlier 6.
17.3.2	Stainless steel wire rope used for the tension regulator of the track overhead equipment snapped in service	Examination of the failed wires showed considerable corrosion of the strands particularly in the zone, which moved back and forth on the pulleys. The so-called stainless steels are not always corrosion resistant. They are affected by sulphurous gases and by saline sprays. Stainless steels, which have not been correctly heat-treated, corrode more rapidly. Abrasion and bending damages the protective oxide coating and accelerates corrosion. The problem was solved by (a) including corrosion tests in specs for wire, (b) replacing the wires periodically, (c) increasing the radius of some pulleys, (d) reducing the radius of rope strands and increasing their number, (e) periodical lubrication of ropes.

Contd.

Case Study Number	Failure Mode	Failure Mechanism and Remedy
17.3.3	The guy wire of an anchor mast in railway overhead wire system was made of galvanised, stranded steel wire. One such wire parted, causing the mast to bend	On investigation it was observed that the distance between the base of the mast and the anchor foundation was (wrongly) provided at only 30 per cent of the standard distance due to the non-availability of space. As shown in Figure 17.1. 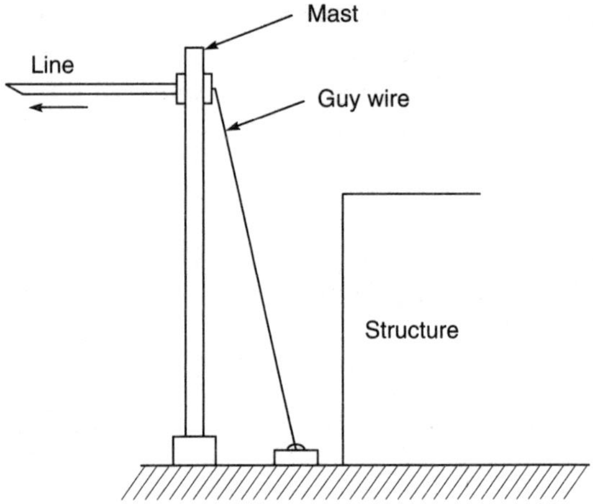 **Figure 17.1** Guy rope anchor too close to mast This resulted in a greatly increased tension in the guy wire. When new, the wire was able to withstand the excessive tension but after a few years, the wire was weakened possibly by corrosion and it parted. The problem was solved by the adoption of the standard method for such locations, comprising a short mast and guy at the correct angle as shown in Figure 17.2. 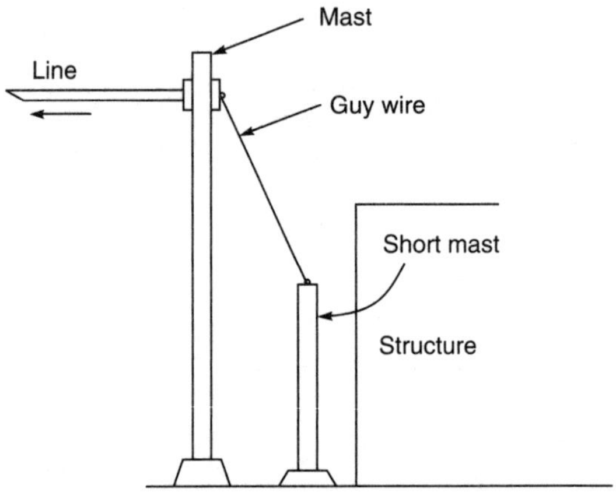 **Figure 17.2** Guy rope angle increased by short mast

— *Contd.* —

Case Study Number	Failure Mode	Failure Mechanism and Remedy
17.3.4	In one workshop a locally designed winch with a stranded steel wire rope was being used to drag locomotives from one assembly station to another. The rope snapped after several years of use	Investigation did not expose any corrosion of the wire. However, tensile tests on five samples taken from the vicinity of the fracture showed a tensile strength varying between 50 and 80 per cent of the original or specified tensile strength. It was then seen that the diameter of a pulley over which the rope took a 90 turn, was only about 20 times the diameter of the rope. See Figure 17.3. **Figure 17.3** The recommended ratio as per the wire manufacturer's manual was at least 60. The rope was replaced and scrapped. A new pulley of diameter three times the original was provided.

17.4 Do's and Don'ts for Preventing Wire Rope Failures

- Use wire ropes manufactured by reputed manufacturers.
- Before acceptance of supplies carry out all the specified tests. In particular the corrosion tests if the ropes are exposed to the elements during use.
- Protect the wire ropes against corrosion, inspect and lubricate them periodically. Carry out periodical proof load tests.
- Ensure that adequate resilience is provided in the connections to heavy masses.
- Ensure that pulleys over which the ropes pass have adequate radii as recommended by the manufacturers of the ropes.
- Where ropes are used in parallel, ensure that they share the total load evenly through equalising devices.
- Refer to the relevant IS and to manufacturers' data sheets for details such as tensile strength, minimum radius of bend, terminations, etc.

Chapter 18

Spring Failures

In this chapter we learn

- *Different types of springs.*
- *Common failure modes and mechanisms of a large variety of types of spring designs.*
- *Failure modes and mechanisms of springs.*
- *Preventive measures at design, manufacture and maintenance stages.*
- *Case studies.*

18.1 Introduction

There are several categories and shapes of springs in use in different types of machines or equipment. These are Helical Tension, Helical Compression, Spiral, Laminated Flat, Torsion Rod or Tube, Belleville or Disc, as shown in Figure 18.1.

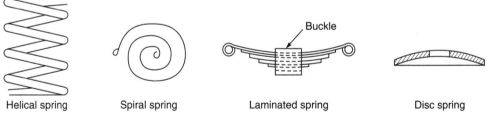

Helical spring   Spiral spring   Laminated spring   Disc spring

Figure 18.1 Some common types of springs

While the shapes and sizes vary widely and the number of designs run into thousands, the basic principles of operation as also the common failure modes and failure mechanisms are very few. The most common mode of failure is fracture through fatigue.

As illustrated in Figure 18.2 S, is the magnitude of the peak stress in the material of the spring and N is the number of stress reversals or fluctuations. As the point (S_2, N_2) falls in the shaded portion outside the S–N curve of the material, fatigue failure is certain after N_2 cycles. The peak stress S_1 is less than the endurance limit, and failure would never take place due to the asymptotic nature of the S–N curve along the N-axis. (For a more detailed

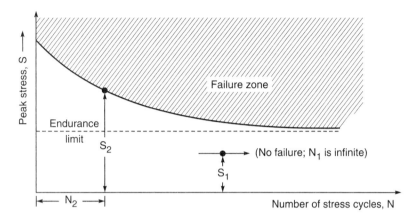

Figure 18.2 S.N. curve for spring steel

explanation of Metal Fatigue, refer to Chapter 13 of the book *Electrical Fires and Failures by A.A. Hattangadi, Tata McGraw Hill.*)

If the initial design of the spring is defective, i.e. the peak stress in normal operation exceeds the endurance limit shown in Figure 18.2, the spring will fracture sooner or later during normal usage. When the fracture takes place will depend on the excess in the alternating component of the peak stress over the endurance limit and the number of stress fluctuations. However, failures due to such gross design deficiencies are rare.

Failures are likely to be due to local increases in stress caused by stress raisers such as nicks, sharp bends, or inclusions in the material, or due to reduction in the endurance limit of the material due to corrosion, poor surface finish or defective heat treatment. Sometimes there may be a general increase in stress as a result of increased amplitude of spring deflections caused by some external defect.

Springs that have a constant or unvarying deflection fail very rarely since fatigue failures require large numbers of reversals or fluctuations of stress.

A few typical cases of spring failures are described below. Although some equipment names like airflow relay, locomotive bogie, etc., are mentioned, these applications are not relevant to the issues here. An identical failure with regard to mode and mechanism could occur in some other, totally different application. The common factor is always the component, its shape, its material and the manner in which it is operated and stressed.

18.2 Spring Failures due to Sharp Bends

CASE STUDY

Figure 18.3 shows the principal parts of an airflow relay. The tension in the spring SP keeps the vane V pressed against the stop ST and the pair of electrical contacts remain open. When the

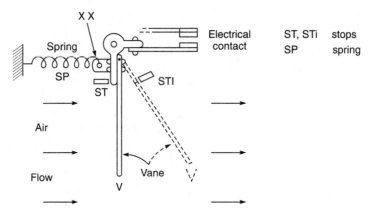

Figure 18.3 Vane type of air flow relay fitted in the air flow duct

air velocity across the vane is high enough, the force on the vane is adequate to overcome the tension in the spring and causes it to move in the direction of the air flow. This results in closing the electrical contacts and completing the circuit for operating the electrical system.

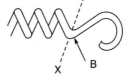

Figure 18.4 Enlarged view of the hook at the end of the spring in the vane type air flow relay

Spring fractures across the line XX were experienced in electric locomotives. Figure 18.4 shows an enlarged view of the hook at the end of the spring SP. Due to the sharp bend at B, micro cracks develop in the material of the spring wire. These lead to further stress concentration, fatigue cracks and eventual fracture at B. Figure 18.5 shows a further enlargement of the bend area. The material in the outer layers at O is stretched by nearly 100 per cent as compared to the centre line of the wire. Figure 18.6 shows the obvious remedy. The bend radius R should be at least four times the radius 'r' of the wire.

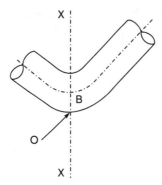

Figure 18.5 The material at O is stretched 100%

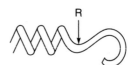

Figure 18.6 Generous radius at bend prevents fatigue fractures

18.3 Spring Failures due to Dents

CASE STUDY

Figure 18.7 shows another type of spring. This is much larger in size, made of a very thick rod, and the spring is in compression. It is the main primary suspension spring in a locomotive bogie. It supports the weight of the locomotive and provides cushioning against the effects of track irregularities. Under normal conditions, there is a gap G between adjacent turns of the spring. When the locomotive is running, the effects of track irregularities and the resultant vertical oscillations of the locomotive causes the gap G to increase and decrease. Sometimes, when the amplitude of the oscillation is too high, the gap can become negligible and the coils of the spring touch each other. It then behaves like a solid mass.

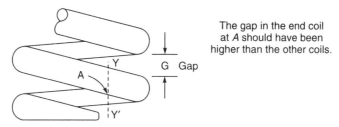

Figure 18.7   End coils of the main spring of a locomotive. Fracture through YY′

Fractures of these springs through Section YY′ in Figure 18.7 were reported in some electric locomotive sheds.

It was found that the ends of the springs were not cropped and ground properly and sharp corners were present. They caused dents in the 1st coil at A in Figure 18.8 when the springs closed solid. These dents acted as stress raisers that developed into fatigue cracks and eventual fractures. The remedy was to grind the ends of the spring as shown in Figure 18.9. If this does not solve the problem, it would be advisable to review the complete design of the spring and the track conditions to see whether the spring closes solid too often, and even when the track is in good condition.

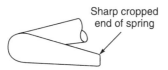

Figure 18.8   Sharp corner at the cropped end caused indentation on the adjacent coil

Figure 18.9   Removing sharp corner on cropped end of spring by grinding prevented fracture of adjacent coil

18.4 Spring Failures due to Wear at Its Terminal Fittings

CASE STUDY

Figure 18.10 shows a type of termination used in certain applications for tension springs. The spring is threaded into the perforated lugs during assembly.

Fractures of the springs at the point of entry into the terminal lugs were experienced in some of the switchgear where tension springs of this type were in use.

Due to constant abrasion under heavy pressure between the spring S and the lug L, in Figure 18.10, there was wear around the holes in the lugs and on the spring coils. This wear became significant only after many years of service. On some of the springs, the wear pattern was such as to form grooves on the spring coils.

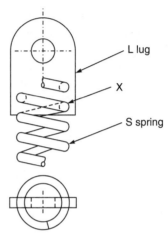

Figure 18.10 Spring failures due to wear at the termination

Due to stress concentration at the edges of these grooves (and not so much due to the reduction in the cross-section), fatigue cracks developed, propagated and caused fractures of the spring wires at X in Figure 18.10.

The remedy, is to replace the worn springs along with the lugs. A temporary reprieve was obtained by threading the springs a little more into the lugs, thereby shifting the highly stressed zones to the un-grooved parts of the springs.

18.5 Spring Failures due to Corrosion and Wear

CASE STUDY

Locomotive pantographs are devices that are used for current collection from the overhead contact wire, which runs over the railway track of electrified railways. The pantograph pans are kept in contact with the wire, with a constant force by means of helical springs. The contact wire level keeps changing. It is low under bridges and high at level crossings. Due to the fluctuations in contact wire height, the pantograph helical springs are subject to changing deflection and consequently to fluctuating stresses.

Failures of these springs were experienced on one railway. On close examination it was observed that the surfaces were badly pitted by corrosion and fatigue cracks were seen to have originated from such pits. The failures are not due to reduction in cross section; they are due to stress concentration at the surface imperfections. These springs are exposed to the weather, being mounted on the roof of the locomotive. The remedies are (a) to replace all springs with signs of corrosion and, (b) to minimise corrosion by the application of a thick oil over the spring surfaces during periodical maintenance.

18.6 Spring Failures due to Abrasion (I)

CASE STUDY

In a 25 kV circuit breaker of one particular make, helical springs are used to open the circuit breaker at high speed. After years of trouble free operation there were cases of spring failures due to fracture at the central part of these springs. See Figure 18.11.

Due to space limitations, the springs were designed, as revealed during investigation of failures, to operate with stresses close to the endurance limit. Under static condition there seemed to be adequate clearance between the spring and the structural member. As the springs were rather long they were subject to some lateral vibrations whenever the circuit breaker was opened and there was some rubbing action at X, in the central part of the spring and Y, on an adjacent structural member. The abraded area with sharp edges acted like stress raisers, and fatigue cracks initiated from such points developed into fractures.

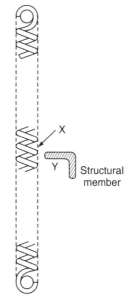

Figure 18.11 Spring failures due to contact and abrasion at X–Y

These failures were completely stopped by:
- Replacing all springs with signs of excessive wear.
- Hand grinding of spring surfaces with slight wear to remove all traces of stress raisers.
- Grinding off some material from the structural member at the critical location to increase the clearance and prevent contact.

18.7 Spring Failures due to Abrasion (2)

CASE STUDY

In Section 18.6 we have seen the effects of inadvertent abrasion on springs due to unexpected interference between moving parts. There are some cases where abrasion is inherent in the design. For example, in the case of laminated springs over rolling stock axle-boxes, there is continuous abrasion between the leaves of the spring as the spring deflection keeps changing. See Figure 18.12.

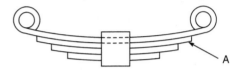

Figure 18.12 Laminated spring. Abrasion between leaves causes groove formation on longer leaf at A

Sharp corners are produced on the longer leaves at the ends of the shorter leaves in contact, as at A in Figure 18.13. These act as stress raisers and fatigue cracks initiated at such points propagate and develop into fractures.
Corrosion at such spots may add to the problem.

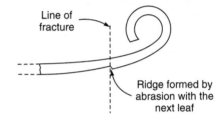

Figure 18.13 Sharp corners produced on longer leaves due to abrasion

The remedies are:

- To apply thick lubricating oils between the leaves of the springs during assembly and periodically during service.
- Replacement (during periodical overhauls) of leaves which have developed visible stress raisers.
- Dye-penetrant tests at vulnerable locations on the leaves during periodical overhauls to detect incipient, hairline cracks.

18.8 Failures of Other Shapes of Springs

We have so far discussed failures of only helical and laminated flat springs. There are other shapes of springs in use but in lesser numbers. These are spiral springs, flat strip springs, torsion rods, etc. The shapes may be different but the principle of operation is the same. They make use of the property of elasticity of metals. The failure modes also are basically similar. Fatigue cracks start from stress raisers in highly stressed areas and then develop into fractures. The stress raisers may be at sharp changes in section either in the original shape of the component or in the worn shape. Possible remedies have to be developed on lines similar to those suggested in Sections 18.2 to 18.7.

18.9 Spring Failures due to Design and Manufacturing Defects

Look for manufacturing and design defects when the fractured spring or other similar ones in service do not show any of the visible defects of the type discussed in Sections 18.2 to 18.7. It is possible, that failures continue to take place even after the visible defects have been eliminated. There are now two possibilities. Either (a) the peak stress in the material of the spring is excessive or (b) the strength of the material is less, due to a deficiency in its composition or in its heat treatment.

Excessive stress (in comparison with the endurance strength of the specified material) can be due to two conditions. The deflection of the spring from its free condition and the variations in the deflections during operation may be in excess compared to the assumptions made during the design of the spring. Alternatively, the selection of the spring wire and coil dimensions may be incorrect. In other words the spring design may be defective.

The endurance strength of spring steel depends not only on the composition of the material but also on its heat treatment. If Chrome-Vanadium spring steel with a very high shear strength is specified and the steel actually used is high carbon steel, or if the required heat treatment is not meticulously given, there is a strong probability that the failures are due to one or more of these deficiencies. In such cases, the cause of failure can be determined only after chemical and metallurgical analyses.

It is necessary to retain the failed components for chemical and metallurgical analysis. A very important part of this is the examination of the fracture aspect. It is possible to learn much about the failure mode, mechanism and even their causes by visual and microscopic examination of the fracture surfaces. Fatigue fracture surfaces can be usually divided into two distinct parts:

- Relatively smooth but dull parts representing the gradual growth of the fatigue crack.
- Brighter and rougher part representing the sudden fracture which takes place when the reduced section is unable to withstand the consequent increase in stress during normal operation.

If one reaches this stage after elimination of the other causes of spring failures, it becomes necessary to refer the case to specialists in spring design, and metallurgy. Fortunately, such cases are relatively rare.

18.10 Spring Failures which do not Look Like Spring Failures

Sometimes the component that has failed does not look like a spring; but its principle of operation and also its failure mechanism are in fact those in the above discussions. Hence, such failures are included in this chapter.

Consider the fixed contact of a knife switch shown in Figure 18.14. When the knife is inserted in the slot, the two wings of the slot deflect outwards because the thickness of the knife is slightly larger than the width of the slot. This deflection has to be an elastic deformation. When the knife is pulled out the wings must revert to their original shape. The elastic deflection is essential for ensuring that there is a constant force between the fixed and the moving contact. As discussed in Chapter 43 on electrical contacts, this force is very important. If it is less than the required value the contact between the knife and the slotted fixed contact will develop a high contact resistance and will get overheated. It can even melt and start a fire.

The wings of the slot have to behave exactly like flat springs. Stress raisers like those shown in Figure 18.15 or defects in the material will cause one of the two wings to fracture followed by either an open circuit or arcing at the knife switch contacts.

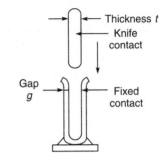

Figure 18.14 The gap in the fixed contact must be less than the thickness of the knife contact

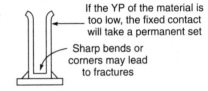

Figure 18.15 Causes of knife switch failures

In case of fracture of the slotted contact, the first thing to look for is the presence of stress raisers illustrated in Figure 18.15. The other important points to check in such an application are (a) the type of fit between the knife and the slotted fixed contact and the limits or tolerances on the dimensions of these components, (b) the corresponding stress limits in relation to the yield point. If the switch is opened and closed frequently, the endurance limit of the material may be considered.

18.11 Spring Failures due to Permanent Deformation

There is another distinctly different failure mode and mechanism for springs. They may take a permanent set or deformation. This happens after installation and commissioning of the equipment but it may remain undetected because there may be no visible sign of this type of failure. It will manifest itself in some other way. For instance, if the spring was meant to provide the contact force between a pair of electrical contacts, there may be overheating and burning of these contacts. In another instance, if a spring meant to hold traction motor field coils in place develops a permanent set, there would be insulation failures due to vibration of the field coils on their cores. In some cases, the permanent set is visible as in springs under compression where the gap between the coils gets visibly reduced. In extreme cases of this type the spring may be closed solid.

Permanent set or deformation of springs is the result of the internal stress exceeding the yield point of the material at some point or stage in its installation, maintenance or operation. Such failures take place soon after the equipment is put into service. This may be due to a design defect, but it is more likely to be due to a deficiency in the material as compared to the design.

There may be no visible damage or change in the spring itself but there would always be adverse effects on the performance of other connected components due to a reduction in the force exerted by the spring. There may be vibration, movement or slackening of other components, and in the case of electrical contacts there may be overheating, burning or arcing.

The obvious remedy is to use spring steel with a yield point, which is higher by an adequate margin than the maximum stress that is likely to occur during normal operation or during assembly.

Two examples of failures caused by spring deformation have been described. Listed below are few more examples:

(a) Failure of plug and socket contacts due to permanent set on the spring elements of the sockets.

(b) Failure of inter-turn insulation of transformer coils due to flattening (through permanent set) of disc springs over the coil assemblies.

(c) Bad riding of railway rolling stock and road vehicles due to permanent set on helical or laminated springs.

Permanent set or deformation of the springs can be detected very easily by comparing the free dimensions with those given in the relevant drawings. New springs should be tested by compressing, stretching or deflecting the springs to the extent indicated in the appropriate drawings. The free dimensions should be measured before and after this load test. In principle, only a single application of the test load should be adequate, but many specifications stipulate a number of such repeated applications in order to detect any gross internal faults, which may cause them to fail by low-cycle fatigue.

18.12 Maintenance of Springs

Springs do not require any special maintenance as there is no wear or deterioration if they are well designed and properly manufactured. It is only necessary to see that there is no corrosion of springs during service and no mishandling during maintenance of other components. Exceptions to this rule are laminated springs where wear is normal.

However, if the design is based on peak stresses that are close to the endurance limits, or if there are inclusions in the material or surface flaws and stress raisers, there are possibilities of fatigue cracks developing in service. In such cases, i.e. if there is a history of spring fractures, periodical visual inspection for surface defects such as dents, sharp bends, etc. and by dye penetrant tests for hair line cracks may help to prevent some failures in service. Care should be taken after such tests to clean and coat the springs with oil for corrosion resistance. During such inspections, the free dimensions may be checked with reference to the working drawings.

In case of laminated flat springs with a history of spring fractures, it is desirable to dismantle the leaves periodically, to clean them thoroughly and to inspect them for hairline cracks particularly at the vulnerable locations at the ends of the leaves.

If hairline cracks or sharp nicks on the surface are detected during periodical inspections it is best to replace the springs or leaves as necessary. There is no dependable way by which springs with hairline cracks can be reclaimed or repaired.

18.13 Do's and Don'ts for Preventing Spring Failures

- Investigate every case of spring fracture or permanent deformation, by looking for the various probable causes of spring failures described in Sections 18.2 to 18.11.
- Consider also cases of failures of components where their elastic deflection is involved in its operation.
- If none of the above causes of spring failures are present, then consider the basic design of the spring to check whether the calculated peak stress exceeds the endurance limit or the yield point of the material.
- When visible defects are not detected and calculated stresses are also within safe limits, consider the possibility of defective material. Have the failed spring material analysed and tested to determine ultimate tensile strength, elongation, hardness, chemical composition, microstructure, etc.
- During routine inspection and maintenance of springs, look for hair-line cracks and permanent set. In critical applications with a history of fatigue fractures, consider the use of magna-flux and dye-penetrant tests.

Chapter 19

GASKET FAILURES

In this chapter we learn

- *Applications of gaskets.*
- *Failure modes and failure mechanisms in gasket failures.*
- *Measures to be taken to prevent gasket failures.*
- *Case studies.*

☐ ☐

19.1 Introduction

Gaskets are used in electrical and mechanical equipment for sealing removable covers, to prevent leakage of compressed air or any other fluids and entry of dust, etc. They are also used in boilers, reactors and other types of pressure vessels. Some examples are gaskets between:

- transformer tank and tank cover;
- pipe flange and compressor output opening;
- traction motor carcass and inspection cover;
- air-flow relay flange and air duct;
- cylinder head and cylinder block of an automobile engine;
- inspection covers, pipe connection flanges etc. and pressure vessels.

Gaskets are generally treated as trivial components. They are inexpensive, replaceable and to be used and discarded. However, failures of gaskets can cause severe and expensive damages.

There is not much technology involved in their installation but it is very important to ensure that certain simple precautions are taken.

19.2 Failure Modes for Gaskets

There are two main failure modes. These are:

- Leakage of fluid through the joint over or under the gasket.
- Fracture and blowing of gasket and leakage of fluid.

The same or similar defects may lead to either of these failure modes. The latter mode can be catastrophic.

19.3 Gasket Failure due to Metal Creep

CASE STUDY

In a large power station, the gasket of a high-pressure receiver blew out. The sudden leakage of high pressure, high temperature water resulted in a few deaths and injuries.

The operating temperature of the receiver was over 700 °C —which was beyond the creep temperature limit for the special steel bolts used to hold down the cover of the receiver. These bolts are subject to metal creep and relaxation of tension in the bolts. The nuts must be re-tightened periodically.

In the case mentioned above, there were two deficiencies (a) the nuts were not re-tightened at the intervals recommended by the manufacturer and (b) the gasket had not been replaced during the previous overhaul when the receiver cover had been opened out. The gasket may have been damaged at that time.

19.4 Gasket Failures due to Shrinkage of Gasket

CASE STUDY

There were many cases of gasket failures in a fleet of locomotives. These were fitted in flanged joints between compressor outlets and delivery pipes. Investigation showed that the gasket material was shrinking in thickness under the effect of heat and pressure. By substituting a different material, which remained stable despite high temperature and pressure, the problem was solved.

19.5 Gasket Failures due to Unevenness of Flange Faces

CASE STUDY

Gaskets fitted on the top face of a large transformer were developing leakage of oil. Investigation showed that the gasket thickness was 10 mm before compression and 8 mm when fully compressed. As against this, there were many locations on the flanges where the variations from straight edges placed across them were 2.5 mm i.e. more than the general compression of the gasket. By tightening of the flange bolts, the flanges and the top plates were bending and giving the appearance of good fits but there were spots where the local flange pressure was not adequate to prevent oil leakage. The gaskets cannot possibly compensate for poor workmanship in the fabrication of the metal parts. Ensuring that the maximum variation from straight edge to the flange faces was less than 0.5 mm solved this problem.

19.5 Do's and Don'ts for Preventing Gasket Failures

Arising out of a number of gasket failure investigations the following common causes, failure mechanisms and precautionary measures may be summarised. These are absolute musts if reliable performance is expected.

Lacunae in any one of the following details relating to the specification, testing of materials and installation can lead to failures. These lacunae are in the nature of seed defects. They may not necessarily cause failures in every case but some of them are sure to cause failures. To ensure reliable service, it is imperative to avoid such defects.

(a) The material of the gasket must not shrink further during service after it is compressed at the specified pressure determined by a specified torque for the bolts/nuts.

(b) If a certain amount of shrinkage is unavoidable in the gasket after applying the specified pressure for a prolonged period, the bolts/nuts over the flange must be re-tightened to the original specified torque until there is no further shrinkage.

(c) The force applied by all the bolts/nuts over the full flange area must be at least 25 per cent higher than the total force due to the specified pressure mentioned in (a) above over the area of the cover.

(d) The surfaces of the flange and of the flange seat must be flat. The maximum variation from a true straight line in any direction shall not exceed 10 per cent of the compression that takes place in the gasket when the flange bolts are fully tightened. See Figure 19.1.

(e) The surfaces of the flange and of the flange seat must be smooth. The maximum depth of valleys or heights of peaks in the surface profile shall not exceed 2 per cent of the compression that takes place in the gasket when the flange bolts are fully tightened.

(f) Before the gasket is assembled, it should be carefully examined visually to see that there are no cracks, tears, flaking, etc. The holes in the flange should be so accurate in diameter and location that the gasket can be inserted over the screws or bolts very easily and without force. If bolts are inserted after placing the gasket in situ, it should be ensured that the holes in the flange seat, the gasket and the flange match perfectly so the bolts or screws can be inserted without damaging the gasket.

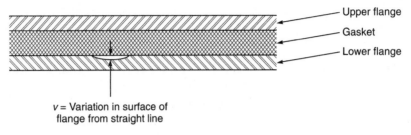

v = Variation in surface of
flange from straight line

Gasket thickness x when free, y when fully compressed
v must not be more than $(x ñ y)/10$

Figure 19.1   Shows how gaskets should be checked (Flange bolts are not shown)

(g) Before assembly of the gasket, the surfaces of the flange, the flange seat and the gasket should be cleaned carefully to remove any foreign matter.

(h) The adhesive or sealant recommended by the manufacturer should be applied in a uniform layer to the flange and the flange-seat surfaces while assembling the gasket.

(i) The nuts/bolts/screws which hold the flange down should be tightened in the specified order in three stages: first with 10 per cent, then with 60 per cent torque and finally with 100 per cent of specified torque while holding the bolt head stationary.

(j) In critical application, gaskets removed from previous use should be discarded and not re-used.

(k) Gaskets should be branded products obtained from approved and reputed suppliers. Gaskets should be inspected carefully for every dimension and material quality. In particular, the compressibility of the material under specified pressure and temperature and further compressibility under augmented pressure should be carefully checked.

(l) If manufactured locally, similar care should be exercised in the procurement of the raw material and the cutting of the gaskets with the help of templates and special tools.

(m) The material of the gasket should withstand contact with the contained fluids without degradation of any of the mechanical properties.

In critical applications, i.e. in cases where the consequential effects of gasket failures can be serious, the precautions enumerated above must be taken. If these measures are ensured, then it can be stated with certainty that there would be no break downs.

The material of the gasket must be as specified by the designer or manufacturer. However, the possibility of the recommended material being unsuitable for the actual working conditions may have to be considered if gasket failures continue to occur. The dimensions and physical properties of the failed gasket should be compared with those given in the relevant drawings. If there is any degradation that can explain the failure mode, it may be necessary to consider other materials.

Chapter 20

O-RING AND U-RING FAILURES

In this chapter we learn

- *The applications of O-rings and U-rings in machinery and plant.*
- *Their failure modes and failure mechanisms.*
- *Measures to be taken for preventing failures of these rings.*
- *Case studies.*

❑ ❑

20.1 Introduction

O-rings became well-known when the Challenger space probe disaster was attributed to the failure of an O-ring. The investigation exposed many engineering, managerial and administrative lacunae in the system. Here we are concerned with the engineering aspects of O-ring and U-ring failures in general.

O-rings are rings of circular cross section, made of synthetic rubber or elastomer, and used for sealing high-pressure fluids in containers and machine components. They are static but sometimes may be used as rolling seals on pistons. See Figure 20.1.

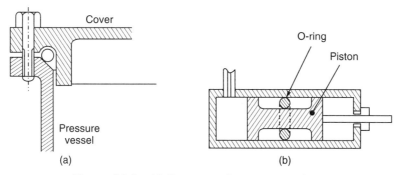

Figure 20.1 (a) O-ring used as a static seal
(b) O-ring used as a dynamic, rolling seal

20.2 O-ring Failure Modes and Causes

The O-ring is held in a rectangular groove or corner space and pressed by the cover or mating component.

The main purpose of an O-ring is to prevent leakage of the contained fluid and its failure mode is the leakage of such fluid. Case histories can be summarised as leakage of fluids due to one or more of the following causes:

Cracks in the O-ring.

Shrinkage of the O-ring.

Distortion of the O-ring.

Permanent set on the O-ring.

Inadequate pressure on the O-ring.

Excessive or inadequate width of the groove.

Incorrect dimensions of the O-ring.

20.3 Cracks, Shrinkage, Distortion and Permanent Set

Many failures are due to cracks, shrinkage, distortion and permanent set of the O-ring. All these defects arise when the material used is unsuitable and procured from new suppliers without a proper evaluation. The procedure include environmental tests, endurance tests and service trials.

Inspite of being obtained from reputed and proven suppliers, it is necessary:

(a) to specify full details of the service conditions and reliability/durability expectations to the suppliers in order to procure the most suitable material.

(b) to carry out certain acceptance tests and measurements on the basis of defined sampling procedures on all supplies received.

(c) to replace the O-rings at recommended intervals.

Some of the tests, which determine the quality of the material, are:

(a) Soaking and pressing of O-rings in hot oil or specified chemical, and measuring changes in dimensions, shore hardness, permanent set and surface condition.

(b) Endurance test in device simulating service conditions and accelerating the wear and tear.

20.4 O-ring Failures due to Unsuitability of Material

CASE STUDY

O-rings are reliable components if the few precautions mentioned in Section 20.3 are taken. However, if any one of these is neglected failure is almost certain. In a refrigeration compressor an O-ring was used to seal a crankcase cover. After it was replaced during periodical overhaul, the new ring developed cracks, caused leakage of refrigerant and failure of the system. On checking it was found that the replaced ring was not suitable for the low operating temperature and perhaps the chemical action of the refrigerant.

There are many such cases where the failure is due to the unsuitability of the material of the ring for the conditions of service such as temperature and fluids in contact. In general, special elastomers are used if the operating temperatures are much below or much above the range 10 to 40° C.

20.5 Failures of O-rings due to Dimensional Errors

Dimensions of the O-rings themselves are as important as those of the grooves which house them; but this is not a common problem if the O-rings are obtained as branded products from reputed suppliers.

If the O-rings are obtained from new and unproven suppliers ring dimensions must be checked with as much care as the material of the ring. It is not easy to measure the dimensions of these components because of their softness and lack of rigidity. The ring diameter can be measured with the help of stepped cylindrical gauges. The step on which the ring fits snugly without being either too loose or too tight, indicates the diameter of the ring.

20.6 O-ring Failures due to Errors in Groove Dimensions

CASE STUDY

Designers of equipment should consult the manufacturers of the O-rings for advice on the dimensions and tolerances of grooves or seats for O-rings.

There was a case of air leakage from the cover of a compressed air motor. It was due to an error in the dimension A in Figure 20.2. The ring did not get compressed sufficiently and air leaked past the ring despite the cover bolts being tightened properly.

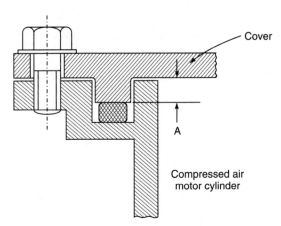

Figure 20.2 Failure of O-ring gasket due to error in dimension A

O-rings are precision components and all the dimensions of the grooves and their tolerances are extremely important for ensuring reliability. Deviations from the recommended dimensions and tolerances do not necessarily or immediately lead to failures; these should be considered as seed defects. Some of them may cause failures long after installation while many may remain in service without causing problems. As it is not possible always to determine the exact failure mechanism for different types of deviations, it is easier and cheaper in the long run to insist on total adherence to the specified dimensions and tolerances.

20.7 Failures of U-rings

U-rings are generally used for rotary or reciprocating sliding action. Some U-rings are provided with steel inserts and steel springs. A wide variety of materials are used for these seals depending upon operating conditions such as speeds, temperatures, fluids in contact etc.

As in the case of O-rings the same holds good for U-rings. The correct material and adherence to dimensions of the U-rings and of the grooves which hold them, and tolerances is of the greatest importance for attaining total reliability of these components. There is an added precaution to be taken in the case of rotary and reciprocating sliding seals. Measures should be taken against any possibility of grit, sand, or such other abrasive material getting in, by appropriate devices such as secondary felt seals, labyrinth seals, covers, filters in the fluid circuit, etc.

20.8 U-ring Failures due to Damage During Testing

CASE STUDY

There was an unusual case of U-rings developing excessive wear and leakage on new refrigeration compressor shafts soon after commissioning. It was observed that when these rings were replaced, they worked satisfactorily.

The cause of the failure was established as damage due to excessive operating temperature during a 100-hour load test in the manufacturer's premises. During this test the compressor was worked as an air compressor and the operating temperatures were higher than the normal operating temperatures when it was used as a refrigeration compressor later in service. The material of the seals was specially selected for operation at low temperatures. Action taken: Temperature data regarding different O-ring materials were obtained from the manufacturers of O-rings. Seals made of a different material suitable for higher temperatures were used during the 100-hour special running in test. At the end of these tests, the seals were replaced by normal low-temperature seals.

20.9 U-ring Failures due to Incorrect Assembly

CASE STUDY

Sometimes failures of hardware occur due to apparently trivial errors. U-rings must always be assembled with the U-cavity facing the higher pressure side of the seal. The pressure causes the lips of the seal to press harder on the metal cavity surfaces and prevents leakage. When a trainee assembled a U-ring in the reverse direction, heavy leakage ensued as soon as the equipment was put into use. It had to be stopped and the seal re-assembled.

20.10 Do's and don'ts for Preventing O-ring and U-ring Failures

- Obtain supplies from approved suppliers and as per proven designs. Any attempt to obtain supplies from new sources or new designs must include type testing and service trials.
- Accept supplies only after carrying out routine and acceptance tests on dimensions, shore hardness, resistance to oils and other specified materials as required, compression set, etc. as specified.

- Ensure checks on the dimensions; surface finishes and material composition of not only the O-rings and U-rings but also the grooves and mating components.
- Ensure freedom of the contained fluids from grit, abrasive matter, dust, etc.
- Ensure that the limits on operating temperatures and pressures are not transgressed.
- Ensure that O-rings and U-rings are replaced periodically at intervals to be determined on the basis of advice from the manufacturers and testing/examination of items removed from service.

Chapter 21

RUBBER HOSE PIPE FAILURES

In this chapter we learn

- *Applications of rubber hose pipes in compressed air systems.*
- *Failure modes of rubber hose pipes viz. bursting, air leaks, slipping out.*
- *Failure mechanism viz. mechanical damage, degradation of rubber (normal or accelerated).*
- *Case studies.* ❏ ❏

21.1 Introduction

ℛubber hose pipes are used in pneumatic systems that use compressed air for the transmission of power and controls. They are used for making flexible connections between pipes, for example between railway coaches, locomotives, and parts of machinery where there is relative movement or vibration, and at locations where it is necessary to connect and disconnect compressed air pipes.

Rubber degrades fast as compared to steel or copper. Moreover, degradation is accelerated by environmental conditions such as abrasion, heat, sunlight, oil and petrol. Therefore, flexible rubber pipes have a much higher failure rate than rigid steel or copper pipes.

Vibration, flexing and movement are the main conditions which necessitate the use of rubber pipes, which have their own adverse effects on the material.

Despite all these factors, it is possible to get reliable service from rubber hose pipes provided these factors are allowed for in design, manufacture, operation and maintenance. And if any of them are ignored, failures are certain to occur.

Rubber hose pipes consist of several layers of fabric, woven in mayflower fashion, impregnated and interleaved with vulcanised rubber either natural or synthetic. Sometimes metal wire layers may also be incorporated.

21.2 Failure Modes

The most common and obvious failure mode is bursting of the pipe. In some cases there may be merely leakage of compressed air. If the hose is not clamped properly on the fitting, slipping out is another possibility. Although there are only two failure modes, their possible causes and mechanisms are many. These are listed below:

(a) Excessive air pressure
(b) Degradation of materials of construction by:
   (i) Heat or temperature.
  (ii) Oil entrained in compressed air.
 (iii) Sunlight.
  (iv) Oil or chemical drips from outside.
   (v) Abrasion or other mechanical damage.
  (vi) Ageing.

The last item in the above list may be considered first as it is an inevitable and perhaps unavoidable reality. Accelerated ageing can be prevented but normal ageing has to be accepted and rubber hose pipes in service are to be replaced and scrapped at suitable intervals.

Very often the rubber hose pipes to be replaced seem to be normal in appearance, but their bursting pressure is much less than when the hose pipes were new. Once the optimum replacement intervals are determined by destructive tests on ageing samples, they should be adhered to. For further research in this regard hose pipes removed for replacement may all be subjected to destructive tests. The records may be reviewed periodically for any changes made in the quality of the hose pipes or for any service failures. The intervals may be modified but the discipline of replacement must be maintained.

21.3 Case Studies

The failure modes, other than those due to normal ageing, might be considered further by looking at a few case studies:

CASE STUDY NUMBER	FAILURE MECHANISM	REMEDIAL MEASURES
21.3.1	The sharp edge of the hose fitting cut into and weakened the hose pipe, which then failed at normal air pressure; See Figure 21.1.	The edge of the hose fitting was given a smooth and curved surface. The edge of the clamp was also curved to prevent damage to the hose pipe, as illustrated in the enlarged view of A in Figure 21.1.

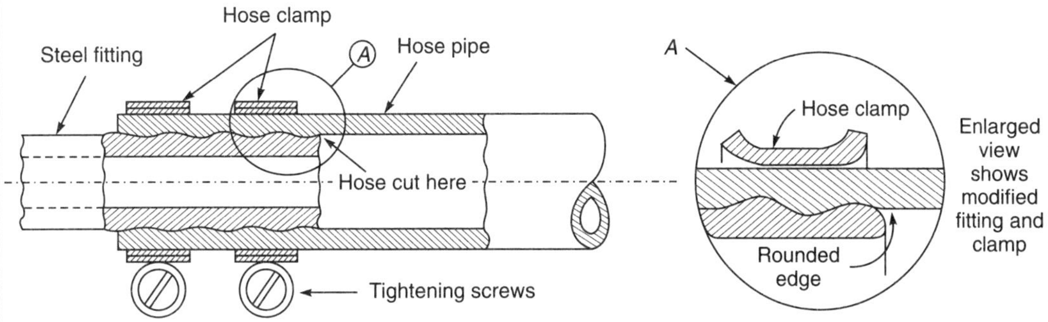

Figure 21.1 Hose pipe failure due to sharp edges on fitting

21.3.2	The inner surface of the hose pipe was affected adversely by the oil entrapped in the compressed air; it got softened and the hose pipe burst at normal air pressure.	All such hose pipes were replaced by new ones with oil resisting, polychloroprene linings and covers.
21.3.3	Hose pipes meant for brake cylinders operating at 3 kg/cm^2 were inadvertently fitted in the main compressed air pipe work operating at pressures up to 8 kg/cm^2; These pipes did not burst immediately due to the large factors of safety in their design. However, they failed prematurely, i.e. before they became due for replacement.	The hose pipes of lower strength were replaced by hose pipes of the appropriate strength. To prevent similar interchange in future, the fitting sizes were changed to make it impossible to use the low strength hose pipe in high-pressure locations. Another possible measure would have been to standardise the high-pressure hose pipe even for the low-pressure application.
21.3.4	Pneumatic hose pipes for main pipe and brake pipe between coaches were rubbing against each other continuously due to the relative motion of the coaches. The abrasion caused wear and weakening of the hose pipes and eventual failures.	The lengths of adjacent pipes were adjusted to ensure that there would be no contact between them.

Contd.

CASE STUDY NUMBER	FAILURE MECHANISM	REMEDIAL MEASURES
21.3.5	In a steel fabrication shop, pneumatic tools and electric welding were in use in the same working area. Damage to pneumatic hose pipes occurred from time to time due to contact with hot components and particles.	Suspension of pneumatic tools and their hose pipes from very light jib cranes not only prevented damage to hose pipes but also made it easier for operators to handle the heavy pneumatic tools.
21.3.6	In one application, no clamp was provided on the hose pipe over the fitting. After a very short time the hose pipe slipped out. The pull out force developed by the air pressure in the pipe is about 49 kg on a hose pipe of ID 2.5 cm at 10 kg/cm^2. This is sufficient to cause the hose pipe to slip out unless the radial force between the hose pipe and the fitting is in excess of 200 kg.	In order to ensure that there is a positive and adequate radial force between the hose pipe and the fitting a hose clip with a screw drive was fitted over the hose pipe at each fitting as shown in Figure 21.1.

21.4 Do's and Don'ts for Preventing Hose Pipe Failures

- Select a type of rubber hose pipe that is suitable for the maximum fluid pressure; service conditions such as vibration, relative movement, resistance to the fluid carried by the pipes, etc. and environmental conditions such as sunlight, temperature, abrasion and chemicals in the air, etc. If needed provide additional protection.
- Provide suitable clamps on the terminal fittings so that the hose pipe is not cut into by the fittings, and not pulled out by the internal air pressure.
- Determine replacement schedules for the hose pipes based on observations, tests and experience. Mark prominently the scheduled replacement date whenever a new hose pipe is installed.
- Maintain records of brands, installation, replacements and failures of hose pipes.
- Investigate every case of hose pipe failure to determine the exact mechanism so preventive measures can be carried out at all similar locations.

Chapter 22

PIPE AND TUBE FAILURES

In this chapter we learn

- *Applications of pipes and tubes in the electrical, mechanical and chemical industries.*
- *Modes and mechanisms of failures of pipes and tubes. Mechanical vibration, corrosion, over-heating and manufacturing defects are the usual causes.*
- *While compressed air pipe/tube failures can usually be eliminated by design modifications, failures of tubes in water tube boilers are usually due to maintenance deficiencies.*
- *Case studies.* ❏ ❏

22.1 Introduction

*P*ipes and tubes are used in electrical and mechanical machines and in chemical industries, for conveying fluids, at high pressure. Pressures of the order of 10 kg/cm^2 for compressed air are common and much higher pressures as in hydraulic systems are also used in machinery like cranes, earth moving machines, etc. There is not much difference between the terms pipe and tube as far as failure modes are concerned. The term pipe is commonly used for larger diameters and the term tube for smaller diameters. In this chapter, the term tube will be used and will include pipes too.

Pipes and tubes may be either seamless or with linear/spiral welded joints.

Failures of equipment utilising compressed air or hydraulic fluids are often due to fracture of tubes and leakage of fluids from joints. Failures due to blockage of tubes are rare.

22.2 Fractures of Tubes at Bends

If the initial selection is defective, i.e. if the tube thickness is too small or the material is too weak for the pressure building up inside the tube, obviously the tube is going to develop cracks or to burst. Such failures are rare, but other errors are made which lead to failures after some period of service.

Tubes are often bent at a number of places and if the bend radius is too small plastic flow and micro-cracks form in the tubes. In order to prevent them from collapsing at the bends they are filled with sand or with a spring before bending, which may be carried out in special bending dies.

In Section 46.7, Chapter 46, the effects of bending solid copper bars are discussed. It is shown that the stress and strain in the bars become excessive if the bend radius is small in relation to the thickness of the bar. Similar calculations apply to tubes as well as solid copper bars as referred in Chapter 46. The minimum inner radius of bends must not, be less than certain limits, which depend on the type of material, and the outer diameter of the tubes. These limits are usually indicated by the manufacturers in their handbooks. They would be, naturally, lower for ductile metals like copper.

The effects of 'sharp' bends (i.e. bends of radius smaller than the recommended minimum limit) may not always be visible; but if the material is overstrained, and if there are constant fluctuations in the internal pressure failures may occur after many years. To avoid all possibility of such problems, ensure that all bend radii are higher than the specified limits.

22.3 Fractures of Tubes due to Other Causes

Tubes can fracture if there are long unsupported lengths and begin to vibrate as a result of machine vibrations or fluctuations in internal pressure.

Corrosion of tubes and consequent tube cracks are not uncommon despite the fact that these are obvious causes of failure. They tend to be neglected because many corroded tubes continue to work since there are adequate margins of safety in design. It is difficult to judge what level of corrosion would lead to certain failure. Hence it is desirable to replace tubes which show any signs of corrosion and to take every possible precaution to prevent corrosion. During periodical maintenance schedules all tubes carrying fluids must be checked for such signs.

22.4 Fractures of and Leakages at Tube Joints

High-pressure tube joints are usually of the flare on cone type as shown in Figure 22.1.

The tube is flared with a special flaring tool and the expanded conical end is held tightly by a nut against the cone of the pipe fitting. If the flaring tool is correctly shaped and sized and if the flaring is done carefully the conical end of the tube gets a shape which fits well with the cone of the pipe fitting and there are no cracks on the flared end. The material must also be of the appropriate ductility. Tubes for such application are usually tested for flaring by a special testing tool.

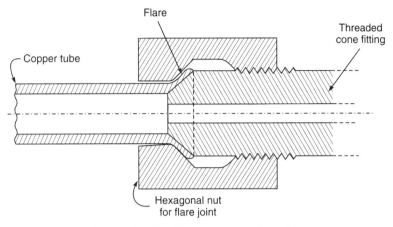

Figure 22.1 Flare on cone tube joint

Machining accuracy of the nut and the pipe fitting must be ensured. The conical surfaces which have to grip the flared end of the tube must be truly circular, and at right angles to the axis of the nut. The pipe fitting should not have sharp corners and angle of the cone should be correct. If any of these requirements are not met, the flared end will not be gripped correctly all over and leakage can occur.

The tube flare must rest evenly on the cone of the pipe fitting *before* the nut is tightened and the nut should not be used to pull the pipe into correct position. The nut should be tightened completely.

Pipes are often provided with threaded joints. The gaps between the threads cut on the pipes and the internal threads of the pipe fittings such as couplings, bends, tees, etc. are then sealed with strings or ropes and putties or compounds. If these joints are badly made fluid leakages are inevitable. There are many types of patented pipe jointing systems. Joint failures can be avoided only by ensuring the dimensional accuracy of the threads and the meticulous observance of the fitting instructions given by the manufacturer of the jointing system.

The most reliable system of jointing is obtained by welding or brazing, and a high level of skill is needed for making such joints.

22.5 Failures due to Thermal Stresses

CASE STUDY

The failure modes and failure mechanisms defined in Section 22.2 to 22.4 are responsible for the vast majority of tube failures. There are a few rare cases and one such is described here:

In one incident there was cracking of a tube and leakage of fluid through the crack. In the control panel of a moulding press there was a straight copper tube of length approximately 125 mm and diameter about 8 mm, across and between two valves as shown in Figure 22.2.

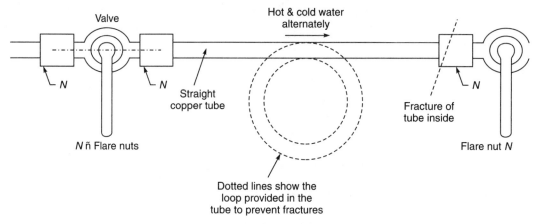

Figure 22.2   Copper tube fractures due to thermal stresses

The pipe cracked at the point where one end of the conical surface of the nut bit into the copper tube. Examination of the fractured surface showed signs of metal fatigue. Since fatigue requires alternating or fluctuating stresses, it was concluded that thermal stresses were developing in the copper tube, which was constrained from expanding and contracting freely by being held at both ends by the valves fixed to the panel. The pipe was used for the flow of hot and cold water alternately for use in the moulding cycles. The action taken was to provide a loop in the copper pipe as illustrated in Figure 22.2 highlighted by dotted lines.

22.6   Failures due to Defective Material

CASE STUDY

In another case of tube leakage through a crack at the flared end, investigation did not reveal any of the possible causes of failure described above. It was observed that the crack was axial or along the tube over the flared end of the tube. In this case the material of the copper tube was defective. The deficiency was revealed on carrying out a flaring test and bend test on the copper tube. Chemical analysis later revealed unacceptable levels of impurities.

22.7 Tube Failures in Water Tube Boilers and Heat Exchangers

Tube failures in water-tube boilers and heat exchangers often have serious consequences. If the failures occur at the points where the tube ends are fitted into the boiler drums or end plates of heat exchanges, they are usually due to the defects in the methods or tools used in fitting the tubes.

If distilled or de-mineralised water is not used in the boiler, hard deposits accumulate on the inner surfaces of the water tubes and if these are neglected they grow to such thicknesses as to interfere with the flow of heat. The steel tubes get over-heated and they eventually fail by bursting. These failures are obviously due to poor maintenance and preventive measures are necessary.

Corrosion of the tubes is unavoidable due to high operating temperatures and pressures and the tubes call for replacement from time to time. Monitoring of the condition of the tubes during annual overhauls and replacement of those damaged beyond permissible limits is as essential a part of maintenance as the control of the quality of water used initially and for make up. As the overall or total cost of replacing tubes in water tube boilers is extremely high, much expertise, effort and care is necessary at every stage of the following activities.

(a) Selection and inspection of the water tubes;
(b) Fitting of water tubes in situ;
(c) Control on the quality of water;
(d) Condition monitoring of water tubes during boiler overhauls;
(e) Maintenance and monitoring of protective devices like low water level alarms, high-pressure alarms and water quality alarms.

While most cases of water tube failures can be traced to one of the above factors, other unusual occurrences have been known to have caused tube failures. Here are a few such cases:

22.8 Failures due to Maintenance Deficiencies

CASE STUDY

- A tube expander tool was left inside a tube after fitting it from inside the lower drum of a boiler. This restricted the flow of water and eventually it overheated and burst within a few days after the re-commissioning of the boiler.

- When a tube inside a cluster of tubes burst, it was decided to cap it at both ends and to re-commission the boiler quickly. Initially, the wrong tube was capped at the lower end. When the error was discovered while filling up the boiler, the correct tube was capped but the cap on the wrong tube (at the lower end only) was not removed. The result was that this tube also burst on re-commissioning the boiler.

22.9 Failures due to Vibration

CASE STUDY

A 30 mm galvanised steel pipe carrying compressed air was fitted on the under-frame of an electric locomotive. The length of the pipe was about 4 metres and it was clamped at two places between two fixed bends. The pipe fractured, while the locomotive was in service, through the roots of the threads at one of the bends, about 22 years after commissioning of the locomotive.

Initially, the cause of this failure was explained as ageing and proposals for replacement of all such pipes were initiated. However when one more similar case occurred on another locomotive in the same age group the matter was re-examined. Inspection of the broken pipe showed some corrosion but its extent did not seem to warrant the conclusion that the failure was due to ageing and corrosion. A hydraulic test showed that the corroded pipe could withstand a pressure of 25 kg/cm^2 which was two and a half times the maximum operating pressure. Examination of the fracture surface showed signs of bending and fatigue.

Further inspection of the pipe in situ showed that the intermediate clamps were a size too big and allowed the pipe to vibrate. See Figure 22.3. Since the failures occurred after nearly 25 years, it seems likely that the magnitude of the resultant stresses was slightly in excess of the endurance limit for steel.

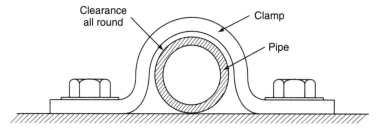

Figure 22.3　Oversize clamps allowed pipes to vibrate and fail through fatigue

It was clear from the aspect of the fracture, that it was caused through fatigue due to vibration of the pipe. The clamps were modified. Similar locations on the whole fleet were checked and modifications made wherever necessary. There were no further fractures of these pipes during the next four years.

22.10 Failures due to Vibration

CASE STUDY

There were several cases of fracture of copper tubes carrying compressed refrigerant from a stationary air conditioning compressor. The tube fractured inside the flare and cone fitting. The compressor was mounted on resilient supports and being a reciprocating compressor was subject to some vibration. It was observed that the tube had a loop and was clamped at C as shown in Figure 22.4.

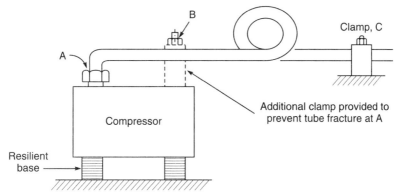

Figure 22.4 Fracture of output tube from compressor, due to vibration

The problem was solved by adding a clamp B to the compressor body, preventing any flexing at the joint, as shown in Figure 22.4 at B in dotted lines.

22.11 Do's and Don'ts for Preventing Tube Failures

- Select the material of the tube and the design of the fittings to suit the working conditions such as pressure, temperature, vibrations, environment, etc.
- Ensure that the pipe joints and connections are made accurately according to the directions of the manufacturer of the fittings.
- Ensure that the bend radii are not less than the specified limits.
- Do not allow tubes to vibrate, by providing effective clamps.
- Ensure that there is no flexing or bending at the fittings by correct location of the clamps and the provision of loops or bellows.
- Where flare joints are provided ensure that the flaring tools are of good quality and that the tubes themselves pass flaring tests.
- If there are possibilities of thermal stresses in the tubes provide loops in the tubes to absorb changes in lengths.
- Prevent failures of water tubes in boilers, by periodical cleaning of insides to remove scales and by control on the quality of water.

Chapter 23

COMPRESSED AIR VALVE FAILURES

In this chapter we learn

- *Different types of compressed air valves.*
- *Common failure modes in these valves.*
- *Failure mechanisms such as degradation of rubber, wear, accumulation of foreign matter.*
- *Measures to prevent failures.*
- *Case studies.*

23.1 Introduction

$\mathcal{T}$here are many types of compressed air valves in any installation that uses compressed air for the transmission of power. Some of these are: non-return valves, isolating valves, drain valves, control valves, electro-valves, pressure regulating valves, safety valves, limiting valves, triple valves, etc.

The number of makes, types and designs of valves in existence would run into hundreds of thousands. A large variety of materials are used in their construction. Different service conditions and system requirements divide the field in many ways. Yet, the basic failure modes, the mechanisms and the causes of failures of all these valves would be very few in number, very similar in nature and mostly identical. Inspection of the failed device shows at once the failure mode. The mechanism of failure and its cause can be determined easily. Here are some usual failure modes:

(a) Leakage to atmosphere.
(b) Leakage through the valve.
(c) Blockage of air flow, partial or complete.
(d) Excessive air flow.
(e) Failure to regulate or limit the pressure or the flow as required.

The failure mechanisms are very simple and obvious and the causes can be generalised into just a few groups as follows:

- Dimensional errors causing bad fits between components.
- Wear of components in contact with each other.
- Degradation of non-metallic components.
- Fracture of stressed components.
- Accumulation of water, oil or particulate matter from compressed air.
- Corrosion of ferrous components.

All these failures are avoidable and if the failure mechanism is clearly and correctly identified, the remedial or preventive measures can be determined. To be able to do so the most important step is to dismantle and to dis-assemble the failed device very carefully while observing the appearance of each component. The following case studies will illustrate some very diverse types of equipment and their failures, some of them due to trivial causes and some to unusual and unexpected causes.

23.2 Failures due to Incorrect Installation

CASE STUDY

A gravity cup type of non-return valve functioned erratically. Sometimes it would fail to block the return airflow. See Figure 23.1.

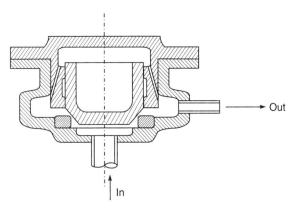

Figure 23.1 Non-return valve, gravity type

It was observed that the valve was installed with the cup axis horizontal. Turning the valve through 90° solved the problem. The weight of the cup initially and the air pressure later ensured the sealing of the valve.

23.3 Failures due to Residual Magnetism

CASE STUDY

In this electro-valve, shown in Figure 23.2 the valve closed on being energised but failed to open on being de-energised. This type of failure occurred sometimes but not always.

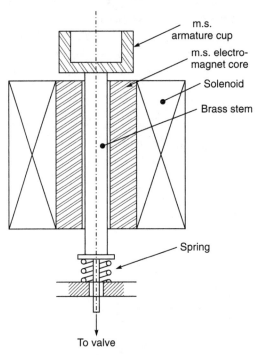

Figure 23.2 Electromagnet system for electro-valve

The dimensions of the valve components were supposed to be such that the magnetic armature cup maintained a small clearance from the face of the magnet. However on some valves, the valve stems were too short and the armature actually touched the magnet face in the closed position. The residual magnetism in the core was then adequate to hold the armature down even when the valve was de-energised. The force exerted by the spring was not adequate to counter-act the weight of the stem and the cup plus the magnetic attraction due to the residual magnetism. The problem was solved by a thin brass shim on the head of the magnetic core.

23.4 Failures due to Accumulation of Dust

CASE STUDY

Figure 23.3 shows the construction of a simple safety valve. Though such simple devices are fool-proof, one unit did fail in a cement plant. The compressor was switched on and off by another pressure regulating control switch. The safety valve failed to open during a routine annual test. This was not a service failure but it exposed a problem.

On examining the safety valve a hard cake comprising dust and cement was found above the valve disc. The compressor was in a room adjacent to the cement packing and loading unit and there was some cement dust in the

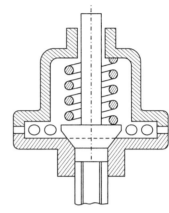

Figure 23.3 Simple fixed setting safety valve

air. This had accumulated in the space above the valve disc and hardened by absorption of water from the atmosphere or some stray splash. Since the room had other vulnerable equipment, the doors and windows of the room were made dust-proof, the room was pressurised and ventilated with a blower and air filter.

23.5 Failures due to Leakage

CASE STUDY

In the electro-pneumatic system of brakes on railway coaches, electro-valves control the charging and discharging of the compressed air brake cylinders. In one such system it was found that some holding valves which should close completely when energised, were allowing compressed air to leak out.

On dismantling the defective valves it was observed that pipe scale was getting lodged in between the valve seat and the valve stem. Improved procedures for de-scaling and cleaning out the pipes were introduced. Improved air filters were also installed.

23.6 Failures due to Damage to Rubber Diaphragm

CASE STUDY

In one design of pressure regulating valve, a rubber diaphragm separated the low pressure and high-pressure chambers. The diaphragm was normally held down by a spring and it operated a small valve, which allowed air flow from the high-pressure side to the low-pressure side. Another similar device operated an electrical contact. See Figure 23.4.

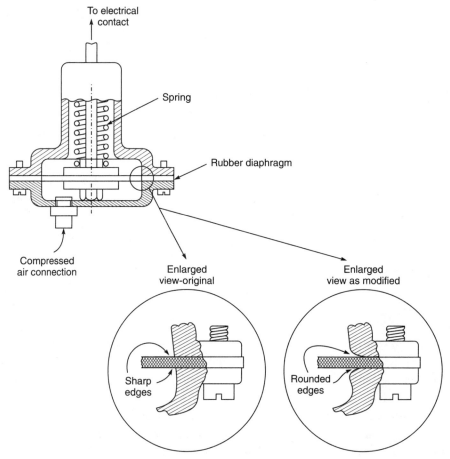

Figure 23.4 Relay failure due to cutting of rubber diaphragm by sharp edges of metal components

The usual failure mode was bursting of diaphragm. Tests on the material of the failed diaphragm showed no deterioration in its strength.

Examination of the location of the damage showed that in every case there was a circumferential cut in line with the corner of the flange holding the diaphragm. Rounding off the corners of the flanges as illustrated in the enlarged view in Figure 23.4 solved the problem.

23.7 Failures due to Design Defects

CASE STUDY

Figure 23.5 shows an automatic drain valve, which is fitted, below a compressed air reservoir. When the water level rises with the accumulation of water, the float rises and opens the drain valve. When the water level drops, the float drops and closes the drain valve.

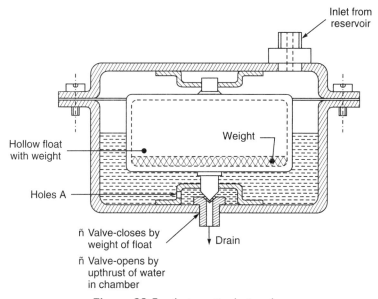

Inlet from reservoir

Weight

Hollow float with weight

Holes A

ñ Valve-closes by weight of float

Drain

ñ Valve-opens by upthrust of water in chamber

Figure 23.5 Automatic drain valve

In a particular case it was observed that a large quantity of accumulated water from the reservoir was discharged on opening a manual drain cock, indicating that an automatic drain valve under trial was not functioning. When the air pressure in the reservoir dropped to about 80 per cent of the normal pressure, the automatic drain cock also started functioning.

There was obviously a design defect. The buoyant force of the float was not adequate to counter the weight of the valve, and the force due to air pressure on the valve. The automatic drain valve started working when the weight of the float was reduced. The valve was adjusted to operate even at 110 per cent of normal operating pressure.

23.7 Do's and Don'ts to Prevent Pneumatic Valve Failures

- To determine the failure mode and mechanism correctly, dismantle a few failed valves very slowly and carefully.
- Look for corrosion, accumulation of scales, dust, grit, incorrectly seated valve discs, worn parts, fractured springs, swollen or damaged rubber components, bent stems, etc.
- Measure pick-up and drop out pressures, clearances, spring deflections, contact forces, etc.
- Compare with a new valve in good working order.
- Check adjustments, where provided.
- Ensure that the fluids (liquid or gaseous) that flow through the valves are free from particulate matter. Verify the effectiveness of filters.
- If any rubber components are either too soft or too hard or cracked, check the quality of the material through appropriate tests.

Chapter 24

FAILURES OF COMPRESSED AIR
OPERATED EQUIPMENT

In this chapter we learn

- *The types of compressed air operated equipment.*
- *Their common failure modes.*
- *The causes and effects of moisture in compressed air.*
- *The causes and effects of dust and particulate matter in compressed air.*
- *The causes and effects of air leakage from pipe-work and within compressed air equipment.*
- *The causes and effects of low air pressure in the compressed air system.*
- *Various case studies.*

24.1 Introduction

Compressed air operated equipment is used in locomotives, electric multiple unit stock and industrial machinery. Also wherever controlled forces have to be developed for various purposes as in the application of brakes, operating tap-changers, pantographs, doors, circuit breakers, isolators, etc. The main advantage of this system is that it is easy to control and transmits the necessary mechanical power to any required position.

This system comprises many devices, some of these are: air compressors, air motors, valves, pressure regulators, safety valves, air filters, pipe-work, pipe joints, reservoirs, inter coolers, after coolers, drain valves, etc.

All these are precision devices generally fabricated out of non-ferrous metals, with elastomeric and plastic components. Each device has its own failure modes and mechanisms. Any errors in their critical dimensions or any degradation of material due to corrosion or other chemical phenomena will cause these devices to fail in service. These failure modes have been discussed in the appropriate chapters separately.

The general causes of failures which occur more frequently are enumerated here in order of importance and frequency of occurrence.

(a) Moisture in compressed air.

(b) Dust or other solid particulate matter in compressed air.

(c) Low air pressure.

(d) Air leakage.

24.2 Moisture in Compressed Air

When air is compressed in air compressors, the compression is adiabatic. The heat developed during compression is not dissipated and it heats up the air. The air coolers that are provided have a very important function. If the compressed air were to be allowed to go directly into the valves and air motors, the hot air could damage these equipment.

As the compressed air gets cooled while passing through the pipes and the equipment, it sheds the moisture in the form of water droplets because the amount of water vapour that can be held in air gets reduced as its temperature drops.

The heating of the equipment as also the entry of water droplets interferes with the working of the devices, which are designed to operate on cool and dry air. It is necessary to provide after-coolers, air reservoirs and drain valves to cool the air and to remove all the water dissolved in it. This is achieved by the arrangement shown in Figure 24.1.

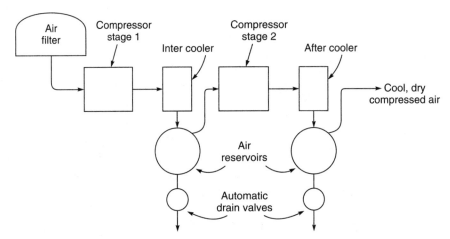

Figure 24.1 Block diagram of compressed air supply system, to ensure supply of cool, dry, compressed air.

The cooling surface area of the after-cooler must be sufficient to bring the air temperature down to that of the air inlet to the compressor. The air precipitates the excess water in the reservoir(s). These must be of adequate capacity and fitted with drain valves that operate automatically or are drained manually at regular intervals.

24.3 Failures due to Moisture in the Air

CASE STUDIES

There are a number of causes of failure of the system mentioned above. Some of them are rather obvious and trivial but the fact is that many equipment failures in actual service are due to such causes, as enumerated below:

CASE STUDY NUMBER	CAUSE OF FAILURE	FAILURE MECHANISM AND REMEDY
24.3.1	Inadequate capacity of after-cooler.	The compressed air did not cool sufficiently to shed all the moisture held in it. Some of it was deposited in the valves and devices. On increasing the after cooler capacity to the required level, the failure rate of equipment dropped by more than 60 per cent.
24.3.2	Automatic drain valves that were provided sometimes failed to operate.	Within a few days the water level in the reservoirs rose beyond the air outlet level and water entered the equipment causing failures. Manual draining of reservoirs at regular intervals was introduced and the automatic drain valves were replaced with new and modified designs.
24.3.3	In one installation there were compressed air operated linear motors, which collected water and failed frequently.	The devices in question were situated at quite some distance from the compressor. The after-cooler was apparently not of adequate capacity. As a result the compressed air output from the after-cooler was hot and moist. Further condensation was taking place in the long pipeline, which was sloping towards the motors. Water entered the motors and caused failures. Reversing the slope of the pipe, fitting a small reservoir at the lowest level of the pipe and draining it regularly solved the problem.
24.3.4	In a cold storage plant, compressed air motors operated the sliding doors of the cold chambers. These motors often collected water and failed.	Although adequate after-coolers and drains were provided for normal use of compressed air, the further condensation, which took place in the cold pipes and devices in the cold chambers, was the cause of the failure. Running a length of the air pipe within the cold chamber and fitting a small reservoir and drain valve at the end of this pipe before connecting the devices for operating the doors solved the problem.

24.4 Failures due to Dust and Solid Particulate Matter in the Air

CASE STUDIES

The best place to eliminate dust from compressed air is even before it is compressed, i.e. at the air inlet to the compressors. It then helps also to minimise wear and tear on the compressors.

Probably some very fine dust escapes these inlet air filters and even passes through the compressors to enter the compressed air system. The reservoirs and some large size pipe-work are often made of steel and their internal surfaces develop rust and scale. The compressed air therefore does contain some dust and particulate matter. If there is vibration as in the case of locomotives and rolling stock, the effect is more pronounced as the rust and scales are dislodged by the vibration. The quantities may be minute but they accumulate over long periods at vulnerable locations and cause failures.

Whenever there are failures of compressed air equipment, the devices should be opened out carefully and signs of dust, particulate matter and water must be looked for. Here are some case studies which illustrate the effect of dust and solid particles in the compressed air.

24.4.1	Some electro-valves used for compressed air control were leaking in the closed position.	Careful examination of the valves showed the presence of grit or scale particles at the conical valve seat. There were several such valves in the control cubicle. The internal piping in this cubicle was made of copper. A small air filter was provided at the compressed air inlet point of the cubicle.
24.4.2	The cylinders of the compressed air motors driving a tap-changer were getting scored.	The lubricating oil from the crankcases below these cylinders was collected, dissolved in petrol and filtered. The filter paper showed particulate matter like scale and grit. Steel pipes were partly replaced by copper pipes and air filters were installed at the air entry point to the copper pipe system.
24.4.3	The control cylinder of an air blast circuit breaker was becoming inoperative due to jamming of the air valve.	While overhauling the defective equipment grit particles were detected in the valves. Air filters were provided at the compressed air entry to the circuit breaker control panel. On opening out these air filters it was observed that they were ineffective due to loose packing of the horse hair filter element. The filter elements were replaced by more effective designs.

24.5 Low Air Pressure

It is obvious that compressed air equipment would fail to operate if the air pressure fell below the appropriate pressure limit. For this reason most installations are provided with pressure switches to detect low pressure and either to sound an alarm or to switch off equipment as considered necessary. The low pressure can occur either due to failure of the air compressor or to excessive leakage of compressed air. These two causes viz. air compressor failure and excessive air leakage have been discussed separately in Chapters 22 and 27. In this section we will discuss an infrequent condition which causes low air pressure resulting in equipment failures.

As discussed in Section 24.3, air filters are provided in the compressed air circuits to remove dust and particulate matter from the compressed air. While they prevent failures due to entry of dust and grit into sensitive equipment, they also introduce an additional point for maintenance. If these filters are not cleaned up at the appropriate intervals they get clogged up and cause drop in air pressure in the equipment.

Such failures were experienced in some installations. The small, almost invisible air filters in the pipe work were completely forgotten in one particular installation and not opened up for several years. Only when failures occurred due to low air pressure, the investigation at that stage revealed the presence of the air filter and its clogged condition.

Under normal conditions if the air filters are partially blocked there would be no measurable reduction in air pressure. If there is a sudden increase in demand for compressed air a momentary reduction in pressure, may affect the performance of the equipment.

24.6 Leakage of Compressed Air

The supply of compressed air is limited in most installations where compressed air is used for operation of air motors, contactors, circuit breakers, etc. The supply is limited by the capacity of the compressor. While the consumption of air for the actual operations of the devices is to be expected, leakage of air is not. Air leakage can occur not only from pipe joints and fittings but also across the seals or pistons in the devices. The only way to limit the overall air leakage to reasonable levels is to aim at 'zero' leakage at every pipe joint and inside every device. The definition of failure of air operated equipment systems should include air leakage. It is not sufficient that the equipment performs its functions. It must neither consume more air during operation nor allow any air leakage. The normal overhaul procedure for any air operated device should include an air consumption test as well as an air leakage test before commencing overhaul of the device, as received from service as also after overhaul before despatch to the assembly shops. The first test helps to reveal service conditions and the need for investigation.

Air leakage through the devices is usually due to degradation of rubber seals and accumulation of grit or dust in valves. Air leakage from pipe joints is almost always due to a manufacturing or installation defects. The most reliable joints are those which do not rely on rubber or similar elastic materials as sealing materials because they are liable to degradation.

24.7 Do's and Don'ts for Preventing Compressed Air Equipment Failures

- Ensure that the compressed air supplied to the equipment is clean and dry, by providing adequate inter-coolers, after-coolers and filters in the air supply lines. Where necessary, chemical air dryers, in which the air passes through columns of hygroscopic chemical pellets, may also be provided.
- Ensure the periodical maintenance of the air filters, drain valves and air dryers.
- As far as possible use copper pipe work. If steel pipe work is used ensure that effective air filters, installed close to the vulnerable equipment, are provided to remove any scales and rust particles from the air.
- Ensure the calibration and testing of pressure regulating devices in order to maintain the air pressure within specified limits.
- Carry out periodical air leakage tests and air consumption tests, attending to them when detected.
- Whenever any air operated equipment fails in service, dismantle the equipment very carefully while looking for various possible causes.

Chapter 25

FLUID FILTER FAILURES

In this chapter we learn

- *Different types and applications of fluid filters.*
- *Their failure modes viz., blockage and through leakage.*
- *Mechanisms of failures and remedial measures.*
- *Case studies.*

□ □

25.1 Introduction

$f$ilters in industrial machinery are mainly of two types: liquid and gas. The most common among these are oil and air filters. There are, other types that are used in the main processes. This chapter will cover air and oil filters which are commonly used in industrial appliances.

Air filters are used for removing atmospheric dust from the air before it is used for cooling or for transmission of power after compressing it. Oil filters are used for removing carbon, grit etc., from lubricating oil before recycling it through the machine.

There are various filter mediums such as cloth, paper, plastic mesh, ceramic, horsehair, etc. Each design of filter and each filter medium has its own peculiar characteristics which determine their suitability for different applications.

Defects in filters lead to different types of equipment failures. The filters by themselves, as devices do not fail but filter defects cause the failures of equipment serviced by them. There are two basic types of defects in filters:

(a) A defective filter allows an excessive flow of the filtrate to escape into the output stream of fluid.

(b) Defective filters block the free flow of fluid being filtered either partially or completely.

(c) Both types of defects (a) and (b) may occur simultaneously.

25.2   Excessive Flow of Filtrate into Output Stream

CASE STUDY

Defects of this type are unlikely to occur by degradation of the filter element. It is more likely that such a defect was overlooked at the stage of installation or maintenance but detected later.

Figure 25.1 illustrates these factors. The fabrication of the filter casing was defective and a gap was present between the filter element and its seat. This allowed the dust-laden air to bypass the filter element and enter the output stream of compressed air. The remedy is simple and apparent.

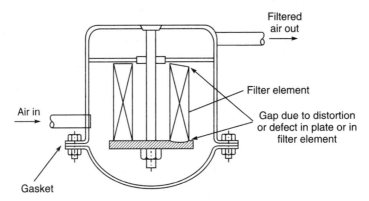

Figure 25.1   Defect in filter unit

A defect in the filter element dimensions could also have resulted in a similar defect. When installing or replacing filter elements, care should be taken to eliminate such defects.

Certain oil filters are provided with bypass valves, which open if the main filter element is blocked by the filtrate substance. This arrangement is provided where the blockage of oil flow is more harmful than the presence of dirt in the oil. Timely cleaning and replacement of filter elements, at required intervals is the solution to prevent such problems.

25.3   Failure due to Blockage and by-passing of Filter Element

CASE STUDY

Air supply to certain air operated mechanisms became restricted due to blockage of filter elements. This led to failures, as the filter elements were not designed for re-use after cleaning. They should have been discarded and destroyed after the specified intervals. These were determined by checking the reduction in fluid flow due to accumulation of the filtrate. A practice of replacement of the filter elements was introduced.

If the filter elements are designed for re-use after cleaning, the recommended cleaning fluids and the process for cleaning should be adopted.

In critical locations the filter is provided with a detector to monitor the pressure differential on the two sides of the filter element. If it is either too low or too high an alarm is sounded.

While blocked filter elements are common, in one exceptional case, when particulate matter was detected in failed valves, the air filters in the line were checked and found to be clean! Further investigation showed that the filter elements were not fitting correctly on their seats due to defects in dimensions and the air was bypassing the filter element altogether. The remedy was obvious.

25.4   Do's and Don'ts for Preventing Filter Failures

- Look out for signs of filter failures—either low air pressure at the equipment, excessive pressure drop across the filter, or dust, grit etc., in the equipment. In large installations it is often desirable to install devices to measure continuously the pressure drop across filters.
- Do not re-use filter elements after cleaning, which are not designed for re-use. Discard and replace them at specified intervals.
- It is desirable to analyse periodically the material collected by the filters. It can give advance warning about defects in other equipment.

Chapter 26

COMPRESSED AIR MOTOR FAILURES

In this chapter we learn

- *Applications of rotary and linear air motors.*
- *Failures of air motors due to component failures.*
- *Failure mechanisms of some special types of failures.*
- *Preventive measures.*
- *Case studies.*

26.1 Introduction

The two most common types of air motors are rotary and linear motors. The former is similar to internal combustion engines with components such as pistons, cylinders, crankshafts, connecting rods and valves. Spark plugs, carburettors and fuel pumps are not included. The source of power is the compressed air.

Linear motors consist simply of long cylinders, pistons, piston rods and inlet/outlet valves.

In the relevant chapter the causes for the failure of common components of these devices have been discussed. This chapter covers other types of failures of compressed air motors, as they are due to some special feature, or to a combination of several defects.

26.2 Failures due to Blocked Air Filters

CASE STUDY

In one application, rotary air-motors were used to drive on-load tap-changers for electric locomotives. The failure mode was a reduction in the operating speed of the motor, which could lead to electrical problems and was therefore unacceptable.

To investigate the cause, a pressure gauge was connected to the air inlet pipe of the air-motor. It was observed that this air pressure dropped as soon as the motor started working while the air pressure was unaffected at the main reservoir. The inlet air pressure returned to normal as soon as the motor stopped working. This indicated the presence of some resistance to airflow somewhere in the pipeline. Careful examination revealed a small in-line air filter that had remained unnoticed and unattended for a long time. On opening and cleaning out the filter element in this filter, the problem was solved. The filter was shifted to a more visible and more accessible location and painted yellow. A metal tag 'CLEAN EVERY THREE MONTHS' was attached to the filter. A survey was made to identify, clean and mark similar filters in other locomotives.

26.3 | Failures due to Incorrect Choke Sizes

CASE STUDY

A linear air-motor used for operating train doors was working too fast. Examination of the airflow-regulating choke revealed that the size of the choke was correct. Further examination of the device showed that a blowhole in the casting was allowing the air to by-pass the choke and causing the air motor to operate too fast.

The blowhole was blocked by brazing. Similar castings in other door mechanisms were checked, and two more similar defects were detected in three hundred sixty door mechanisms. The manufacturer was alerted to this problem and advised to investigate the designs of the moulds and the casting process. Further supplies were tested at 50 per cent higher air pressure.

26.4 | Failures due to Excessive Airflow

CASE STUDY

A double acting air motor was provided for the operation of a two-position rotary drum switch used as a reversing switch for an electric locomotive. Compressed air entry to and discharge from the cylinders in opposition was controlled by electro-valves. Linear cams and auxiliary switches controlled the electric supply to the electro-valves. Similar, traction/braking, rotary drum switches were provided with identical air motors. Figure 26.1 shows the general arrangement of these switches.

It was observed that the reversing drum switches were operating faster and with more vibration than the traction/braking drum switches. The minimum operating pressure for the former was about 2.3 kg/cm^2 while for the latter it was 3.4 kg/cm^2 as compared to the normal

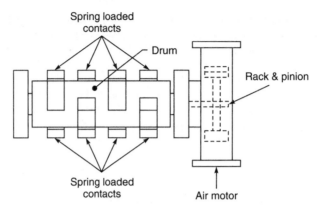

Figure 26.1   Air motor operated cam switch

air pressure of 5 kg/cm^2. This difference was obviously due to the longer drum of the latter needing a greater operating force.

The performance of the smaller braking drum switches was improved and the vibration eliminated when the cams controlling the electro-valves were modified to cut off air supply to the cylinders a little earlier in the stroke.

26.5   Failures due to Rubber Degradation

CASE STUDY

Compressed air motors are used to raise and lower pantographs of electric locomotives and motor coaches. Rubber piston seals are used on the large pistons to make them airtight. Figure 26.2 shows the general arrangement of these linear air motors.

These air motors have practically no air leakage but at one stage excessive air leakage was observed on several locomotives. It was noticed that all these defective air motors were supplied during the same and most recent period. In other words, other air motors, which had been in service much longer, were not prone to these defects.

Examination of the piston seals revealed that their shore hardness was more on one side than on the opposite side indicating hardening in some zones. It was verified that this was on the upper side of the seals. Similar examination on seals taken from other locomotives with much longer service did not show any such problem. It was suspected that the air motors, which are exposed, to the sun on the roofs of the locomotives became much hotter on the side exposed to the sun and the quality of rubber was not suitable for higher temperatures. The manufacturer of the rubber seals was cautioned about this observation. Suitable corrections in

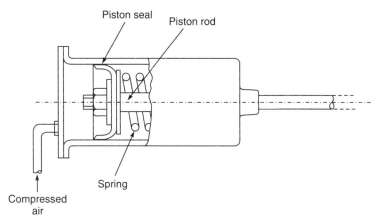

Figure 26.2   Linear air motor for pantograph operation

the quality of the rubber eliminated the problem. The specification in the drawing for the rubber seal was also suitably amended by stipulating that the rubber seal shall be suitable for operation at a temperature of 85 °C.

This investigation highlighted the usefulness of tabulating data about failed components. It showed that the problem was restricted to a relatively small group of locomotives of recent manufacture and the comparative measurements on piston seals from different manufacturers helped to convince the manufacturer of defective seals sufficiently to make him withdraw and replace all the defective seals.

26.6   Do's and Don'ts for Preventing Air Motor Failures

- In case of any problem regarding operating speeds, measure the inlet air pressure during operation.
- Refer to the relevant chapters in case of failures of components like seals, valves, pipes, filters, etc.
- Chapter 24 specifically covers compressed air equipment in general.
- Consider also the possibilities of defects, such as increased resistance, in the driven equipment.

Chapter 27

AIR COMPRESSOR FAILURES

In this chapter we learn

- *The application of air compressor in industry.*
- *The types of compressor failures due to component flows.*
- *Failure types discussed in this chapter viz. carbonisation of valve plates, breakage of valve plates.*
- *Excessive wear in cylinder liners, pistons, piston rings, etc.*
- *Case studies.*   ❏ ❏

27.1 Introduction

Air compressors come in all capacities and pressures from very small units, of capacity only a few litres per minute and pressures of the order of a few kg per square cm, to very large ones of capacity in thousands of litres per minute and pressures of the order of several hundred kg per square cm. The former are used mainly for operating small devices like brakes, doors, switches, etc. whereas the latter are used in large chemical and metallurgical industries where the compressed air is used in the main processes.

This chapter is based on experience of compressors at the lower end of the scale and mainly on reciprocating compressors. As far as failure modes and mechanisms are concerned, there is little difference between the large and the small machines. The same basic phenomena like fatigue, wear, creep, thermal expansion and contraction, corrosion, condensation, etc. are involved. There are very few physical or chemical effects that are relevant only for large or only for small machines.

There are many components in compressors which have been dealt with in other chapters. The failure modes, mechanisms and their causes are the same whether such components are in compressors or in motors or in some other type of machinery.

A few failure modes that are specific to air compressors are discussed in this chapter. They are:

(a) Carbonisation of inlet and outlet valves.

(b) Breakage of flat valve springs.

27.2 Carbonisation of Inlet and Outlet Valve Plates

The most common design of compressor valves is shown in Figure 27.1. The valves consist of thin and flat spring steel strips, which rest on valve plate openings. The valve strips either bend/deflect to allow the air to flow, or sit flat on the valve plates to block the air from flowing in the reverse direction. In some designs separate springs may be used to act on rigid valve plates.

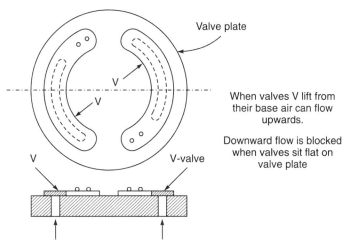

Figure 27.1 Valve plate with 2 flat valves

Carbonisation of the springs or plates is perhaps a misnomer. It is not the steel components that get carbonised. It is the lubricating oil in the cylinder that gets splashed on to the valve plates and gets carbonised and deposited on the valve plates.

When air is compressed it gets heated because part of the work done in compressing the air is converted into heat and this raises the temperature of the air unless the heat is removed as fast as it is developed during compression. This is not practicable as the compression takes place in a fraction of a second. This process, called adiabatic compression, is an inexorable and immutable natural phenomenon. The following table gives the temperatures attained after adiabatic compression:

Final pressure kg/cm^2	Absolute temperature ratio	Air temperature after compression if initial temperature was 27 °C and initial pressure was 1 kg/cm^2
4	1.5	177 °C
8	1.8	267 °C
12	2.05	332 °C
16	2.23	386 °C

At temperatures above 150 °C, lubricating oil starts decomposing and carbonising. Initially the rate of decomposition is slow but as the temperature rises, the rate of decomposition and carbonisation rises exponentially.

In order to limit this temperature rise of the compressed air, compression is usually done in several stages. A compressor with an output pressure of 10 kg/cm^2 first compresses the air to about 3 kg/cm^2 in the first stage. This compressed air at a temperature of about 150 °C is then cooled in an inter-cooler back to about 30 °C. It is then compressed in a second stage to the required 10 kg/cm^2 and the temperature again rises to about 150 °C (instead of 300 °C, that would have been reached if the compression had been done in a single stage).

With this background of the failure mechanism, the following case studies will illustrate how failures due to carbonisation occur.

CASE STUDY NUMBER	DEFECT	FAILURE MECHANISM AND REMEDY
27.2.1	In one type of compressor, the inter-cooler was inside the base casting; the cooling air outlet of this inter-cooler was blocked inadvertently	The temperature of the compressed air leaving the inter-cooler, i.e. the temperature at the inlet to the second stage became higher than normal. The increase in temperature of the compressed air after the second stage compression was even higher. This led to the gradual carbonisation of the lubricating oil entrained in the compressed air. The valve plates in the cylinder head became covered with a carbonaceous deposit in the second stage cylinder. This led, further, to reverse leakage of air during the suction stroke and to reduced efficiency and output of the compressor. The remedy was to remove the blockage of cooling air
27.2.2	The lubricant used in a compressor was changed to a new brand for trial	After about one years service, carbon deposits were observed on the valve plates of the trial compressor while the other compressors still continuing with the original oil were not affected. The trial with the new oil was abandoned

27.3 Failures due to Breakage of Valve Spring Plates

CASE STUDY

The mechanisms of failures and the causes of failures of springs as discussed in Section 27.2 are largely applicable here too although these springs are not similar in appearance to normal helical springs.

The original supply of compressors of one particular make did not have any problem with regard to the valve plates. However, when they were due to be replaced after many years of service spares could not be obtained from the original manufacturer due to obsolescence of that design. Substitutes obtained locally worked satisfactorily for some time but after several years, breakage of valve plates were experienced.

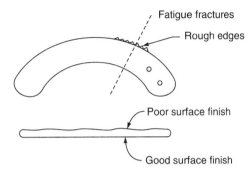

Figure 27.2 Fractures of valves in compressor heads

The fractures were examined carefully and it was observed that the fractures were due to fatigue. It was also seen that the surface finish on the edges and corners of the plates was not as good as on the flat faces as shown in Figure 27.2.

The manufacturer of the valve plates was advised to make further supplies with improved surface finish not only on the flat faces but also on the edges and corners. All earlier supplies were removed from service and replaced. There were no further cases of plate fractures.

27.4 Failure due to Breakage of Valve Plates

CASE STUDY

In another case, similar to that described in Section 27.3, investigation showed the quality of spring steel used was not up to the required standard with regard to fatigue strength. The surface finish also needed improvement.

The valve plates used in this application have a very severe duty of operating temperature, alternating stresses and wear on the contacting faces. It is essential to obtain the valve plates from the original manufacturers of compressors. The material, dimensions and the surface finish of the plates should be carefully specified and inspected during purchase from manufacturers who have been pre-qualified with emphasis on quality.

27.5 Normal Wear and Tear

As in the case of automobile engines, reciprocating compressors also suffer normal wear and tear on the pistons, piston rings and cylinder liners. The effects of excessive wear are (a) reduction in capacity; (b) increase in oil content of compressed air. The capacity can be measured very easily by measuring the time taken to raise the pressure in a reservoir of known capacity from 90 per cent to 110 per cent of the rated pressure. Norms should be established by making tests on new compressors. The compressor speed should be measured in these tests and corrections made if necessary.

A certain amount of oil in the compressed air is expected. All rubber components in the compressed air circuit should be oil resistant. Oil consumption based on the quantity to be added to the crankcase periodically for topping up should be noted and consumption per thousand hours calculated. If the consumption rate rises by more than 50 per cent as compared to the level when new, the pistons, piston-rings and cylinder liners may need to be replaced.

27.6 Compressor Speed Control

CASE STUDY

Wear and tear of reciprocating machines is more severe during starting from rest. Similarly, the starting of electric motors is equally onerous. When induction motors are switched off there are switching surges and standing waves produced in the windings. Contactors used for switching of electric motors suffer more wear and tear during every starting or stopping operation.

Air compressors are used to fill up compressed air reservoirs. As the air consumption rate is not constant, compressors have to be switched on or off when the air pressure falls below or rises above the specified pressure range. This constant on/off operation causes increased wear and tear and fluctuations in the electrical load.

This problem was solved with the availability of Variable Voltage Variable Frequency (VVVF) convertor drives for induction motors. With this device the compressor motor is not turned on and off, but its speed is regulated so as to maintain the reservoir pressure within close limits, despite fluctuations in the compressed air consumption. The same device is used for starting the set initially. With this arrangement there is considerable improvement in the reliability, durability of components as well as in cost of maintenance.

27.7 Do's and Don'ts for Preventing Compressor Failures

- In case of component failures refer to the appropriate chapters.
- In case of carbonisation of cylinder heads and valve plates, measure temperatures at different stages in the compression and consider whether inter-cooler performance is satisfactory. If not, consider the need for cleaning or modifications.

- If temperatures are satisfactory, examine the suitability of the lubricating oil.
- Consider the suitability of the intervals between cleaning schedules for removal of carbon deposits.
- In case of fatigue fractures of valve plates or valve springs consider factors like stress raisers and endurance limits of material.

Chapter 28

VACUUM BRAKE FAILURES
ON DC EMUS

In this chapter we learn

- *Failures of vacuum brakes on DC EMU stock.*
- *Definition of brake failure.*
- *Failure modes and failure mechanisms viz. general brake binding.*
- *Remedial measures determined through statistical analysis of failure data.* ❑ ❑

28.1 Introduction

Before 1951, on Indian Railways, the Electric Multiple Unit Stock on the Bombay Suburban Sections were fitted with vacuum brakes. These brakes operated at a vacuum, which was 50 cm of mercury column below atmospheric pressure. Exhausters driven by DC motors created the vacuum.

The vacuum brake system is inherently a fail-safe system. Brake failures involving loss of brake power are very rare. However, brake binding cases, i.e. brake failure where the brakes could not be released either on all the coaches or on one coach only, were common. It is these types of failures, called brake binding, which are discussed in this chapter.

Brake binding was very rare in the initial years but as the traffic increased and the loading on the coaches, on the substations and the overhead track system increased, brake binding was experienced from time to time. By about 1954/55, the number of such incidents increased to unacceptable levels.

28.2 Principle of Operation of Vacuum Brake System

Figure 28.1 shows a schematic of the braking system. Only the relevant components on one coach are shown to explain first the principle of operation and then the mechanism of failures.

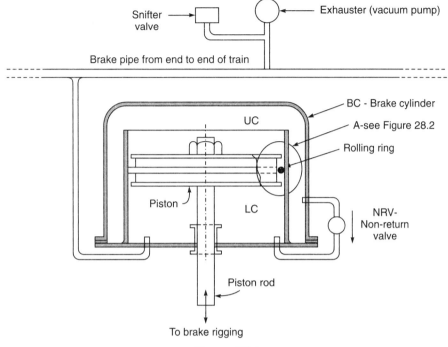

Figure 28.1

When vacuum is created in the brake pipe by the exhauster in the motor coach, air is evacuated from the spaces UC and LC in the brake cylinder BC both above and below the piston. The non-return valve NRV allows the air to flow from the upper chamber UC to the lower chamber LC, and then to the brake pipe. The pressure or the vacuum on both sides of the piston is equalised. The heavy piston drops down to its lowest position, thereby releasing the brakes.

When air is re-admitted to the brake pipe by the driver on applying the brake, air enters the lower chamber LC but not the upper chamber UC because the non-return valve NRV blocks the air from flowing from the lower chamber LC to the upper chamber UC. The vacuum in the upper chamber remains intact and the difference in pressure between the lower and the upper chambers LC and UC, causes a force to be developed on the piston which rises up and applies the brakes.

If vacuum is recreated in the brake pipe, the air flows from the upper chamber to the lower chamber LC and thence to the exhauster until it is equalised on the two sides of the piston and the piston drops down thereby releasing the brakes. The system is now ready for re-application of the brakes.

Brakes can be applied, released and reapplied any number of times. If however, the vacuum that is re-created after any application of the brake is less than the vacuum created initially, a small difference would remain between the two sides of the piston and this would prevent it from dropping fully and releasing the brakes. The defect known as brake binding would occur.

This re-creation of vacuum in the brake pipe to a lower level, after initial creation of vacuum to a higher level, is the root cause of brake binding. Such an inequality can and did occur due to differences in the line voltage, which caused changes in the speed of the DC motor driving the exhauster.

28.3 Failures due to Snifter Valve Defects

CASE STUDY

When such failures occurred, it was observed that when the driver took the brake handle to the release position and tried to start the train the vacuum brakes did not get released completely. The brake blocks would be rubbing on the tyres and increasing the train resistance, thereby preventing the train from picking up speed. Sometimes, after a little run the brakes would release fully and the train would then run smoothly for the rest of the journey.

At times the problem would be so severe that the motorman or the maintenance staff had to release the brakes manually on each coach.

The effects of these failures were manifold: trains were delayed, by a few minutes or even longer periods depending upon the time, the place and the intensity of the failure. Sometimes the starting resistors became defective and there would be disorganisation of service.

Since differences in line voltages along the line are not uncommon in traction systems, a valve known as a snifter valve was provided near the exhauster as illustrated in Figure 28.1. This was a safety valve in reverse, it allowed air to enter the brake pipe whenever the vacuum created was higher than a prescribed limit, due to a high line voltage.

In the case under study, statistical analysis of the failures on lines similar to those described in Section 12.3, Chapter 12, showed that the failures occurred at mid-points between substations, during the peak traffic hours, when the line voltage was the lowest. It was also apparent that certain motor coaches were significantly more likely to figure in such failures. This directed attention to the snifter valves in the motor coaches. On testing them it was seen that the settings had drifted and were too high on some motor coaches and generally higher than necessary.

After carrying out certain experiments and tests, new lower settings were decided upon and all snifter valves were adjusted accordingly. As against seven to ten such failures every month, the failure rate dropped to zero as soon as these adjustments were completed on all the motor coaches.

This case study demonstrated the usefulness of statistical analysis of failure data and that when remedial measures are taken after determining correctly the failure mechanism, the failure rate drops to zero.

28.4 Brake Binding due to Defective Rolling Rings

CASE STUDY

We have discussed above cases of general brake binding that occurs on most of the coaches. There are other cases where the brake binding may occur on only one or two brake cylinders.

Such cases of brake binding on individual cylinders were due to defects in the brake cylinders and the most common defect was due to degradation of the rolling ring seal between the cylinder and the piston shown in Figure 28.2.

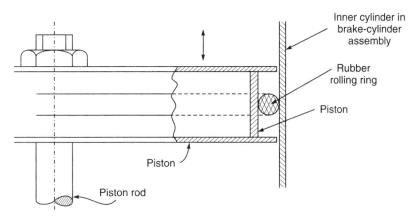

Figure 28.2   Enlarged view of part A from Figure 28.1 shows how the rolling ring provides an air tight seal between the two sides of the piston

The ring rolls on the groove of the piston between the cylinder and the piston. The dimensions of the rubber ring are very important. If the ring is too tight it will not roll easily and the resultant sliding friction can prevent the piston from dropping freely under its own weight to release the brakes. On the other hand if the ring is too loose, it will take a wavy shape and get twisted. Again the ring will not roll freely and prevent the piston from dropping freely when brakes are released.

Cases of this type were eliminated as soon as the dimensional accuracy of the rolling rings and the quality of rubber used as regards swelling under the influence of oil were controlled through stringent inspection of purchased spares. Refer to Chapter 20.

28.5 Brake Binding due to Bending of Piston Rods

CASE STUDY

A few cases of brake binding were traced to bending of the piston rods, due to mishandling of spare brake cylinders in the railway yards. These were eliminated by the introduction of simple tests with low vacuum testing in which the minimum vacuum needed to move the piston was determined. All cylinders that underwent repair or overhaul were subjected to this simple test. All defective brake cylinders were thus detected and repaired.

The number of cases of brake binding due to defects in individual cylinders came down sharply after introduction of this test.

28.6 Do's and Don'ts for Preventing Failures due to Brake Binding

- Ensure the quality of the rubber components such as rolling rings, valve seats, gaskets, etc., with reference to the drawings and specifications.
- Ensure the dimensional accuracy of the cylinders, pistons and valves.
- Ensure the constancy of vacuum created at all times and locations along the track.
- Test all assembled brake cylinders by checking at minimum vacuum/pressure needed for operation.
- Since brake binding can occur due to a combination of various factors, carry out statistical analysis of a number of such failures before reaching any conclusions.
- To facilitate statistical analysis, maintain data relating to each case of failure.

Chapter 29

Fan and Blower Failures

In this chapter we learn

- *Applications of fans and blowers.*
- *Effects and types of fan/blower failures.*
- *Failure modes viz. fractures, vibration and reduced output.*
- *Failure mechanisms and preventive measures.*
- *Case studies.* ❑ ❑

29.1 Introduction

Fans and blowers are used in many industries for main processes and also as auxiliary machines for enhancing the output of other equipment. This chapter is about fans and blowers of the latter kind, examples of which are enumerated below:

(a) Cooling of internal combustion engines;

(b) Cooling of electric motors;

(c) Cooling of transformers and other static equipment.

Some fans are mounted directly on or inside these machines and driven directly. Some fans or blowers may be mounted and driven separately. In some cases fluids are used for the cooling of the equipment and fans are used for the cooling of the fluid. The choice of the system depends, in the ultimate analysis, on the overall cost.

This chapter deals with fans or blowers by themselves. Failures of fans and blowers cause failures of the main equipment that they are supposed to serve. Since failures of the main equipment are inevitable if there is inadequate ventilation, effective protective devices are provided to stop their operation in the event of any reduction in the output of the fan or blower.

Failures of fans and blowers are relatively rare as they consist, of nothing more than a cast or welded impellers inside steel casings. Yet, despite this simplicity, failures do occur from time to time mainly due to design and manufacturing deficiencies. Failures due to operating or maintenance deficiencies are rare. Three common failure modes of fan and blower impellers are enumerated below:

(a) Excessive vibration.

(b) Cracks and fracture.

(c) Reduced output.

29.2 Excessive Vibration

CASE STUDY

Figure 29.1 illustrates the impeller of a blower fitted on the shaft and held in place by a nut. Within a few days after installation it developed severe vibration and had to be stopped. It was found that the impeller had become loose on the tapered end of the shaft although the nut was secure in its place.

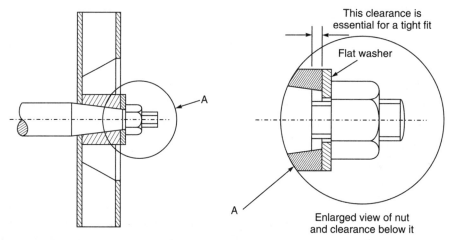

Figure 29.1 Impeller working loose due to machining error

Close examination showed that due to incorrect machining of the tapered bore on the impeller, it was going too far on the shaft end and the nut was not effective in holding it in place. There should be a small clearance between the end of the shaft and the face of the impeller as shown at A in Figure 29.1 to ensure that the nut bears down hard on the impeller when fully tightened. Checking with the working drawings, in the manufacturers works, confirmed the machining defect.

There were other cases of vibration in blowers but these were due mainly to cracks and distortion of the impeller plate. This type of defect is discussed in the next section.

There are other possibilities of impeller vibration such as those due to unbalanced impellers but such manufacturing defects can be detected during normal initial inspection. We will discuss in this chapter only those failures or defects which are not present on new machines and which develop after some time.

29.3 | Failures due to Cracks and Fractures of Impellers

CASE STUDY

Figure 29.2 shows the hub of an impeller and the crack emanating from the root of a fillet weld between the hub and the impeller plate. Fillet welds are the starting points of cracks and fractures. Another design with rivets had similar problems starting from the rivet holes. In both cases changing the design of the impeller to use a butt weld between the hub and the impeller plate solved the problem. After welding the joint was ground on both sides to give a smooth face as shown in Figure 29.3.

It is not always possible to replace fillet welds by butt welds. In such cases it becomes necessary to explore other methods of reducing or eliminating such stress raisers. One solution is to replace the fabricated design by a casting; another solution would be to use a lighter but stronger material such as aluminium alloy. Increasing the thickness of the impeller does not solve the problem because such increase leads to higher centrifugal forces and the stress remains the same. Yet another method would be to re-design the blower to operate at a lower speed.

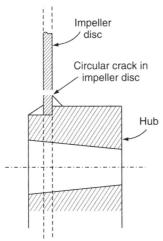

Figure 29.2 Blower impeller crack near fillet weld

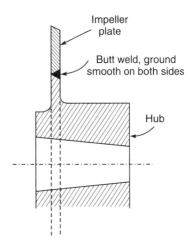

Figure 29.3 Blower impeller with butt weld

The failed welds or rivet holes from which fatigue cracks have emanated should be examined very carefully by metallurgists. If undercutting of welds, incompatibility of electrode and impeller material or structural damage due to welding is detected suitable corrective action can be taken.

All such defects start as invisible micro-cracks which develop into visible cracks and then into fractures. Cracks and fractures are due to stresses caused basically by centrifugal forces and these forces are unavoidable.

The concentration or augmentation of stress at fillet welds, rivet holes and changes in section can be minimised. If the augmented stress at any point exceeds the endurance limit and if there is fluctuation in this stress, cracks are initiated by metal fatigue. In calculating the stress it is necessary to take into account the effects of stress raisers, which are often the root cause of failures.

29.4 Reduced Output from Blower

CASE STUDY

In an unusual case of a blower with vertical axis provided for cooling a rectifier, there were incidents of air flow relay operation resulting in tripping of the main circuit breaker. This prevented damage to the rectifier but interrupted operation.

Investigation showed that the airflow output from the blower was less than normal and the cause was accumulation of dust deposits in the impeller. Oily fumes and dust in the environment caused the deposits to accumulate in the form of a caked substance which grew not only on the wire mesh in air passages but also on the impeller surfaces. Increase in the frequency of cleaning up of the air passages and the impellers was the solution. See Figure 29.4.

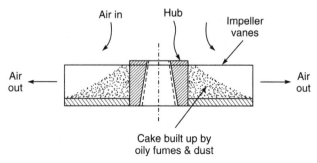

Figure 29.4 Reduction in air output due to accumulation of oily fume and dust on impeller

29.5 Do's and Don'ts for Preventing Fan and Blower Failures

- As vibration is one of the main causes of fan impeller failures, the vibration levels should be measured initially after installation. If any observations of noise or palpable vibration are made, repairs should be carried out accordingly.

- Impellers should be dynamically balanced, if necessary.

- During routine maintenance, welds (fillet welds in particular) discs and vanes should be carefully examined for hairline cracks. Stresses in rotating discs are usually highest near the inner radius where the disc may be welded to a hub. In case of a history of weld failures, arrange dye penetrant tests on the vulnerable areas, during overhauls of the equipment.

- Look for signs of corrosion and for deposits and accumulation of dust in corners of vanes and discs.

- Ensure that clearances between the fixed and rotating parts are within the limits specified in the design.

- Ensure that the fits/limits and fasteners used in the assembly of the impellers on the shaft are as prescribed in the design.

- Check the quality of the material, i.e. the ultimate tensile strength and the endurance limit with reference to the specification.

- If all the above checks reveal no defects, the basic design of the impeller has to be examined and the stresses in the zones where cracks are developing have to be calculated. This may require help from the manufacturers' design engineers.

Chapter 30

THERMAL CRACKS ON MOTOR COACH TYRES

In this chapter we learn ·

- *The real possibility of thermal cracks developing on the tyres of electric multiple unit stock.*
- *Special conditions of operation that are conducive to the formation of thermal cracks on tyres.*
- *Failure mechanism in the formation of thermal cracks.*
- *Preventive measures.*

30.1 Introduction

*T*his chapter is mainly relevant to railwaymen in charge of maintenance of electric multiple units. The failure mechanism and the manner in which the problems were controlled could be applicable to other machines where heavy rotating masses are braked frequently. A similar problem was reported in a steel rolling mill, which reversed every few minutes.

Modern high performance electric rolling stock which operate at high speeds and high retardation rates are usually fitted with disc brakes, so that thermal cracks on tyres are not a problem. Many older motor coaches still in service are fitted with cast iron or composition brake blocks which act directly on the tyres. These motor coaches face the problem of cracks developing on the surfaces of the tyres. If allowed to continue in service these cracks propagate through the full depth of the tyre and cause a tyre fracture.

This type of failure has the potential of causing a derailment and must be tackled immediately and effectively. Where such brakes are in service, the maintenance staff must check the tyre surface periodically for the appearance of cracks and should a crack be detected, the coach should be withdrawn from service for necessary repairs.

30.2 Mechanism of Failure (Cracks on Tyres)

Figure 30.1 shows the schematic of a motor coach wheel with clasp brakes on the tyre treads.

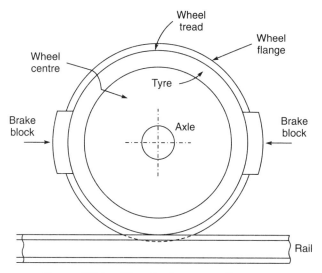

Figure 30.1 Clasp brakes on a railway tyre

Consider a typical motor coach in Electric Multiple Unit operation. Assume the following operating conditions:

- Speed at the instant of application of brakes: 60 km/hr = 16.7 m/sec
- Length of the brake block = 30 cm; Wheel diameter = 1 m
- Tyre speed 16.7/π = 5.3 rps

With the above conditions any point on the tyre passes from one end of the brake block to the other in 0.3/16.7 second, i.e. 18 milliseconds. Due to the braking force and the speed a great deal of heat is generated at the interface between the tyre and the brake shoe. This raises the temperature of the surface film to about 500 °C within 18 milliseconds. For about 80 millisecond the tyre is exposed to the air until it reaches the opposite brake block. In this interval the surface temperature drops to about 120 °C, by radiation to the atmosphere and conduction of heat into the thickness of the tyre. The tyre surface again passes under the second brake block and in the next 18 millisecond its temperature again rises to about 500 °C. In this manner as every point on the surface of the tyre tread passes under the two brake blocks alternately, the temperature of the surface fluctuates between 500 °C and 120 °C at a frequency of about 10 cycles per second.

Only the surface of the tyre undergoes these severe temperature variations. As the depth in the tyre increases, the amplitude of the temperature variation drops rapidly. Due to the differences in the temperatures of the different layers of the tyre, and the effects of thermal expansion and contraction alternating tensile and compressive stresses are produced in the tyre steel. And these stresses have the same frequency as the temperature cycle.

This phenomenon takes place in all tyres with brake blocks operating on tyre treads. The maximum stress developed at the surface of the tyre does not ordinarily exceed the endurance limit of the steel. Moreover the stressed steel is worn out continuously. Unless the intensity of this phenomenon is exceptionally high or the fatigue strength of the tyre material is exceptionally low, cracks do not develop in the surface of the tyre. When the following factors combine to make the peak stress exceed the endurance limit, fatigue cracks develop at the surface of the tyre.

(a) High speed at brake application.

(b) High unit pressures between brake blocks and tyres.

(c) Low fatigue endurance limit of tyre steel.

(d) High initial stress due to shrink fit of tyre on wheel centre.

Tyre cracks that appear by this process are called thermal cracks but basically they are fatigue cracks.

30.3   Preventive Measures

The above list suggests the measures to be taken to contain the problem when these thermal cracks appear in any railway system. There are limits to which these measures can be applied and when these are reached the only way is to give up this system of braking and to adopt disc brakes.

(a) The speed of the train at the instant of application of brakes can be reduced only at the cost of increasing the running time. By increased coasting this can be done with the least effect on the running time.

(b) Reducing the air pressure in the brake cylinder, at the cost of increasing the braking distance can reduce the unit pressure between the brake block and the tyre. If the initial pressure was excessive, the increase in the braking distance is marginal.

(c) The type of steel used for the tyre can be changed to use a steel with a higher fatigue endurance limit. If the steel used initially was of inferior quality this is a viable solution.

(d) The initial stress in the tyre due to its shrink fit can be reduced if it was excessive initially.

It will be seen that in every possible line of action there is a limiting condition due to either economics, availability or safety. If these limits, which differ from case to case, are reached the solution for new systems is to adopt disc brakes. The solution for existing systems where thermal cracks have started appearing is to determine the optimum practicable limits through service trials and to respect them even if it means reducing schedule speeds. On the other hand if it is found, that there was unnecessary or inadvertent excess in regard to speed, brake pressure, shrink fit stress, or deficiency in regard to fatigue endurance limit of the tyre steel, the necessary corrective steps can be taken. The problem can be solved without operational repercussions.

30.4 Emergency Measures

If thermal tyre cracks start appearing in an existing system in commercial operation, it is necessary to take emergency steps to prevent any accidents from occurring before the final solutions can be implemented. These steps are as follows:

(a) Withdraw the coach with thermal cracks on its tyres immediately on detection. Machine the tyres with a light cut sufficient to remove the surface cracks and only then return the coach to service.

(b) Intensify the periodical checking of tyres to detect thermal tyre cracks.

(c) If cases of the above type are coming up frequently, check their ages since last machining. If they are above a convenient and economical machining interval for coach tyres, introduce this for periodical machining of tyres. If not, then

(d) Reduce the brake cylinder pressure marginally, while verifying that the braking distance limit is not transgressed.

(e) Verify and ensure that coasting practices are being practised correctly.

(f) Introduce an upper speed limit for the trains.

30.5 Thermal Cracks on EMU Tyres

CASE STUDY

In one railway, the EMU stock was running normally without any incidence of thermal cracks. However, after many years of trouble free operation, thermal cracks started appearing on the tyres. The problem was detected when one tyre fractured across its full width. On careful examination of other tyres in service, a few were detected with hair line cracks on the tyre treads.

The action taken was: to review the conditions discussed in Sections 30.2 to 30.4; to withdraw from service wheel-sets with hair line cracks; to reduce the brake cylinder pressures by 10 per cent on the whole fleet of EMUs; to intensify the routine inspection of tyre surface; to reduce the interval between machining of tyres. There was no significant increase in the braking distance and in the running time. The thermal cracks were eliminated.

30.6 Do's and Don'ts for Preventing Thermal Cracks in Steel Tyres

- In case of thermal cracks appearing in railway tyres, take emergency steps enumerated in Section 30.4.
- Check brake cylinder pressures with calibrated instruments, reduce it marginally if practicable.
- Reduce the speed at instant of brake application by control on coasting.
- Check chemical composition and mechanical properties of cracked or fractured tyres.
- Retain samples of tyre steel around the cracks or fractures until the investigation is complete.

STRUCTURAL MEMBER FAILURES

In this chapter we learn

- *Failures of structural members of machinery and plant.*
- *Principal failure modes are fractures, cracks and deformation.*
- *Mechanisms of failures are usually metal fatigue, corrosion, and yield.*
- *Measures to prevent such failures are usually in regard to design improvements or manufacturing process improvements.*
- *Case studies.*

□ □

31.1 Generalities

We will discuss here, failures of equipment components whose role or function is only to support or hold in place other operating components of the equipment. They are subject to forces caused by gravity or by reaction forces produced during the operation of the equipment. Examples of this category of components, given the name structural members, are:

(a) The end shields of an electric motor. They support the weight of the armature and add rigidity to the whole carcass of the motor.

(b) The suspension bolts of a transformer, that support the weight of the core and the coils of a transformer from its top cover.

(c) The side frames of a printing press. They support all the active components of the printing press. They are subject to weights and also internal reactions of the moving parts of the press.

(d) The main pillars and the platens of a hydraulic press. They are subjected to the weights of the components and, the reactions due to the forces generated within the press during its operation.

(e) The sheet-metal bracket of a small electromagnetic relay. It supports the weight of the solenoid.

Failures of such structural members though relatively rare, do occur sometimes. The failure mode is usually cracking, fracture, and deformation. The causes are generally *design deficiencies*, and in the case of castings and welds, manufacturing defects. Failures due to over-loading and bad maintenance are very rare.

There are few common factors between the variety of failure types; however the failure mechanisms are very few in number: metal fatigue, corrosion, thermal expansion and contraction, yield under stress. These may be explained by a few case studies.

31.2 Failures due to Defects in Material and in Assembly

CASE STUDY

Figure 31.1 shows the manner in which the active part of a *locomotive transformer* is suspended from the cover plate of the transformer tank. The weight of the core and the coils is suspended by four screws welded to the top cover. When the core is lifted, these bolts carry the full weight of the core but when assembled in the tank the load is relieved by contact between the core and projections on the bottom plate of the tank. When the locomotive is in motion, inertia forces caused by accelerations of the order of 1 g add to the weight of the core, due to the vibration of the locomotive.

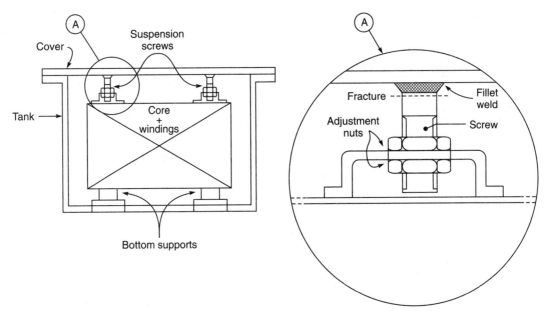

Figure 31.1 Suspension screws of a traction transformer core + windings

In one case the steel bolts were found to develop cracks and fractures. The location of the fractures was from the root of the welds as shown in the inset A in Figure 31.1. Investigation revealed the following defects:

(a) The steel used for the screws was not weldable.

(b) There was some undercutting in the screws at the roots of the welds.

(c) The fractures were of the fatigue type. The area of the fatigue zone was nearly double the area of the sudden fracture zone.

(d) The assembly process for adjustment of the nuts did not include adequate checks on the dimensions to ensure that the core rested properly on the bottom supports and did not hang solely from the suspension screws.

The following steps were taken to prevent recurrence of screw fractures.

(i) A weldable steel of slightly lower tensile strength replaced the material of the screw.

(ii) The welding quality was improved to ensure that there would be no undercutting at the toes of the welds on the screws.

(iii) A special measuring jig was used to ensure that the depth of the core bottom seats was at the specified distance from the top cover of the transformer, while adjusting the nuts on the screws. A similar measuring jig was used to check the corresponding dimension on the tank.

31.3 Failures due to Sharp Bends in Plates

CASE STUDY

The electromagnet of a relay was supported on an L-shaped mild steel bracket as shown in Figure 31.2. These relays were mounted on a platform, which was subject to heavy vibration caused by the operation of the machine.

The L-shaped bracket was found to have fractured at the bend in one case. Examination of similar relays on other machines revealed one with a small crack in the same location. It was also seen that the inner radius of the bend on the 2-mm steel sheet bracket was less than 1 mm. The steel used for the brackets did not pass an 180° bend test.

The following steps were taken to prevent further recurrence of these fractures.

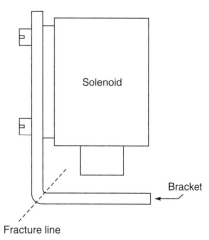

Figure 31.2 Fracture of relay bracket

(i) All relays of the same type installed in various locations were examined. Those with hairline cracks at the bend were replaced.

(ii) The L-shaped steel brackets of the relays were replaced by new brackets with an inner radius of 4 mm at the bend of the L-shaped brackets. See Figure 31.3. Priority was given to relays subject to vibration.

(iii) The steel sheet used for manufacturing these brackets was first tested for compliance with a simple 180°, flat bend test.

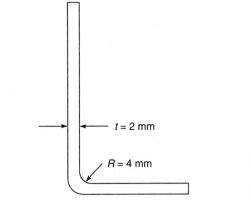

Figure 31.3 Modified bend of relay bracket

31.4 Cracks in Castings

CASE STUDY

Steel castings were used for locomotive bogie frames. Cracks were detected in some of the bogies as shown in Figure 31.4.

X-ray examination of several similar uncracked bogies revealed hot tears in the region. These are cracks in the steel members

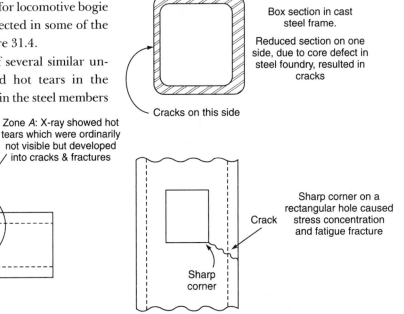

Figure 31.4 Cracks in box sections of cast steel bogie frames in locomotives

produced during solidification and cooling of the castings and caused by the contraction of the solidified steel from about 1400 to 35 °C . These effects have to be prevented by the correct design of the moulds, cores and patterns.

Control on the quality of steel castings for critical applications was ensured through the use of X-ray or gamma ray equipment to inspect the quality of the castings in vulnerable zones.

31.5 Failures due to Excessive Stress Caused by Vibration

CASE STUDY

A silica gel breather was fitted on the side of a conservator tank of a locomotive transformer. The breather consisted of an aluminium alloy casting with a glass container as shown in Figure 31.5.

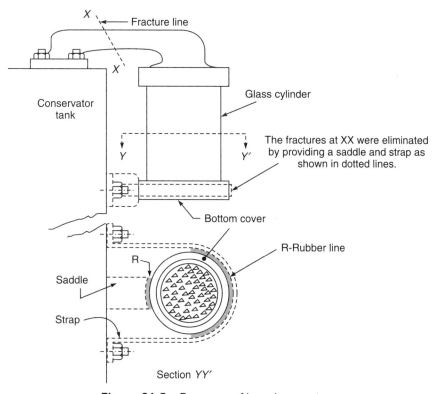

Figure 31.5 Fractures of breather castings

There were several cases of cracks and fractures in the top casting due to inertia forces caused by the vibration of the locomotive. Replacement of that casting by a stronger or improved design could probably have solved the problem. But it was decided to adopt a quicker and probably less expensive solution by providing a lateral support comprising a saddle and a strap at the lower end, to reduce the bending load on the upper casting, as shown in Figure 31.5 highlighted by dotted lines.

31.6 Do's and Don'ts for Preventing Structural Failures

- Although structural failures in machinery are rare, every case must be investigated because the consequences of such failures are very expensive.
- If it is a steel casting, presence of hot tears, blow holes and reduced section should be investigated.
- If it is a fabricated structure possibility of weld fractures and fractures at sharp bends and toes of fillet welds should be investigated.
- In all cases dimensions of machined surfaces should be checked.
- The possibilities of stress concentration should be examined in cases of hairline cracks or fractures.
- The possibilities of excessive stresses due to vibration and inertia forces should be examined.

About Chapters 32 to 65

Chapters 32 to 65 deal with failures of electrical equipment used generally in power stations, transmission and distribution networks, industrial establishments and transportation systems like the railways, airlines and shipping. Many of the components discussed here are used even in commercial and domestic electrical installations.

The types of equipment discussed in Chapters 32 to 64 are classified as electrical equipment. Not all failure modes of electrical equipment involve electrical stresses. Failures like shaft failures, bearing failures and conductor fractures involve only mechanical stresses and mechanical failure mechanisms. These have been discussed in Chapters 7 to 31. While they may be discussed here in brief, the relevant chapters should be referred for more details.

Chapters 32 to 65 deal mainly with electrical mechanisms of failures; i.e. processes where electrical phenomena are also involved in addition sometimes to mechanical processes.

CRIMPED SOCKET FAILURES

In this chapter we learn

- *A brief description of crimped sockets used for terminating cables.*
- *Tests that can be made to verify their reliability.*
- *Failure modes and failure mechanisms.*
    - *(i) Overheating;*
    - *(ii) Slipping out of conductor from socket;*
    - *(iii) Fracture of the palm of the socket;*
    - *(iv) Fracture of the conductor;*
- *Case studies involving the above failure modes.*
- *Remedial or preventive methods.*   ❏ ❏

32.1 Introduction

Crimped sockets are now used universally for the termination of wires. They have replaced soldered sockets and other forms for terminating wires. Crimped sockets come in a wide range of sizes from the smallest for wires of cross-section of the order of 1 square mm to large wires of size around 300 square mm. Figure 32.1 shows a typical crimped socket and Figure 32.2 shows the same after fitting on a stranded wire. A special tool is used to compress or crimp the barrel of the socket around the bare end of the wire. For small wires up to about 5 square mm cross-section, the tools are hand operated. For larger wires hand operated hydraulic tools are used.

The cross-section of the socket is reduced by the tools, at the narrowest section of the crimp zone, to about 85 per cent of the combined original cross-section of the socket and the wire. The resulting deformation of the copper or aluminium produces adequate contact forces to ensure good electrical contact between the wire and the socket. It also secures the wire firmly in the socket and prevents it from coming out. The force needed to pull the wire out is of the order of 40 to 80 per cent of the tensile strength of the wire.

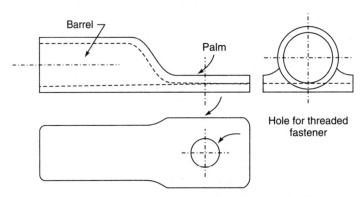

Figure 32.1 Crimp socket before fitting on cable

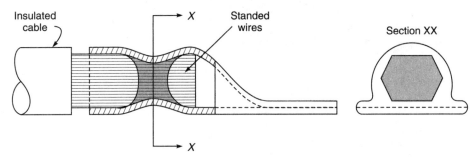

Figure 32.2 Crimped socket after crimping onto cable

If the crimped socket is sectioned through its narrowest zone after crimping on a wire, the appearance is that of a solid wire provided the crimp is satisfactory. If the crimp is inadequate due to some defect in the crimping die or the tool, the cross section will show voids.

The three important tests that can be made on crimped sockets to verify their reliability are:

    Resistance test

    Pull out test

    Endurance test

All the three tests mentioned above are type tests, which are to be made on the prototypes as well as on random samples drawn from lots when specified. They are meant to prove the design of the crimped sockets and also the crimping tools.

The detailed procedures for these tests are given in BS 6360 and IS 8337. There are no routine tests for verifying the compliance with the standard. The only routine check that can be made is with regard to external dimensions of the barrel of the socket and narrowest portion of the crimp zone. It should be ensured that it is the same as of the crimped joints subjected to the type tests.

32.2 Failure Modes and Failure Mechanisms

The four failure modes enumerated below are generally in order of decreasing probability of occurrence.

Over-heating.

Slipping out of conductor from the socket.

Fracture of the palm of the socket.

Fracture of the conductor.

If the failure occurs while the socket is carrying an electric current, arcing and melting may follow, making it difficult to determine the true failure mode.

There are several possible causes and failure mechanisms associated with each of these failure modes. It is, desirable to determine first the failure mode before thinking of probable causes. If there is extensive damage around the socket, it may not be possible to do this. In that case it is more difficult to determine the true cause of failure.

When the damage at the seat of the failure is extensive and the original components are charred or burnt/melted, one possible method which is successful in identifying the cause, is to examine the same zone in other similar systems. If the failure is due to a type defect it is possible to see either design and workmanship defects or early signs of impending failure, such as cracks, loose fasteners, tracking, low clearances, overheating, etc.

32.3 Overheating of Crimped Sockets

The starting point of overheating of crimped sockets is bad contact between the wire and the socket, or between the socket and the terminal, or between one socket and another in contact with it. The cause could be either an inadequately pressed crimp due to a defect in the design of the tool or inadequate tightening of the threaded fastener.

Either of the two defects mentioned above can lead to increase in contact resistance and consequent rise in temperature. If it rises above 90 °C, the contact surfaces begin to oxidise rapidly and this starts a vicious cycle of increasing contact resistance, increase in temperature and increase in oxidation rate. Eventually, there is melting, arcing, and possibly a fire.

32.4 Failures due to Inadequate Thickness of Socket Barrel

CASE STUDY

In one case, an electrical fire was caused by a crimped socket failure where a cable carrying about 300 amperes melted at the termination. There was extensive melting of socket, cable and terminal. It was impossible at first to determine which joint had failed first, (a) the joint between the socket and the cable? or (b) the joint between the socket and the terminal?

The failed socket was compared with other sockets and terminals in the same panel and in other similar panels. It was observed that the wall thickness of the socket barrel of the failed socket was visibly less than that of others currently in use by the same manufacturer and by other manufacturers. A few of the cables with these suspect sockets were taken out and tested for the pull out test and the resistance test. It was found that the pull out force was only about 30 per cent of the minimum specified in the relevant BS 6360. The electrical resistance was about 40 per cent to 60 per cent higher than the permissible limit. On comparing the temperature rises of these sockets with new standard sockets, it was found that their temperature rise was also 50 to 120 per cent higher.

The above evidence was considered sufficient to conclude that the sockets and/or the crimps were defective, and to undertake a major program of replacement of over 200 such sockets identified by measuring the barrel diameters. Priority was given to those sockets, which seemed to be running hotter than the rest. Temperature crayons for 80 °C were used for this purpose. Some of the cables had to be replaced.

32.5 Consequential Damage After Terminal Failure

CASE STUDY

In another similar case, there were several electrical fires involving cable sockets connected to smoothing reactors. While there was similar, extensive damage at and around a terminal, there were a few cases in which the damage was restricted to the terminal only. It was considered possible that all the fires started in the terminal. Further investigation revealed, a design defect in the terminal confirming the hypothesis that all the fires started in the terminal. This is discussed in greater detail in Chapter 39 on Terminals. This case is mentioned here to emphasise that some socket failures could be due to terminal failures.

32.6 Causes of Overheating of and Working out of Cables from Sockets

There are several causes of failures of crimped sockets. Some of these taken from many case histories, may be summarised as follows:

(a) Barrel thickness inadequate (Case Study 32.4).
(b) Defect in terminal board (Case Study 32.5):
   Case studies for the following items have not been described in detail here, but figures are given in two cases.

(c) Cable worked out of socket due to inadequate penetration of the cable into socket. See figure 32.3.

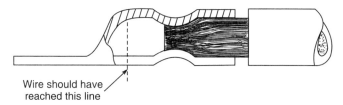

Figure 32.3   Inadequate penetration of the cable into the socket

(d) Cable worked out of socket due to incorrect location of crimp. See figure 32.4.

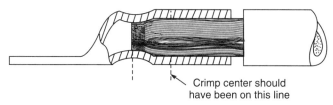

Figure 32.4   Incorrect location of crimp on the socket

(e) Cable worked out of socket due to incorrect selection of socket and/or crimping tool.
(f) Inadequate crimp due to defect in crimping tool, deficient crimping force or due to poor workmanship.

32.7   Causes of Fractures of Sockets and Wires

Fractures of sockets or wires and consequent electrical failures involving, arcing, melting or even fires, are experienced generally on machinery subject to severe vibration. Electric and diesel electric rolling locomotives and multiple unit stock, stone crushers, power hammers, machine tools with reciprocating parts, etc. are some of these machinery subject to vibration.

The first obvious method of preventing such failures is to eliminate the vibration. This could have many other benefits apart from preventing failures of electrical equipment and their connections. However, it is often not possible to eliminate the vibration altogether. In such cases, special precautions have to be taken to prevent electrical failures.

The most effective method is to eliminate the relative vibration or movement at the termination and to transfer the same to the stranded cable. Figure 32.5 (a and b) shows one common example of such an arrangement.

Vibration of the cable end and the crimped socket as a whole is not harmful. It is only when the cable flexes and vibrates relative to the socket that severe alternating stresses are produced in the socket and also in the wire at the point where it enters the socket. This could

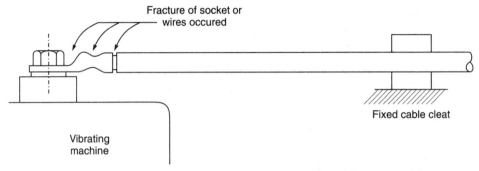

Figure 32.5 (a) Fractures of crimped connections due to vibration and flexing

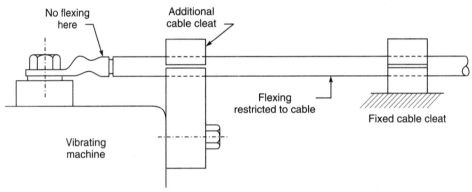

Figure 32.5 (b) Additional cleat fixed to vibrating machine prevents cable and socket failures

lead to fatigue fracture sooner or later depending upon the frequency of vibration and the magnitude of the stress in relation to the endurance limit of the material.

32.8 Failures due to Sharp Bends and Corners

CASE STUDY

In one case crimped socket fractures occurred within two years after commissioning. Examination of the sockets showed sharp bends on corners as shown in Figure 32.6. The stress concentration at

Figure 32.6 Sharp corner in crimp socket can be the starting point of fatigue fractures

such sharp bends raised the stress level and accelerated the fatigue fractures of the sockets. Sockets of the type shown in Figure 32.1, in which the sharp corners were eliminated, were used to replace these sockets.

32.9 Failures due to Defective Material

CASE STUDY

In another similar case it was discovered that the sockets had hairline cracks on the edges of the flattened portions forming the palms. This was due to the material of the tubes used for manufacture of the sockets not being electrolytic copper of the required ductility. The action taken was to replace all such sockets of defective material and to ensure greater care during inspection of new sockets.

32.10 Other Causes of Socket and Wire Fractures

Several other types of defects which have led to fracture of sockets and wires in service are enumerated below:

(a) Excessive crimp deformation due to the use of wrong crimping tool. See Figure 32.7.

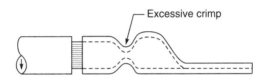

Figure 32.7 Excessive crimp due to use of wrong crimping tool

(b) Sharp edge on the inner bore of the socket cutting into strands. See Figure 32.8.

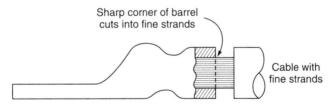

Figure 32.8 Fine strands cut by sharp corner of barrel

(c) Flexing in bare portion of the insulated cable between the socket and the insulation. See Figure 32.9.

Using heat shrunk plastic sleeves over the cable socket and the insulation as shown in Figure 32.10 prevented fractures of the wires.

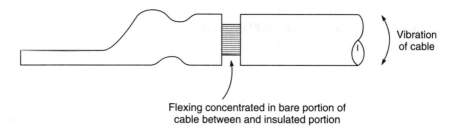

Flexing concentrated in bare portion of
cable between and insulated portion

Figure 32.9 Strand failures, due to fatigue, in bare portion of wire at entry to socket

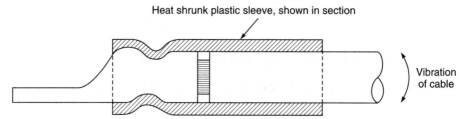

Heat shrunk plastic sleeve, shown in section

Figure 32.10 Heat shrunk plastic sleeve prevents strand failures due to vibration

32.11 Do's and Don'ts for Preventing Failures of Crimped Sockets

The following Do's and Don'ts have been reproduced from Reference (a) given at the end of this chapter.

- Ensure that the socket is made of annealed electrolytic copper/aluminium. Verify the electrical conductivity, the Brinell hardness and the bend test with reference to the relevant specification.
- Do not heat either the raw material or the socket in a reducing flame or reducing atmosphere, at any stage during manufacture, specially if the copper has some dissolved oxygen. Such heating causes hydrogen embrittlement.
- Ensure that the crimping dies do not have any sharp changes of section or sharp corners or edges.
- Ensure that the barrel of the socket has adequate thickness and that the dimensions of the socket and the design of the crimping tool are proved by endurance tests on crimped joints.
- Ensure that the crimp depth and the crimping force are always adequate, i.e. not less than that used during the proving tests by using proven tools and dies.
- Ensure that no change is effected in the design of the socket, the design of the crimping tool or the size of the wire after any particular combination is proved by pull out and endurance tests.

- Ensure that the crimping tools and the crimping dies are verified periodically to ensure that there are no defects due to wear and tear.
- The design of the terminal board should also be checked if the over-heating and burning seem to have originated in the terminal board.
- Ensure that the wire penetrates the barrel fully and that the crimp is located in the centre. If more than one crimp is provided on one socket, refer to the relevant drawing for guidance regarding their spacing and location.
- Ensure that the sockets are inspected carefully at the time of procurement. The dimensions, the quality of the material and the trial crimps should be checked carefully.
- Ensure that no vibratory stresses are produced on the sockets, by clamping the cables in cleats in such a way that there is no relative movement between the cleat and the terminals where the sockets are bolted on.
- In case of small control cables less than 10 square mm, provide a heat shrinkable sleeve to grip the end of the socket and the wire insulation.
- Round off the inner corner of the barrel of the socket to prevent it from cutting into the strands.
- Check the crimp and the crimping tool to ascertain whether the strands are getting damaged during crimping.
- Ensure that the staff who inspect crimping sockets or use crimping tools are trained fully in handling them.

References

(a) Hattangadi, A.A., *Electrical Fires and Failures,* Chapters 5 to 9, Tata McGraw Hill, 1998.
(b) British Standard 6360.
(c) Indian Standard 8337.

Chapter 33

WIRE AND CABLE FAILURES

In this chapter we learn

- *Applications of bare copper wires and cables.*
- *Common causes of failures of bare copper wires.*
    - *(i) Overheating at joints.*
    - *(ii) Short circuits due to stray wires or foreign material.*
    - *(iii) Excessive current through the line.*
    - *(iv) Excessive tension or incorrect sagging during installation.*
    - *(v) Mechanical failure of joints or splices.*
- *Case studies to illustrate the above.*
- *Precautions and preventive measures to prevent failures of brazed and butt welded joints.*
- *Fatigue failures of insulated wires due to vibration.*
- *Aluminium wires.* ❑ ❑

33.1 Introduction

Wires and cables are used extensively to connect different electrical components and equipment. The word 'wire' is used when the cross-section is small and it is not insulated. The word 'cable' is used when cross-section is larger and the wire is insulated. This definition is not very rigid and sometimes these two words are used interchangeably. In this chapter, failures of bare or un-insulated wires and cables will be considered first. A few cases of strand fractures in insulated wires will also be dealt with later.

Failures of insulation on insulated wires and cables will be discussed in Chapter 34.

The main purpose of a wire or cable is to conduct electric current through it. It must always be insulated so that it does not make contact with any other conductor except where it is supposed to. Even bare or apparently un-insulated wires are in reality insulated by air and the supporting insulators fixed at intervals.

Failure of wire can be of two types. It can break or part thereby breaking or interrupting the flow of current; or it may touch or develop a short-circuit to another conductor thereby diverting the flow of current into unintended paths.

33.2 Fracture or Breakage of Wires

This type of failure is very rare in insulated cables or wires, which are usually supported in pipes, conduits, trays, or cable trenches. Fracture is a common failure mode in un-insulated overhead lines, where the bare wires are strung between poles and supported by insulators.

Some of the more common causes of fractures of overhead wires in order of probability of occurrence are:

(a) Overheating at joints or connections such as parallel clamps, tee joints, splices, etc.
(b) Short-circuits and arcing due to contact with stray wires or foreign material or due to failures of supporting insulators.
(c) Excessive current through the line.
(d) Excessive tension at low temperatures, or contact/low clearance at high temperature due to incorrect adjustment of sag at the time of installation.

33.3 Line Failures due to Parallel Clamp Defects

There have been many cases of parallel clamp failures causing line fractures. These are discussed in detail in Chapter 38 and briefly in this section.

In some cases there was visible damage due to arcing and melting with bead formation at the end, as shown in Figure 33.1, when the failure occurred while carrying current.

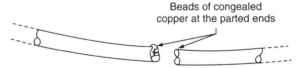

Beads of congealed
copper at the parted ends

Figure 33.1 Aspect of parted ends of line when fracture occurs while carrying electric current

In some rare cases there was local reduction in tensile strength due to overheating and the wire gave way soon after the protective systems had interrupted the current. In this case the reduction in section or necking of the wire strands was visible at the fracture as shown in Figure 33.2.

A connection to an overhead line either at the end of the line or at some intermediate point is often made with a parallel clamp as shown in Figure 33.3.

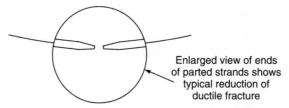

Figure 33.2 Aspect of parted conductor strands when there was no current through them at time of fracture

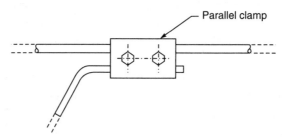

Figure 33.3 Defect in the design, manufacture or installation of parallel clamp can lead to line parting (See Chapter 38)

If the parallel clamp is not installed correctly, or if there is a design/manufacturing defect in it, it may get overheated and reach temperatures high enough to reduce the tensile strength of the copper wire to a level below the operating stress in it. In this case the line failure is a consequence of a defect in the connection and not in the wire itself.

If the parallel clamp is correctly designed and installed its operating temperature should at all times be slightly less, i.e. a few degrees below the temperature of the wire itself. However if there are any lacunae in the design or in the installation it is possible that the operating temperature of the parallel clamp is significantly higher than the temperature of the line itself. In such cases the failure may occur when there is a fault somewhere else, which causes a very high current to be drawn for a short time. In such cases it would not be correct to treat the line fracture at the parallel clamp as consequential damage of the distant fault. The parallel clamp would have failed sooner or later even under normal operating conditions.

33.4 Stray Wires Causing Line Fractures

Where railway traction overhead lines pass under bridges or 11 kV or 22 kV distribution lines pass close to buildings, there are possibilities of wires carried by birds leading to short-circuits. Even in the short time it takes for the protective gear to operate and disconnect the source of power there is often enough damage to the wire due to the arcing, to weaken the wires and cause them to snap under the normal operating tension.

The only effective remedy is to avoid running overhead lines close to buildings or structures. As far as railway traction lines are concerned their passage under bridges is unavoidable. As clearances under bridges are generally reduced for economic reasons, this problem is accentuated. The only effective remedy is to apply several layers of plastic tape to form an insulating sheath over the bare conductor in the vicinity of earthed structures. The sheath thickness should be adequate to withstand double the maximum operating voltage for 1 minute.

33.5 Failures due to Excessive Current

CASE STUDY

In a traction overhead line the current is shared by the catenary, the contact wire and the auxiliary catenary if there is one. There is a number of paralleling connectors between the lines at intervals along the line. The vulnerable point is usually the point connecting the line to the feeder line from the substation where the full current for the whole line passes through the connection.

In one case the feeder from the substation was connected to the lines as shown on Figure 33.4.

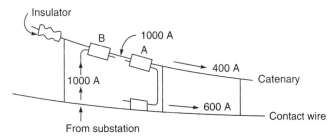

Figure 33.4   The small section AB of the catenary carries 1000 A

Due to the relative position of the feeder wire and the paralleling connector, the short length of the catenary between A and B carried a very high current, much higher in fact than the capacity of the single wire. This turned out to be the most vulnerable point and in the case in question the failure occurred there. The position was corrected by merely exchanging the positions of the feeder clamp and the paralleling clamp as shown in Figure 33.5.

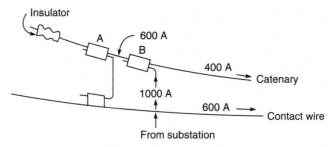

Figure 33.5 Interchange of positions of A and B reduces current in section AB to 600 A

It will be seen that the revised arrangement has reduced the current in the Section AB from 1000 amps to 600 amps and the heating effect in the ratio 1000^2 to 600^2 i.e. 2.77 to 1.

33.6 Failures of Joints in Solid Copper Conductors

CASE STUDY

Contact wire in railway traction overhead equipment is usually a solid conductor of section as shown in Figure 33.6. This wire is drawn from wire rods and the length of each piece may be about 500 metres. As the lengths of the railway track sections are much longer, the pieces have to be joined together either by brazing or by butt-welding. This is done in one of the stages before the final wire-drawing and the joints are not visible. There have been cases of contact wire parting at these joints. The cause is obviously a defective joint because a properly made joint is as strong as the wire itself.

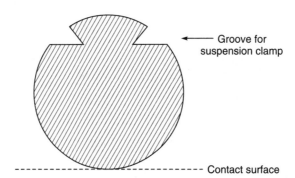

Figure 33.6 Cross-section of solid copper conductor

In one particular case the cause of failure was determined to be an unauthorised change in one of the details of the brazing process, viz. chemical composition of the brazing flux. This led to poor quality of the brazed joint. The parted surfaces showed large areas of copper un-wetted by brazing alloy.

There are several brazing process variables such as composition of the brazing alloy, quality of the brazing flux, the surface preparation, and the temperature to which the copper is heated. All these must be proven together by carrying out tensile, fatigue and bend tests on brazed samples. And once these are determined, they must not be changed without carrying out the same tests. The important variables to be so determined and then maintained without change are as follows:

(a) The chemical composition and the thickness of the brazing foil or sheet;
(b) The chemical composition and the quantity of the brazing flux;
(c) The process for heating the joint;
(d) The heating time and the final temperature of the copper during brazing;
(e) The splice angle for the joint and the surface finish of the faces to be joined. The horizontality of the surface to be brazed;
(f) The method of mechanical cleaning of the faces to be joined before the application of the flux;
(g) The force with which the heated ends of the copper wires are held together;
(h) The cooling time allowed and the final temperature to be reached without any movement or vibration after brazing.

If the joints are made by butt welding, there are different but equally important process details, which must be proved by tensile, fatigue and bend tests on samples and then maintained without change. These are:

(a) Shape and alignment of the machined ends of the wires to be brought together.
(b) The currents and times for which they are allowed to flow during the welding process.
(c) The forces applied between the conductors to be joined at different stages of the heating and welding cycles.
(d) The cooling time allowed and the final temperature to be reached without any movement or vibration.

In the type tests on ten samples of each to be carried out for proving the brazing or the welding process, the average tensile strength of the finished and jointed conductor shall not be less than 95 per cent of the average strength of the conductor itself. The standard deviation for the jointed conductor shall not be more than 105 per cent of the standard deviation for the conductor.

It must be emphasised that there is little that can be done to prevent failures after the installation of the wire at site. Routine ultrasonic tests on finished joints can help to screen out defective joints but these are not foolproof. It is necessary to take the steps to control the manufacturing process as already mentioned.

33.7 Overhead Line Failures due to Excessive Tension

CASE STUDY

In un-regulated track over head equipment on electrified railways, it is very important to adjust the tension in the wires with reference to the wire temperature at the time of installation. As it is not easy to determine the average wire temperature when the sun is shining on the wire it is desirable to do this adjustment a few hours after sunset. It is then possible to take the wire temperature to be the same as the ambient air temperature.

There have been some cases of contact wire parting at low ambient temperatures due to failure to adjust the tension correctly at the time of installation. More often than not, the wires would be carrying current at the time of parting. The resultant arcing causes melting and bead formation at the parted ends. Occasionally, there is no current and the typical reduction in section and necking is observed at the point of parting.

Even in the case of regulated track over head equipment, failures of this type can and do occur if the regulating equipment gets jammed when the ambient temperature is high. In fact, the rise in tension of regulated equipment under these conditions is higher than that in un-regulated equipment. Failures of regulating equipment can lead to other problems too and there are many possible reasons for regulating equipment failures. These are discussed separately in Chapter 60.

Parting of contact wire may remain undetected until the pantograph of a train runs into it and causes a major panto/OHE entanglement. The parted contact wire remains supported by the droppers from the catenary and does not touch the ground. There are many other ways in which these accidents are caused. They are discussed in detail in Chapter 65.

33.8 Other Causes of Contact Wire or Catenary Parting

There are several other ways in which contact wires or catenary wires may part. These are enumerated below:

(a) Failure of splice fitting or end fitting due to slipping of wire or fracture of fitting.
(b) Stationary pantograph being lowered while drawing heavy current.
(c) Overheating of contact wire at stationary pantograph due to a defect in the pantograph.

33.9 Failure of Contact Wire due to Bad Contact at the Pantograph

CASE STUDY

In a very unusual and freak incident, the contact wire parted due to overheating at the contact between the pantograph pan and the contact wire above a stationary locomotive. The current drawn by the locomotive for the magnetisation of the transformer and the small load of auxiliary machines was estimated at less than 40 amps and ordinarily there should have been no overheating at this small current.

Investigation showed that a 6.5-kg weight, which had been suspended from the pan for the purpose of measuring the pantograph force after overhaul, had been forgotten by the maintenance staff. As a result, the actual contact force between the pan and the contact wire was probably a few hundred grams instead of the normal 7 kg +/–0.4 kg. Due to this low contact force and the normal presence of lubricant/dust etc., the contact resistance was so high that it caused overheating of the contact wire, loss of tensile strength and finally parting under the normal tension in the wire.

Many failures are often caused by maintenance tools and spare materials, which are left behind inside or on machines, which have been subjected to inspection, maintenance or repair.

33.10 Traction Motor Lead Failures

CASE STUDY

Mechanical failure of insulated wires are rare because they are usually well-supported and are subject to little tension. However there are exceptions, as in this case study.

Nose suspended, axle mounted traction motors are subject to very heavy vibration. The vibration level on these motors is probably the highest among all the components of electric locomotives or coaches. The leads to these traction motors are usually flexible, stranded and insulated cables fitted with crimped sockets as shown in Figure 33.6.

There were some cases of fracture, arcing and melting in the region of the crimped sockets, leading to shorts, earth-faults and locomotive failures.

Investigation showed that these were new cables fitted only a few months earlier. The first reaction was to assume a defect in the quality of the cables which was cleared up on re-testing the cables. It was, noticed that the outer diameter of the cables was less than the earlier supplies due to some improvements in the materials used for the insulation and for the sheaths. The wooden cleats at the entry to the junction box had not been modified to suit the new reduced

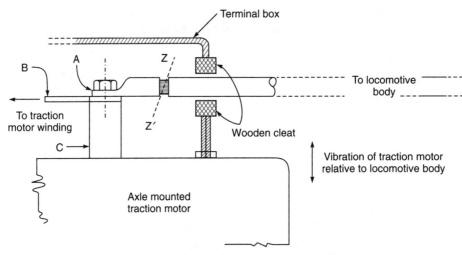

Figure 33.6 Gap in wooden cleat allowed cable to vibrate and flex inside terminal box and caused fractures at ZZ'

diameter of the cables. There was a small but significant clearance between the cables and the cleats. This allowed the vibration of the cables to produce stresses on the terminations (either the crimped sockets or the wires themselves) and then to cause fatigue fractures.

The failures of terminations on the traction motor leads were completely stopped as the wooden cleats were modified to fit the outer diameter of the cable.

33.11 Fracture of Leads to Electrical Apparatus

Cable leads to other electrical apparatus, which are subject to vibration, are prone to fracture if there is relative vibration between a cable support and the apparatus. The fracture may take place at the socket or in the cable itself. Some examples of such cases are shown in Figures 33.7 and 33.8.

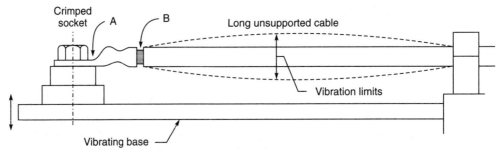

Figure 33.7 Shows how a long unsupported cable will vibrate when the base itself is vibrating. Fracture may take place either at A or B

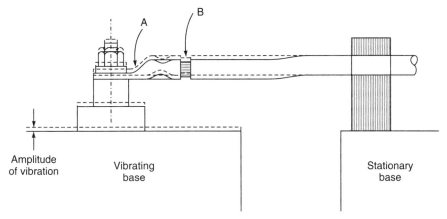

Figure 33.8 Shows conditions when one end of cable is vibrating but the other is stationary. Fracture may take place either at A or B

The solution is generally the provision of a cable clamp on the vibrating apparatus so that the flexing of the cable resulting from the vibration does not take place at the termination; instead, the flexing is transferred to a part of the flexible cable between two clamps. See Figure 33.9.

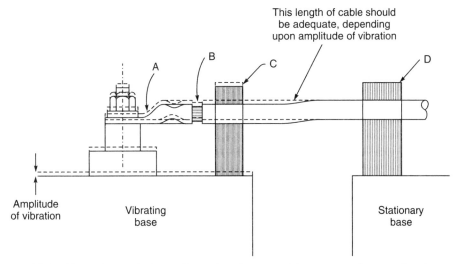

Figure 33.9 The additional clamp C, on the vibrating table prevents flexing of socket and cable at A or B. This prevents fracture. Flexure of stranded cable between C and D does not lead to fracture, if length CD is adequate

Another solution, in the case of control cables of small size is to provide heat-shrinkable sleeves between the sockets and the cables as shown in Figure 33.10.

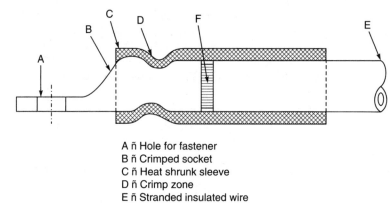

A ñ Hole for fastener
B ñ Crimped socket
C ñ Heat shrunk sleeve
D ñ Crimp zone
E ñ Stranded insulated wire

Figure 33.10 Heat shrinkable sleeve on thin stranded wire and socket provide mechanical stiffness to prevent flexing of strands at F

33.12 Failures of Copper Strands, Inside Insulation of Multicore Cables

CASE STUDY

Fractures of copper strands inside the insulation of the cores of multi-core cables between railway coaches are due to metal fatigue. The causes are often mystifying at first because there are no signs of external damage on the cables. The insulation is intact and only the strands have a clean break as shown in Figure 33.11.

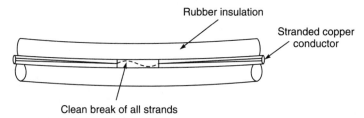

Rubber insulation

Stranded copper conductor

Clean break of all strands

Figure 33.11 Shows unusual type of fracture of all strands inside insulating sheath

When multi-core cables are used to provide control connections between coaches, these are subject to relative vertical and horizontal movements between their anchoring points on the coaches. This causes the outer cores in the main sheath to be subjected to alternating tension and compression especially when the cores are firmly held inside the sheaths. See Figure 33.12.

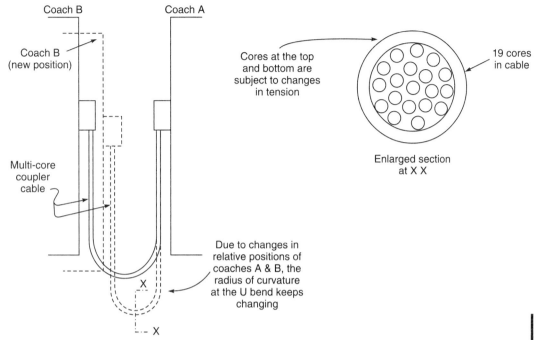

Figure 33.12 Multicore complers between railway coaches

These effects are magnified if the end clamps are tightened before the cable is bent in a U-turn.

33.13 Failures of Splices or End Fittings

All overhead lines or wires have necessarily to be fitted with splices and end fittings and they must have tensile strengths, which are significantly higher than the tensile strength of the wire itself. This margin is necessary because the variance of the strength of cast or forged fittings after being fitted manually is likely to be much higher than the variance of the strength of drawn wires. In general, failures of splices and end fittings are rare but occasionally failures may be experienced. These would be generally due to workmanship errors in fitting or to manufacturing defects in the fittings.

In case of failures of fittings the points to be checked are:

(a) Whether clear, illustrated instructions are provided for each type of fitting, indicating the order of assembly, the tightening torques and clearances to be checked after tightening.

(b) Whether the dimensions of the fittings are within the limits of tolerances provided in the working drawings.

(c) In case of fractures whether there are any flaws at the fractures and whether the material complies with the specification.

(d) Whether appropriate tools have been provided to the staff for the fitting and checking of the fittings.

CASE STUDY

In one case, the contact wire had slipped out of the groove on the splice. On checking the depth of the indentation it was felt that the screws had not been tightened fully. On checking some of the screws already tightened on other splices in the vicinity, it was found that some of them could be tightened further by 20 to 30 degrees of angle at the specified torques. This indicated the need for further training of the concerned staff.

Since splices and end fittings are often fitted on the line at heights and on unstable platforms, special care has to be taken to use double spanners in opposition to ensure proper tightening.

33.14 Aluminium Wires

All the failure modes and mechanisms as also the suggestions for improving reliability, mentioned above with special reference to copper wires are equally applicable to aluminium wires also. A few additional comments relevant to aluminium wires only are discussed here.

Physical properties such as conductivity, tensile strength and hardness of aluminium are different from those of copper and naturally, the dimensions of aluminium wires and fittings are different from those of copper wires and fittings of the same current rating. The relevant data can be seen from wire data books. Failure modes and failure mechanisms are very similar, but there are a few physical properties of aluminium, which are so different from those of copper that a few additional failure mechanisms have to be considered. These distinctive properties are:

(a) The melting point of aluminium is much lower than that of copper and as a result it is subject to metal creep at room temperature whereas copper creeps only at temperatures above 140 °C.

(b) Aluminium oxidises more rapidly and the oxide film is harder when compared to copper.

(c) Aluminium and copper, when joined together, are subject to bi-metal emfs at the junction.

Special precautions have to be taken to prevent failures due to overheating at joints when aluminium wires and fittings are used. These are as follows:

- Since pressure joints lose contact force due to metal creep, spring loaded threaded fasteners must be used. Alternatively, such contacts must be re-tightened periodically—almost every few months.
- Contact faces should be coated with Vaseline and this coating should be scraped in with wire brushes.
- If the design calls for the use of bi-metal washers, they must be used as stipulated in the drawings.
- Pressure contacts should be designed to operate with a temperature rise not exceeding 30 °C.

Where high reliability is desired it is best to avoid the use of aluminium wires. Copper wires should be used.

33.15 Do's and Don'ts for Preventing Wire Failures

The following list of Do's and Don'ts for preventing wire failures is reproduced from Chapters 9 and 12 of the book referred to at (a) at the end of this chapter.

- Fix the wire or cable directly on the vibrating machine by means of a suitable cleat.
- In case of small control cables, provide a heat-shrinkable sleeve to grip the end of the socket and the wire insulation.
- Ensure that multicore cables are clamped in their coupler heads after they are bent to their final shape.
- Ensure that the cores in the multicore cables are free to slide in their sheaths.
- See Chapter 38 regarding wire parting due to parallel clamp failures.
- Ensure adequate clearances on overhead lines from earthed structures under highest wire temperature conditions. Adjust sag correctly with reference to wire temperature.
- Ensure safe tension in overhead lines at lowest temperatures.
- Provide stay insulators where overhead wires can swing and touch earthed structures.
- Where adequate clearances cannot be provided apply sufficient insulation on wire to prevent stray wires from causing shorts.
- Ensure that the current in the overhead lines is always and everywhere within its capacity.
- Ensure that the dimensions and tensile strengths of splices are proven by tests.

- Ensure that the splices are fitted correctly. Provide special training to staff for this purpose.

For Butt Welding

- Ensure that the following parameters are adjusted correctly.

    (a) Duration and magnitude of welding current
    (b) Duration and magnitude of butting force
    (c) Sequence of (a) and (b) above

- Verify the manufacturer's recommendations for the process parameters by carrying out prototype tests on samples.
- Make a daily butt welding test on a test piece and subject it to the bending test.

For Brazing

- Ensure that brazing alloy and flux are obtained from reputed. Manufacturers or tested as per the relevant IS.
- Ensure that the brazing current and time are adjusted to values, which have been proven by tests.
- Check the brazed joints visually to verify proper penetration of the brazing metal and proper wetting of the surfaces.
- Avoid gas flame brazing by using electric brazing tongs.
- If gas flame brazing cannot be avoided, use neutral flame and apply only tip of the flame to the work. Test for hydrogen embrittlement of sample test pieces.
- If aluminium wires and fittings are used take the precautions mentioned in Section 33.14. If possible, avoid the use of aluminium wires by using copper instead.

References

(a) Hattangadi, A.A., *Electrical Fires and Failures,* Chapters 9 and 12, Tata McGraw-Hill, 1998.
(b) Chapter 35 ibid Auxiliary Motor Cable Failures.

Chapter 34

INSULATED WIRE FAILURES

In this chapter we learn

- *Causes of cable insulation failures.*
- *Degradation of cable insulation due to:*
  - *Abrasion.*
  - *Sunlight.*
  - *Water, oil, chemicals.*
  - *Electrical stress.*
  - *Heat.*
  - *Ageing.*
- *Case studies involving some of the above causes.*

34.1 Introduction

A very large number of breakdowns in electrical machines and installations are due to failures of insulation on insulated wires. There are many possible reasons and it is necessary to determine the root cause correctly whenever insulation failures occur. If it is not eliminated effectively, failures will continue to take place.

Insulation failures have the potential of causing fires if the protection system is not effective. Most insulating materials are usually the only combustible items in any electrical equipment. In any case, even if a fire does not ensue, electrical insulation failures will almost always lead to interruptions in service or production. It is, very important to ensure every possible precaution to prevent insulation failures.

34.2 Causes of Insulation Failures

Ageing may be considered first only because it is the rarest. In principle, all insulating materials, except a few like ceramic or pure mica, have a limited life and lose their insulating properties with the passage of time. But this degradation is very slow and most electrical machines, if designed correctly and operated within their specified temperature limits, work for forty years or even more. However, electrical machines do fail due to insulation failures but in almost every case it is due to ageing accelerated by one of the factors mentioned below.

Insulating materials are susceptible to accelerated degradation by abrasion, sunlight, water, oil, chemicals, electrical stress and heat. Naturally, these are the factors responsible for most insulation failures.

It is not necessary for degradation to take place in the entire cable. More often than not, it is localised and the degraded portion is destroyed by arcing and burning which follows a short-circuit. Only in the case of normal ageing or general overheating, the degradation would be over larger areas, being more severe in zones that are generally at higher temperatures.

The last three factors viz. electrical stress, heat and ageing are unavoidable and inter-related. Heat is unavoidable because the passage of current always generates heat in electrical conductors. The degradation by ageing is accelerated at higher operating temperatures. For every 8 °C rise in temperature, the life of the insulation is reduced by half.

If the electrical stress is above a certain limit, the insulation gets weakened continuously as a result of partial discharges that take place in voids inside the insulation. Hence, if the manufacturing process leaves a number of voids in the insulation, premature failure of insulation due to accelerated ageing is to be expected.

The maximum electrical stress and operating temperature have to be controlled carefully within appropriate limits by the designer and manufacturer of the cables while declaring the current and voltage ratings. Similarly, these limits must be kept in mind by the user while selecting cables of appropriate ratings and while investigating failures of insulation.

34.3 Failures due to Overheating by Eddy Currents in Steel Plates

CASE STUDY

There were several cases of electrical fires in electric locomotives of a certain class. In every case, the fire had originated in a cable trough. There were no terminations of cables in the vicinity and there were short-circuits between adjacent cables.

On investigation, it was seen that the ratings of these cables were adequate in comparison with the maximum currents to be carried by them. Examination of the cables in similar locations on other locomotives of the same class revealed softening of the insulation. On a locomotive, which had just returned from service, it was noted that not only the cables but also the steel bottom plate below the cables were abnormally hot. A few tests were made and it was noticed that the steel plates were getting heated while the cables were at normal temperatures in locations where they were not close to the steel plates. It was concluded that the whole problem was due to the heating of the steel plates by eddy currents produced in the vicinity of the cables.

Comparison with other locomotives of different classes with similar arrangements of cables running in steel troughs showed certain differences. In the class in question the motor voltage was lower and the currents were higher. Secondly, the distance between the cables of opposite polarities was higher as shown in Figure 34.1.

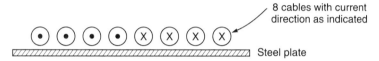

Figure 34.1   Overheating of cables carrying heavy alternating currents due to proximity of steel

The action taken was to rearrange the cables as shown in Figure 34.2, thereby reducing the distance between cables of opposite polarity.

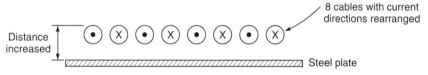

Figure 34.2   Action taken to prevent failures described in Figure 34.1

The distance between the cables and the steel plate was also increased to the extent possible within the available space. After making these modifications the temperature of the steel plate dropped significantly, to safe levels. There were no further cable failures and their temperatures in the cable troughs were also well within the specified limits.

Similar effects have been observed in the case of three phase cables. If they pass through steel plates, there is much less heating than if three widely spaced single phase cables are used. In the former case, the magnetic fields cancel each other and eddy currents are reduced.

34.4   Failures of Insulation due to Abrasion

CASE STUDY

A cable developed a short-circuit with a structural member in its vicinity. Examination of similar locations in other equipment of the same type showed that the cables were actually in hard contact with the structural member in some machines. These machines were also subject to vibration. Closer examination showed that the cable insulation was visibly damaged by abrasion in some of the locations where there was contact between the cable and the structural member. In many locations there was no visible damage despite contact between the cable and the structural member as shown in Figure 34.3.

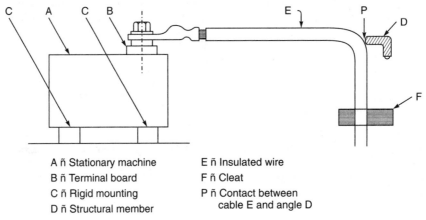

A ñ Stationary machine   E ñ Insulated wire
B ñ Terminal board   F ñ Cleat
C ñ Rigid mounting   P ñ Contact between
D ñ Structural member   cable E and angle D

Figure 34.3 Insulation failure due to rubbing on structural member

The action taken was to adjust the cable layout so that there was adequate clearance between the cable and the structural members.

This was an example of a major fire resulting from a trivial and avoidable defect in installation.

34.5 Insulation Failures due to Overheating

CASE STUDY

In a large thermal power station a number of cables were running in cable trenches. A fire started in one such trench due to cable insulation failure. Investigation showed that these cable trenches were getting filled with fly ash and coal dust. Over the years, the cables got completely covered with this mixture, which prevented the dissipation of heat from the cables, rise in temperature and ultimate failure.

The remedial action was: immediate cleaning of all cable trenches with vacuum cleaners; introduction of periodical trench cleaning practices; review and adjustment of protection relays; inspection of condition of insulation on all cables in similar locations; and replacement of a few cables which were suspect due to overheating.

34.6 Other Failure Mechanisms

As stated earlier there are several degradation processes, which can affect or afflict cables. And there are even more ways in which failure mechanisms get initiated. Here is a selection of such failure mechanisms, all taken from actual case histories:

CASE STUDY

■ Lubricating oil leaking from the crankcase damaged some cables passing under a compressor. There was a short circuit between cables with softened insulation. Action taken was as follows:

• Leakage was stopped;
• Tray with drainpipe fitted under the compressor;
• Cables with oil resisting polychloroprene sheath over the insulation were used under the compressor tray.

■ A cable was passing through a hole in a steel plate. During the initial installation, adequate clearance between the steel plate and the cable was maintained. However, after a few years of service, while carrying out repair works, the cables were dismantled and re-laid. This time there was direct contact between the steel plate and the cable. Several months later, there was a short-circuit between the cable and the steel plate. It was presumed that the cable sheath and insulation got progressively damaged due to vibration and the forces produced by thermal expansion and contraction. The action taken was:

• To provide wooden cleats as shown in Figure 34.4 to prevent all possibility of damage to the sheath and insulation of the cable.

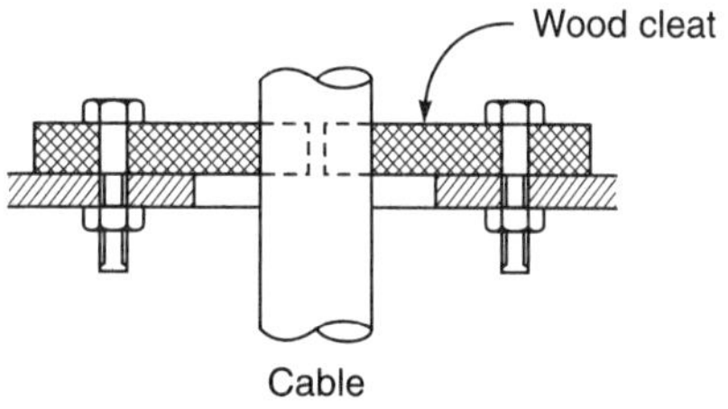

■ Underground cables from a 1500 volt DC substation to the over-head lines on the track developed earth faults. These cables were over 19 years old. On cutting off and testing of an undamaged portion of the

Figure 34.4 Provision of wooden cleat to prevent contact between steel plate and cable

same cable it was found that the breakdown voltage for the insulation was more than 85 per cent of the specified B.D.V. and more than 250 per cent of the operating voltage. Ageing of the insulation could not be considered as the cause of failure. On checking the cable layout it was observed that the failure had occurred at a bend in the cable and, the radius of the bend was about 50 per cent of the minimum bend radius recommended by the manufacturer of the cable. This was the most probable cause of the premature failure of the cable. The insulation and the sheath would have developed cracks, which would then have degraded further in service over the years. Action taken was as follows:

- The cable trench was modified while replacing the cable, to permit a proper bend radius for the cable.

34.7 Do's and Don'ts for Preventing Insulation Failures

The following list is reproduced from the book referred to at the end of this chapter.

- Look into the history or records of failures in the past to see whether there are any common factors or locations.
- Look for the presence of factors that accelerate degradation of insulation in the vicinity.
- Ensure that there is no direct contact between the insulation of any cable and any metallic structural member. Where necessary provide cable cleats.
- Ensure that the cable cleat grooves are dimensioned in such a way as to grip the cable firmly, but not too tightly. Confirm at the same time that there are no spaces left to permit cable vibration within the cleat.
- Ensure that the edges of the cleats are rounded so as to prevent cutting into the cable insulation. Provide elastomeric liners if necessary.
- Ensure that there is no localised overheating of the cables due to external heat sources, such as steel plate heated by eddy currents, resistors, steam pipes etc.
- Ensure that dissipation of heat from the conductors is not impeded by the accumulation of dust, fly-ash etc.
- Ensure that lubricating oil or other chemicals are not allowed to come in contact with cable insulation.
- Select cable ratings after taking into account local ambient conditions and proximity of other heat generating devices.
- Consider the possibility of overheating of terminations when investigating cable insulation failures.

References

(a) A.A., Hattangadi, *Electrical Failures and Fires,* Chapters 10 and 17, Tata McGraw-Hill.

AUXILIARY MOTOR
CABLE FAILURES

In this chapter we learn

- *The supply system for 400 volt auxiliary motors used for driving cooling equipment and air compressors on electric locomotives.*
- *Causes and effects of shorts between phases on the cables to these motors, some of which lead to fires.*
- *Possible methods for preventing such failures and for preventing fires.* □ □

35.1 Introduction

*I*n electric locomotives, there are a number of 400-volt three-phase auxiliary motors that drive machines like air blowers, compressors and oil pumps. In several classes of 25000-volt AC locomotives, the three-phase supply is obtained from ARNO converters, which are actually single-phase motors cum three phase generators. These auxiliary motors are distributed all over the locomotive and a number of three-core cables or single core cables in groups of three run from the control panel to these auxiliary motors.

35.2 Failure Mechanism

The following case illustrates cable failure due to degradation caused by excessive electrical stress.

CASE STUDY

In some classes of locomotives the system design provides protection against earth faults but not against shorts between phases. There were many cases of fires, which originated from the 3-core cables between the control panel and the auxiliary machines. Any short between these cores led to arcing and melting of conductors and the starting of fires if there was any delay in a consequential ground fault.

A review of the insulation on these cables showed that it provided adequate margin of safety and there was hardly any degradation of cables. Breakdown voltage tests were carried out on some of the cables of the same make and age after removing them from the locomotives. These tests showed no degradation of the dielectric in the cables.

In Section 64.3, Chapter 64, the generation of high voltage standing waves in the stator windings of asynchronous motors has been discussed. Just as these waves cause partial discharges and permanent damage to the motor insulation, the same standing waves can and sometimes do damage the insulation on the cables between the contactors and the asynchronous machines. These damages occur every time, when the machine is switched off by the opening of the contactor. See Figure 35.1.

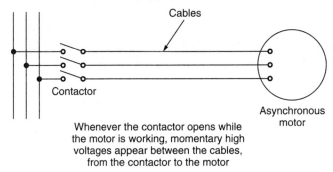

Figure 35.1 Auxiliary motor cable failures

When the cumulative damage is sufficient to cause failure at normal voltage, the short-circuit occurs and there is a failure and probably even a fire.

Evidence of such high voltages is seen also on the bus-bars and contactors in the form of globules of molten metal created by discharges through air.

35.3 Remedial Measures

An effective remedy for three-phase motors of rating up to about 400 volt, 200 HP, is to connect 4 microfarad, 750-volt capacitors between the phases at the motor terminals of those machines, which are started and stopped frequently, e.g. compressor motors. This reduces the peak voltage of the standing waves to less than 500 volts.

Tests can be made with storage oscilloscopes to determine the peak voltages with different capacitors to ascertain the optimum capacitance. In order to avoid new problems due to capacitor failures the voltage rating of the capacitors should be double the nominal voltage. The capacitors may also be provided with built in fuses.

The insulation level of cables leading from the control panel to the asynchronous motors should be increased by 100 per cent for motors, which do not start and stop frequently. It is not necessary to provide capacitors for such motors. It is desirable to run such cables in steel conduits.

35.4 Protection Systems

Prevention of fires starting from short-circuits anywhere in the system calls for provision of adequate protection systems. When there is a short-circuit either in the motor or in the cable between the contactor and the motor, the main circuit breaker must trip. As all these motors are direct-on-line-starting type and there are ten or more such motors of different sizes, many of which start at the same time, it is not easy to devise a common over-current relay which will be immune to starting currents but sensitive to short-circuit currents. On the other hand, provision of individual over-current relays for each motor will increase the number of relays in each locomotive with consequent maintenance and reliability problems.

It is a more economical solution to put all such cables (between the ARNO converter and the contactor cubicle and between the contactors and the motors) in steel conduits. Any short-circuit between phases will get converted very soon into ground faults and the existing ground fault system will operate and trip the locomotive main circuit breaker, thereby preventing fires.

35.5 Do's and Don'ts for Preventing Auxiliary Motor Cable Failures

- If there is no protection against short-circuits between phase cables and only earth fault protection is available, group them in steel conduits separately for each motor.
- Provide 4 micro-farad, 750 volt capacitors between phases at the terminals of motors of ratings up to 400 volt three-phase, 200 HP, that start and stop frequently in order to minimise switching surges.
- Refer also to the Do's and Don'ts at the end of Chapter 34.
- Reference: See also Chapter 33, Wire and Cable Failures.

MULTICORE COUPLER FAILURES

> ### In this chapter we learn
>
> - *The applications of single and multi-core couplers.*
> - *Three common failure modes of multi-core couplers.*
> - *Failures that start with overheating and end in melting of copper conductors and short-circuits.*
> - *Failures due to shorts between cores.*
> - *Failures due to open circuits.*
> - *Mechanisms, causes and prevention of the above types of failures.* ❑ ❑

36.1 Introduction

$\mathcal{A}$ multi-core coupler is essentially an assembly of plug and socket joints. Single plugs and sockets are commonly used for the connections to measuring instruments like multi-meters, electronic devices like oscilloscopes and radios. Two pin couplers are used for small electrical appliances and three pin couplers are used for larger electrical appliances. Multicore couplers with more than three connections—usually 19—in each coupler are used between railway coaches and for many other applications where connections have to be made and unmade quickly. In industrial electronic equipment, computers and many other devices that use microprocessors, low voltage, low current multiple pin couplers are used in large numbers. These are often provided with gold plated contacts and are flat or linear in design instead of round, but the failure modes and mechanisms are the same as those of the heavy duty, circular, multi-core couplers. There are just three failure modes in these couplers, as compiled from a large number of case studies:

(a) Overheating, charring, arcing, melting of contacts

(b) Open circuits in cores or contacts

(c) Short circuits between adjacent cores

Any two or even all three of these failure modes may be present at the same time but there is usually one, which is the original cause, and the others may be consequential. In order to prevent recurrence we have to determine the original causes or the failure modes, which occurred first.

36.2 Overheating, Charring, Arcing, Melting of Contacts

This type of failure usually starts with overheating at the contact between the plug and the socket in any one core. The overheating could be due to current in excess of the rating of the plug/socket or due to inadequacy of contact force as discussed in Section 36.3. The critical point is usually around 90 °C. Once this temperature is exceeded, the degradation starts and continues at an increasing rate.

Such a defect remains undetected. Even when the voltage drop in the fitting, which normally would be only a few millivolts, increases ten-fold with a corresponding ten-fold increase in its temperature rise, there is no adverse or noticeable effect on operation. The increased voltage drop continues to be negligible as compared to the operating voltage. The defective fitting continues undetected in service with progressive deterioration in the vicious cycle shown in Figure 36.1.

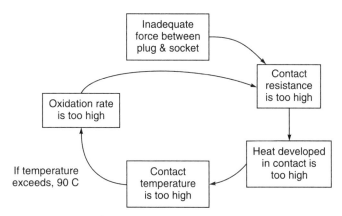

Figure 36.1 Vicious cycle that starts with low contact tone and ends in overheating

The temperature keeps rising, slowly perhaps, but relentlessly. Eventually, there is charring of insulation around the plug/socket, arcing between the plug and the socket, and finally melting of the contacts.

At some stage, a short-circuit may develop to an adjacent core or the core may develop an open circuit or very high resistance and this may cause some malfunction of the main equipment. On the other hand it is possible that a fire may start and spread until it gets detected and the power is switched off either automatically or manually by the operator. The fire may however continue until it is extinguished. The outcome of all these possibilities, cannot be predicted. It is a matter of chance depending upon the operating conditions, the physical and electrical location of the fault, the flammability of the surrounding materials, and so on.

36.3 Mechanism of Failures Starting with Low Contact Force

The starting point of this type of failure is inadequate force between the plug and the socket. The contact resistance is an inverse function of the contact force and the heat generation in the contact is directly proportional to the contact resistance. The temperature rise is, again, directly proportional to the heat dissipation. The rate of oxidation is an exponential function of the temperature. For every eight degrees rise in temperature, the rate of oxidation gets doubled. As the thickness of the oxide film increases, the contact resistance rises. We are back to the starting point of this cycle, which keeps repeating, in a rising spiral as shown in Figure 36.1 above.

36.4 Prevention of Failures due to Low Contact Resistance and Overheating

Since failures are activated by low contact force, the remedy is to ensure that it is adequate.

This force is developed by the deflection of the spring element in the socket. There are various types of springs. Two common arrangements are illustrated in Figure 36.2.

In system (a), the socket itself is made of springy metal and it is split into four quarters. The plug diameter is a little larger than the bore of the socket. The socket quarters have to deflect outward to accommodate the plug and this provides the contact force. In system (b), the contact forces are provided by steel springs acting on contact leaves. The latter system is used for plug/socket connections for high currents where very low contact resistance is required and consequently very high contact forces are needed.

The way to ensure adequate contact force is to control the dimensions of the plugs and the sockets such that the deflection of the spring provides the required contact force. This is a design function. Compliance with the fits and limits specified in the design is the manufacturing function. As far as the purchaser or the user is concerned, he can satisfy himself about the design and the manufacture by measuring the contact forces and the temperature rise at the rated current.

It is difficult to measure directly the contact forces in small sockets and plugs. An easy way out is to measure it indirectly by measuring the pull out forces since the latter are roughly 25 per cent of the former, the ratio being the coefficient of friction.

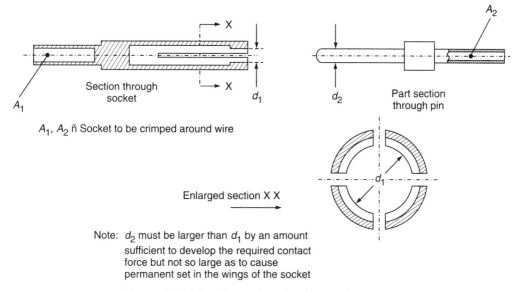

A_1, A_2 ñ Socket to be crimped around wire

Enlarged section X X

Note: d_2 must be larger than d_1 by an amount
sufficient to develop the required contact
force but not so large as to cause
permanent set in the wings of the socket

Figure 36.2(a)   Plug and socket for small currents

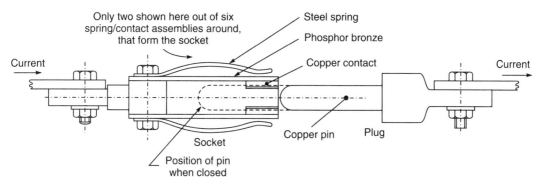

Figure 36.2(b)   Plug and socket for large currents

It may be noted that in multi-core couplers the force should be measured on individual plug and socket pairs and not on the whole coupler.

Tin plating and silver plating of the contact surfaces in the plugs and the sockets can increase the rating of the contacts as they can operate at higher temperatures to start with.

If the stress in the spring element is excessive, i.e. greater than the yield point of the material, the spring will take a permanent set and the contact force will get reduced and may activate the process mentioned earlier. It is necessary to check the contact force after the plugs and sockets have been assembled and pulled apart a number of times.

36.5 Failures of Multi-core Couplers due to Shorts Between Cores

In order to reduce the size and the cost of couplers, the insulating material provided between the cores is made as thin as possible on the assumption of the best manufacturing quality. If there is any error in the manufacturing quality in regard to the electrical quality of the material or in the dimensions of the coupler holders, failures are bound to occur sooner or later.

To verify that the overall design and manufacture is of the required standard certain electrical tests are specified. These are the measurement of insulation resistance and the breakdown voltage between different cores. The breakdown voltage test is a type test to be carried out on samples. In addition a proof test is carried out on all the couplers. The proof test voltage is typically twice the rated voltage plus 2000 Volts.

While the intrinsic dielectric strength of the coupler as designed and manufactured may be adequate, it is possible that this may deteriorate in service due to entry of water. It is necessary to ensure that the couplers are waterproof especially in the case of those mounted outside, e.g. inter-coach couplers. The electrical tests viz. insulation resistance test, breakdown voltage test and proof voltage test should be done after a rain test, in which water is sprayed on to the couplers as specified, to simulate heavy rain.

36.6 Open Circuits

Open circuits in couplers are relatively rare. Defects of this type occur where the manufacturing quality of the plugs, sockets and the crimped joints to the plugs and the sockets is very poor. They can be prevented from occurring in service by careful inspection of the design, manufacture and installation of the individual core couplers and carrying out pull out tests separately on each individual sockets and plugs.

As regards the crimped connections between the sockets or the plugs and the connecting wires, a reference may be made to Chapter 32 for a discussion on the defects and deficiencies in crimped joints. The causes of open circuits due to working out of the wires from the sockets are mainly due to dimensional errors in the sockets and/or in the crimping tools.

36.7 Plug Pull Out Test

This is a very important test on all plug/socket couplers. It is very easy to carry out and should be done on each individual socket or on random sample sockets from lots offered for inspection. This test will eliminate all cases of overheating and of open circuits. All it needs is a test-plug which must have the specified weight and the specified diameter and length of the plug. The specification for the test plug depends on the size and rating of the socket. Figure 36.3 shows a typical test-plug for the current carrying sockets of a 3-pin, 16-Ampere domestic plug and socket to IS 1293.

The test is carried out by holding the socket up-side down and inserting the test plug vertically into the socket and then releasing the test plug carefully. The requirement is that *the test plug should not slide out under its own weight*. The successful completion of this simple test ensures that the contact force exerted by the socket on the plug is adequate to prevent over-heating and open circuits. To ensure that the springs are not taking a permanent set, the test may be made after repeated insertion and with-drawal of the test plug, the specified number of times.

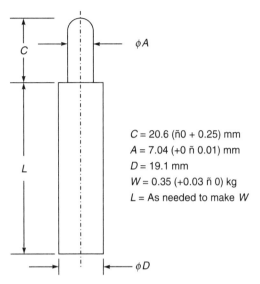

$C = 20.6\ (\tilde{n}0 + 0.25)$ mm
$A = 7.04\ (+0\ \tilde{n}\ 0.01)$ mm
$D = 19.1$ mm
$W = 0.35\ (+0.03\ \tilde{n}\ 0)$ kg
$L =$ As needed to make W

This test pin shall not slide out of the socket under its own weight

Figure 36.3   Test pin for gravity pull out test on current carrying sockets of domestic 16 A plug/socket unit

36.8   Conclusion

Almost all the failures that occur on multi-core couplers are due either to design/manufacturing defects or to installation defects. To prevent these from occurring the following acceptance tests must be carried out on all new supplies in accordance with the relevant specifications:

   (a) Rain or spray test followed by resistance and breakdown voltage test between cores on samples and proof voltage tests on bulk supplies.

   (b) Plug pull-out test on each socket or random sample sockets depending on the numbers to be checked.

   (c) Proving the crimping tools and plug/socket designs by carrying out wire pull-out tests on samples of wires crimped into the plugs and into the sockets.

If failures are experienced on existing installations the first step for investigation is to carry out the above tests on the plugs and the sockets. Breakdown voltage tests and millivolt drop tests necessitate special test equipment but the plug pull out tests can be done easily if the test plug and a sensitive spring balance are available.

36.9 Do's and Don'ts for Preventing Coupler Failures

- Ensure that the contact force between the plugs and the sockets are specified.
- Verify the contact force by carrying out pull out tests and millivolt drop tests at normal current.
- Ensure that the maximum temperature anywhere in the plug socket assembly at normal current is well below 90 °C.
- Carry out break-down voltage tests between sockets and pins on sample couplers.
- Carry out proof voltage tests between sockets and between pins on all couplers.
- Check the diameters of plugs.
- Check the sockets by gravity pull out tests using specified test pins.

Chapter 37

Knife Switch Failures

In this chapter we learn
■ *Applications where knife switches are in use at present.*
■ *Failure modes, failure mechanisms.*
■ *Causes of failures and preventive measures.*
■ *Measurement of contact force.*
■ *Special forms of knife switches.*

37.1 Introduction

The hand operated knife switch was probably the first electrical device used for controlling the flow of electric current. It was a common piece of equipment on electrical control panels of power stations and sub-stations seventy five years ago. Knife switches have been replaced by remote controlled contactors and circuit breakers. However, they are still seen from time to time in equipment for the purpose of off-load isolation of electrical circuits. See Figure 37.1 for examples of a few of these devices. In some they are integrated with other devices.

37.2 Failure Modes

The most common failure mode is overheating, arcing and possibly causing a fire if the knife switch is mounted on a combustible insulating board.

In fuse carriers with built in knife switches as shown in Figure 37.1a, the failure takes the form of a fuse failure in which the fuse blows even when there is neither over-load nor short-circuit. This type of failure is discussed also in Chapter 45, Section 45.8. Due to overheating in the knife contact, the conducted heat to the fuse wire causes the fuse characteristic to get modified and to blow at normal currents. If the fuse is merely replaced and if the knife contact problem is not resolved, the fuse will blow again.

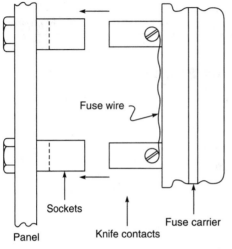

Notes

(i) The sockets are either made of springy material or provided with spring inserts, to provide the required contact force.

(ii) The knife thickness and the socket gap are controlled to ensure required force.

(iii) Any dificiency in contact force will lead to over-heating and failure.

Figure 37.1(a)   Fuse carrier with knife contacts

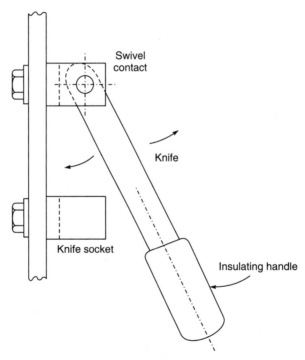

Figure 37.1(b)   Low voltage hand operated knife switch

Sometimes, knife switches are incorporated in sophisticated circuit breakers for the purpose of connection or disconnection when these circuit breakers are inserted into or withdrawn from their cubicles. If the types of defects discussed later in this chapter are present, failures will occur in these equipment exactly as in simple knife switches.

These failures are not detected by electrical protective equipment, (except when the knife switch is integrated with the fuse), because there is neither any increase in current nor any leakage of current to earth, in the initial stages. Short-circuits and over currents may arise in later stages of the failure when it may be too late for preventing the failure but in time to prevent a fire accident. The only way these defects can be detected is by looking at the devices while they are still in operation, through an infra-red camera or viewing device.

37.3 Failure Mechanisms and Causes of Failures

The contacts at either end of the knife switchblade are vulnerable. At one end it is a sliding contact and at the other end it is a swivelling contact, but basically they are identical in principle of operation. They both depend on the normal force between the knife blade and the wings of the contact. It is necessary to ensure that the contact resistance between each of the two fixed contacts and the knife blade is low and the heat developed by the passage of current is not excessive.

As explained in detail in Chapter 43 about contacts, the contact resistance is a function, for all practical purposes, of only one parameter and that is the contact force between the fixed contact and the mobile contact. If this force is within a certain range, the contact resistance and the temperature rise remain under control and there is no overheating. On the other hand, if the contact force falls below the critical level the contact is bound to overheat sooner or later and enter a vicious cycle of increasing temperature and increasing electrical resistance which culminates in overheating, arcing, melting, fire etc.

The contact force is provided by the elasticity of the leaves of the fixed contact. In very high current knife switches, helical springs may be provided as shown in Figure 37.2 to ensure the required contact force. To get the required force there are two pre-requisites:

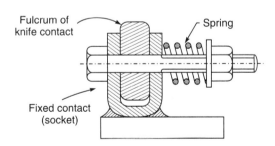

Figure 37.2 Spring loading to augment contact force

(a) The dimensions of the knife blade and the free gap between the leaves of the fixed contact should be such that inserting the blade in the fixed contact produces the deflection necessary for developing the required contact force.

(b) The stress in the elastic leaves of the fixed contact should be well below the elastic limit of the material so that the gap returns to the original free value as soon as the knife blade is removed from the fixed contact. This can be verified by checking that the pull out force does not get reduced after repeated operation.

If either of these two conditions is not fulfilled there is every possibility of the contact failing in service.

The overall dimensions of the knife blade of the fixed contacts should be such that the maximum operating temperature, under worst-case conditions is below 90 °C, (preferably below 75 °C). If the temperature goes beyond 90 °C, the contact surfaces begin to oxidise rapidly and the contact resistance begins to rise. This starts a vicious cycle, which leads to a failure of the switch, and is very similar to that shown in Figure 46.2 in connection with bus-bar failures due to inadequate tightening of the nuts and bolts at their joints.

The overall dimensions must be accurate. The fit between the knife blade and the slot of the fixed contact should be such that the deflection of the spring is adequate to provide the required contact force between the moving and the fixed contacts. And this should be ensured at both ends of the knife blade.

The causes of the above types of failures are mostly due to defective design or defective manufacture. The dimensions and the material of the components of the knife switch will determine the pull out forces and the temperature rise of the contacts. The properties to be checked are the resistivity of the material of components carrying current and the elasticity of the springy components.

37.4 Measurement of Contact Force

Since the contact force between the moving contact (knife blade) and the fixed contacts is so important for reliability, it is desirable to measure or at least estimate this force. As it is not possible to measure this force directly it can be estimated indirectly by measuring the pull out force for the knife blade. The frictional force between the knife blade and the fixed contacts when disconnecting the switch is a good measure of the contact force. It is about 25 per cent of the contact force. A spring balance should be used for acceptance tests. In the case of small switches, the feel of the force when opening or closing the switch is adequate for routine maintenance. The pull out force should be measured after the switch is opened and closed at least ten times.

The limits for the pull out force should be obtained from the manufacturers on the basis of type tests.

The pull out force should be measured on a few switches, selected at random according to an appropriate sampling plan after passing a current, which is at least 50 per cent higher than the rated current for one hour. This force shall not be less than 90 per cent of the pull out force limit specified for routine tests on cold switches. For checking the swivelling contact, the fulcrum pin or screw should be removed to measure the pull out force.

If helical springs are provided to augment the contact force in high-current applications, the deflection of the spring should be determined by measuring the free length and the assembled length of the spring.

For the temperature rise tests on knife switches, they should be mounted on panels and external connections should be made in the normal manner.

37.5 | Special Forms of Knife Switches

Sometimes knife switches are in forms, which are somewhat different from the usual form shown in Figure 37.1. One example is the roof bar on electric locomotives. These bars connect the pantographs with the circuit breaker. They are used also to isolate defective pantographs. The tubular bars are provided with knife shaped end fittings, which fit into spring loaded fork fittings as shown in Figure 37.3.

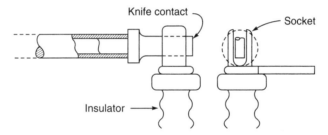

Figure 37.3   Knife and socket connector on locomotive roof

These roof bars carry currents of the order of several thousand amperes in 1500 volt dc locomotives and several hundred amperes on 25 kV ac locomotives. As in the case of any ordinary knife switch, the contact force exerted by the fixed contact on the knife fitting of the roof-bar is very important and must always be within the specified limits. These limits are, much higher for 1500 volt locomotives.

37.6 | Other Failure Modes

There are a few other failure modes but they are rare in comparison with that already discussed above. These are:

(a) Tracking between the knife switch fixed contacts and other terminals on the switch-board panels.
(b) Fracture of the fixed contact leaves due to brittleness of the material.
(c) Overheating due to bad contact with attached terminals. In such cases the failures of the knife switches are consequential effects and not true failures of knife switches.

Careful examination of the failed components and the surroundings is necessary to reach the true cause of failures of the above three types. The remedies would be obvious.

37.7   Do's and Don'ts for Preventing Knife Switch Failures

- Ensure that the contact force between the fixed contact wings and the knife blade is adequate at both ends, i.e. in the swivelling end and in the other sliding end.
- Ensure that the material of the fixed contact has adequate elasticity and static deflection to develop and maintain the required contact force. This can be determined usually by measuring the pull out force.
- If failures through tracking are experienced, increase the tracking distance by providing flash barriers; use anti-tracking varnishes; increase cleaning frequency; change position to make tracking surfaces vertical.
- Where knife contacts are provided on fuse carriers, look for contact force deficiency when the fuse blows even at normal currents.
- Ensure that the operating temperature of knife switch contacts does not rise above 75 °C under highest current and highest ambient temperature conditions.
- If helical springs are provided for developing the contact force, check their deflections in working condition.
- As knife switches are used rarely, for isolation of circuits under off-load conditions, they are likely to get neglected. Special efforts should be made, during overhauls to clean them thoroughly and to operate them a few times.

Chapter 38

PARALLEL CLAMP FAILURES

In this chapter we learn

- *Uses of parallel clamps for making electrical connections.*
- *Mechanism of parallel clamp failures—due to inadequate contact force.*
- *Causes of inadequate contact force.*
    - *(i) Bolt size is too small.*
    - *(ii) Nuts not tightened fully.*
    - *(iii) Clamp plates too weak.*
    - *(iv) Clamp dimensions incorrect.*
    - *(v) Metal creep in clamp plates or wires.*
- *Case studies based on above causes.*
- *Check list for parallel clamps.*

□ □

38.1 Introduction

$f$igure 38.1 shows a parallel clamp used for connecting copper wires used in overhead lines. They are used for connecting feeder lines from sub-stations to catenary lines and for paralleling the catenary and the contact wires. A parallel clamp may also be used to take off a connection from the catenary to an auxiliary transformer for local supplies at a station.

Although these are fairly simple pieces of hardware, many failures, such as parting of the overhead wires are caused by apparently trivial defects in the design, manufacture or installation of these items. These defects can easily be avoided and it is possible to expect reliable performance if a few simple precautions are taken. On the other hand if any of these defects are present in an installation, wire or clamp failure is sure to occur sooner or later.

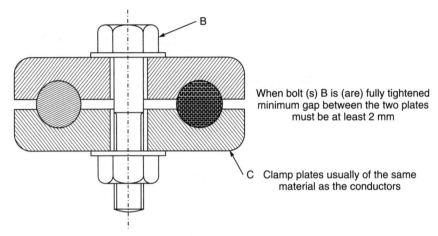

When bolt (s) B is (are) fully tightened minimum gap between the two plates must be at least 2 mm

C Clamp plates usually of the same material as the conductors

Figure 38.1 Typical parallel clamp for connecting two wires

38.2 Causes of Line Failures due to Defects in Parallel Clamps

There is basically only one failure mechanism that manifests itself in different ways depending upon the nature of the defect. This failure mechanism is depicted in Figure 38.2.

The temperature finally reaches such a level that the tensile strength of the wire falls below the tension in the wire. Even if there is no tension in the wire, failure is sure to occur, a little later, due to melting of the wire and the clamp.

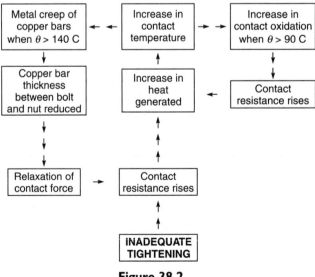

Figure 38.2

The contact force can fall below the optimum level due to any one or more of the following reasons—all taken from actual case histories.

(a) The bolt size was too small and the total force exerted by the two bolts was inadequate.

(b) The nuts on the bolts were not tightened fully.

(c) The clamp plates were too weak, they bent under the force exerted by the bolts, and eventually touched each other. See Figure 38.3.

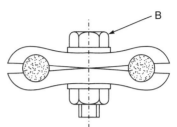

Figure 38.3   Clamp plates too weak to withstand force exerted by the bolt(s) B

(d) The dimensions of the clamp were incorrect and the plates touched as in Figure 38.3.

(e) The clamp plates were subject to metal creep as they were made of aluminium alloy, a material of relatively low melting point. The deflection of the plates kept on increasing and eventually the plates touched as in Figure 38.3 although the original dimensions were correct and the deflections were not excessive.

38.3   Failures due to Inadequate Tightening of Bolts

CASE STUDY

In one case, the clamps and the wires had been badly damaged by arcing and melting. It was not possible to find out whether the fasteners were tightened correctly or completely.

The maintenance staff were asked to install a few parallel clamps in the same manner as they were doing in the past. The specified tightening torque was then applied to these test clamps and also to a number of other installed clamps in the same maintenance zone. It was observed that in almost every case the nuts turned by 10 to 40 degrees.

Inspection of the tools used by the staff showed that some of the spanners were rather loose on the nuts. On checking the dimensions, they were found to be outside the specified limits. Some of the staff complained that if they tried to tighten the nuts fully, the spanners would slip on the nuts.

It was also observed that most staff were using single spanners for tightening the nuts on the clamps assembled on the lines. In this case there was a tendency not to tighten the nuts fully as the wires deflected visibly. It was demonstrated that by using two spanners in opposition on the nut and on the bolt head they could be further tightened by 30 to 50° if they had been first tightened with single spanners.

The action taken was—to explain the need for full tightening to the staff; to provide them with tools which did not tend to slip on the nuts; and to supply two spanners for tightening the nuts/bolts in opposition on clamps installed on overhead lines.

There were no further cases of parallel clamp failures.

38.4 Failures due to Metal Creep

CASE STUDY

Parallel clamps provided on aluminium conductors used for the return wire conductors in one-division had a high failure rate. The clamps were also made of aluminium. Careful examination of old clamps in service for many years showed that they had all bent as shown in Figure 38.4(b).

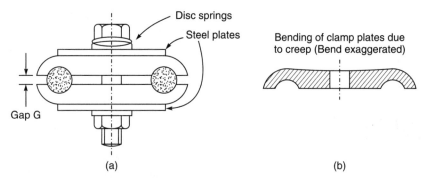

Figure 38.4 Method used to reclaim weak clamp plates of aluminium alloy

It was clear that the clamp plates were getting deformed by creep strain and this was causing the contact forces to get reduced by relaxation of the elastic strain in the bolts, due to the occurrence of the creep strain in the clamp plates. This would lead to increase in contact resistance, overheating of the clamps and eventual failure.

The action taken was—to provide steel plates, longer bolts and disc springs as shown in Figure 38.4(a) and to establish a maintenance practice of re-tightening the nuts every year and checking the gap G between the plates. This gap should be at least 2 mm.

38.5 Check List for Parallel Clamps

The following points should be examined when investigating overheating or melting of clamps and wires under the clamps.

(a) Check whether the material of the clamp is subject to metal creep at temperatures of the order of 100 °C. If yes, steel plates and disc springs should be used around the parallel clamps. Periodic re-tightening of the bolts/nuts must be a part of the maintenance schedule.

(b) Check whether the temperature of the clamps is a few degrees lower than the temperature of the wire when carrying the maximum expected current, by carrying out a low voltage, high current test.

(c) Check whether there is a clear residual gap of at least 2 mm between the clamp plates when the bolts/nuts on them are fully tightened.

(d) Check whether the thickness of the clamps is adequate to ensure that the plates do not take a permanent set under a force which is about 50 per cent above the normal force exerted by the bolts/nuts.

(e) Check whether the nuts/bolts are being tightened fully during the installation. This can be verified by applying the specified torque and checking the angle through which the nut turns.

(f) Use an infra-red temperature sensing device to check the operating temperatures of the parallel clamps when carrying full load current. This is a very expensive piece of equipment but there is no cheaper alternative that can be used safely on live lines.

(g) Use two spanners in opposition, one on the head and one on the nut, while tightening them on overhead lines.

The first four items in the above list depend only on the intrinsic design and manufacturing quality of the clamp. The fifth item depends on the quality of installation and maintenance. If any of these requirements are not complied with, clamp failure involving parting of the wires is likely to occur sooner or later. On the other hand if they are complied with, reliable performance can be expected.

38.6 Do's and Don'ts for Preventing Parallel Clamp Failures

- If there is any case of overhead line parting, check whether it has occurred under a parallel clamp connector. If yes, then it can be assumed that the failure was due to a defect in the clamp.
- Follow the checklist given in Section 38.5 above.
- If you have taken over an installation including overhead lines and parallel clamps, carry out random checks as above at a few locations.
- If you have access to an infra-red temperature sensing device use it periodically when the lines are carrying full load currents.

Chapter 39

TERMINAL BOARD FAILURES

In this chapter we learn

- *Common designs of terminal boards.*
- *Case studies of terminal failures to illustrate failure mechanisms and causes of terminal board failures as listed below:*
  - *(i)   High ambient temperature.*
  - *(ii) Shrinkage of insulating boards.*
  - *(iii) Metal creep of aluminium bars.*
  - *(iv) Incorrect assembly of spring washers.*
  - *(v)   Tracking between terminals.*
  - *(vi) Passage of current through fastener shanks.*
  - *(vii) Moisture entry through cut edges of laminated insulating boards.*

❑ ❑

39.1 Introduction

*I*n all elsectrical installations and equipment there are many separate units or cubicles which are interconnected by cables. Even within such units or cubicles there may be many different components, which may be similarly connected together. These divisions and subdivisions are created in order to facilitate manufacture, installation and also maintenance. The inter-connections are usually made between terminal boards on which different connections are grouped together suitably to facilitate troubleshooting and repairs.

The terminal boards can be of many different designs but essentially they consist of insulating boards with terminals and threaded fasteners. Figure 39.1 shows a view of one terminal at which three cables are joined together.

There may be many such terminals either on the same insulating board or on different boards. The sizes of the terminals depend on the maximum currents to be handled. The spacing between the terminals depends on the anticipated maximum voltages between the terminals.

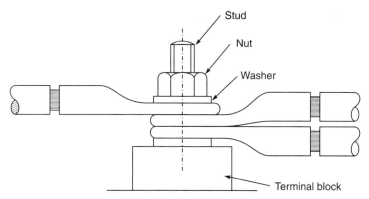

Figure 39.1   Typical terminal block with threaded fastener

The deceptive simplicity of these terminals is probably the cause of adequate attention not being paid to their design. As a result of apparently trivial defects or deficiencies in the design or manufacture, there are many cases of electrical failures originating on terminal boards and often culminating in expensive fires.

39.2   Failure of Bolted Terminal due to High Ambient Temperature

CASE STUDY

In one particular design, the insulating panel used for the terminal board was suitable for temperatures of the order of 300 °C. The sizes of the terminals and bare copper bars used for connections, were selected to operate at a temperature of about 120 °C in order to effect savings in the costs of copper bars and terminals. As the equipment was used for heating air, the heat generated was not considered a disadvantage. The designer had apparently overlooked the fact that pressure contacts such as those between sockets, or between sockets and terminals must not be allowed to rise beyond a temperature of about 90 °C. Above this limiting temperature, the copper terminals and sockets get oxidised rapidly. This increases the contact resistance and starts a vicious cycle as shown in Figure 39.2.

There were consequently many failures of these terminals. They would get heated up to the point of melting and damaging the terminal boards. The action taken was to replace the bolted joints by brazed joints wherever possible and to use larger bars and to locate the terminals, where bolted joints were considered essential, in ventilated compartments at room temperatures.

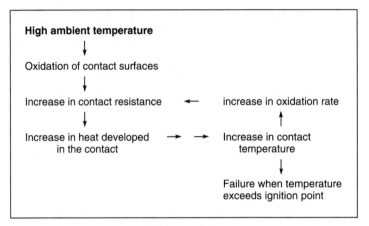

Figure 39.2

CASE STUDY

In another case of terminal board for cables carrying up to 1000 amperes, the design of the terminal was as shown in Figure 39.3.

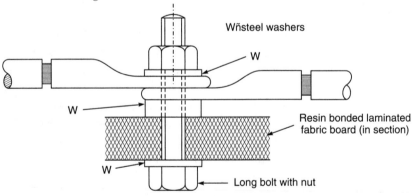

Figure 39.3 Terminal of defective design

The terminal sizes were adequate and the normal temperature rise was only about 35 °C. Yet these terminals became overheated after some months of service to such an extent as to cause charring of the insulating panel, short-circuits and eventually fires.

The cause of this defect was determined after investigation as shrinkage of the insulating board under the influence of pressure, heat and time. This caused relaxation of the force exerted by the threaded fasteners and initiated a vicious cycle as shown in Figure 39.4

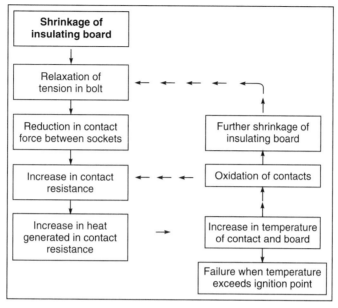

Figure 39.4

The temperature eventually rose to levels higher than the ignition temperature of the insulating panels.

The design of the terminals was basically defective in principle. The force circuit of the fasteners included the insulating material, which had the property of shrinking under pressure and heat. The action taken was to redesign the terminals as shown in Figure 39.5. The shrinkage of the board now had no adverse effect on the contact force between the sockets and the terminals.

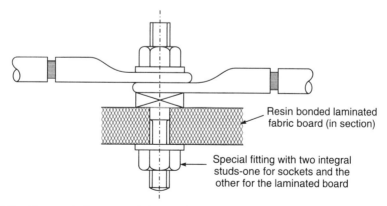

Resin bonded laminated fabric board (in section)

Special fitting with two integral studs-one for sockets and the other for the laminated board

Figure 39.5   Terminal of improved design. Shrinkage of board does not affect contact force

39.4 Failure of Terminal Board due to Metal Creep

CASE STUDY

Another case of terminal board failure occurred in a large reactor in which the main winding was made of aluminium bars. The bars were connected to terminals with threaded fasteners as shown in Figure 39.6(a).

Several months after commissioning, the terminal got overheated and eventually the aluminium bar melted at the terminal. It was thought at first that this was due to bi-metallic action between the aluminium bar and the copper terminal. Use of bi-metallic washers did not solve the problem. It was observed after two months of service that the temperature rise of the terminal had

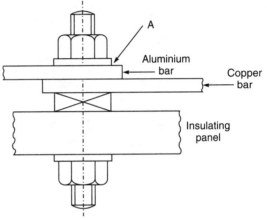

Figure 39.6(a) Terminal board with aluminium bar and copper bar

increased from the initial 25 °C to about 60 °C. On applying the same tightening torque as at first installation, it was seen that the nut turned through about 30°. It was realised at this stage that there was relaxation of contact force due to metal creep of the aluminium bar. Reference to creep data showed that aluminium creeps at temperatures above 7 °C.

The action taken was to remove the bi-metallic washers and to provide disc-spring washers. The disc springs were selected to get nearly flat at the tightening force and two disc springs were used in series as shown in Figure 39.6(b). A maintenance

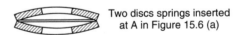
Two discs springs inserted at A in Figure 15.6 (a)

Figure 39.6(b) Disc springs, in series, in open condition. Nut to be tightened until they flatten

practice of opening the terminals, cleaning the bars, applying Vaseline, checking the disc springs and re-tightening the nuts every 3 years was also introduced. This modification was introduced at all similar locations where aluminium bars were connected with threaded fasteners. There were no further cases of aluminium bar joint failures.

39.5 Failure of Bolted Joint due to Assembly Error

CASE STUDY

There was a minor fire at the terminal board of a motor in continuous use. It was observed that the socket of the main power lead had melted and parted at the hole. The lead had then got disconnected from the terminal.

Careful examination of the terminal showed that a spring washer had been inserted between the internal and external cable sockets at the terminal, instead of just under the nut. See Figures 39.7(a) and 39.7(b).

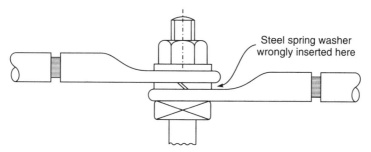

Steel spring washer wrongly inserted here

Figure 39.7(a) Terminal as wrongly assembled

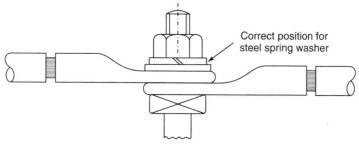

Correct position for steel spring washer

Figure 39.7(b) Terminal as correctly assembled

This resulted in the current passing through the spring washer instead of directly from the incoming to the outgoing cable sockets. The higher resistance of the steel washer led to a higher voltage drop, a higher heat generation and higher temperature rise and gradual degradation of the terminal. Eventually the contact resistance became so high as to cause melting of the socket.

This is an example to show how apparently trivial errors can lead to major failures.

39.6 Other Causes of Terminal Board Failures

Some of the other and more obvious and common failure mechanisms are enumerated below as a checklist for investigators. It is based on actual case studies.

(a) Terminal boards, particularly those that have horizontal surfaces, collect dust between the terminals. If the environment is salty and moist, tracking between the terminals over the surface of the terminal board may be followed by short circuits. If the protection system is not adequately sensitive, fires may result from such failures. The remedies are:

   (i) increasing the creepage or tracking distance between terminals on critical surfaces;

   (ii) periodical cleaning of the surface between terminals;

   (iii) use of vertical terminal boards wherever possible;

   (iv) use of anti-tracking varnishes on critical surfaces;

   (v) air-conditioning or pressurisation with dust free air of rooms housing the equipment.

(b) Threaded fasteners made of copper are used sometimes at the terminals in the belief that they provide greater conductivity. Actually, terminals of this type are more prone to failures. Because of their lower tensile strength, the nuts on copper fasteners are tightened much less than steel fasteners. The contact force is less and the contact resistance is higher. The temperature rise is consequently higher. It is safer always to use steel threaded fasteners and to provide direct face-to-face contact between the sockets or between the socket and the terminal. There is very little current through the steel fasteners. The steel fasteners can be galvanised or cadmium plated to prevent corrosion.

(c) Laminated boards made of synthetic resin bonded paper or fabric are used sometimes for terminal boards. While they have a smooth and tracking-resistant surface, the cut edges of these boards are often not as good in this regard. If the quality of the lamination is not very good, edge-wise entry of moisture is also possible. It is, therefore, desirable to seal the cut ends of such boards with anti-tracking varnish as soon as possible after cutting and preferably after heating them in an oven to 120 °C.

39.7 Terminal Designs

The main features of a good terminal design are:

(a) The threaded fastener is made of galvanised or cadmium plated steel.

(b) There is direct face to face contact between the cables or terminals to be connected.

(c) Little or no current passes through the bolt or screw.

(d) The temperature rise of the terminal is less than 30 °C when carrying maximum current.

(e) Adequate tracking distance is provided between terminals.

(f) The cut edges of laminated insulating boards are sealed with anti-tracking varnish.

(g) If aluminium sockets or bus bars are connected at the terminals, disc springs should be inserted below the fastener nuts.

(h) The force circuit for the main contact force does not include non-metallic insulating boards.

39.8   Do's and Don'ts for Preventing Terminal Failures

- Ensure that the terminal design has the features enumerated in Section 39.7.
- Where large numbers of terminals carrying high currents are in use try to check them periodically with infra-red viewers.
- Ensure that the fasteners are fully tightened every time. Train the staff, particularly the new entrants, to ensure full tightening of fasteners.
- Look for signs of over-heating, such as softening or discoloration of insulation, during scheduled maintenance. If such signs are seen, investigate fully to determine the causes.
- Look for deposits of dust and other foreign matter on all tracking surfaces. All vulnerable locations should be thoroughly cleaned during scheduled maintenance.

Chapter 40

CAPACITOR FAILURES

In this chapter we learn

- *Applications of capacitors in electrical equipment.*
- *Failure modes, failure mechanisms and causes of failures of capacitors.*
- *Case studies involving:*
  - *High ambient temperature.*
  - *Over voltage.*
  - *Over-age.*
  - *Poor manufacturing quality.*
  - *Open circuits.*

40.1 Introduction

Capacitors are used in electrical equipment for:

(a) Improving power factor.

(b) Absorbing or suppressing high-voltage, high-frequency surges or transients.

(c) Timing circuits.

(d) Tuned oscillator circuits.

(e) For storage of energy for fast discharge.

The intrinsic reliability of commercially available capacitors is lower than that of many other circuit elements like conductors, insulators, inductors, resistors, diodes, solenoids, coils, etc. Moreover great care must be exercised in their selection and utilisation, as they are sensitive to variations in design parameters.

There are just two failure modes and failure mechanisms (short circuits and open circuits), but many causes of failures such as over-temperature, over-voltage, over-age, poor quality of manufacture, etc.

When dealing with capacitor failures, it is often useful to compare the actual failure rates with those indicated by reliability data furnished by the manufacturers of the capacitors.

Care should be taken, however, to take into account all the factors which affect reliability such as temperature, de-rating factor, vibration, etc.

40.2 Capacitor Failures due to Excessive Temperature

The reliability of capacitors depends on the ambient temperature. In fact, curves depicting the relationship between the failure rates and the ambient temperature are usually published by the manufacturers of capacitors and by those who test or evaluate the reliability of capacitors. One such curve for particular makes of capacitors is shown in Figure 40.1:

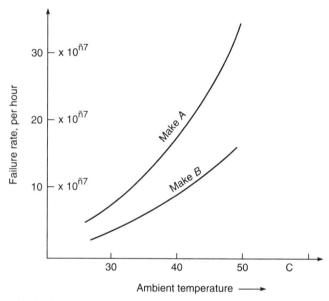

Figure 40.1 Variation of capacitor reliability with ambient temperature

Comparison between the failure rates of different devices used in electronic equipment shows that some types of capacitors are not only more failure prone but also more sensitive to increase in temperature than other devices.

It is clear that there is a significant drop in the reliability of capacitors when the ambient temperature rises. Conversely, there is an increase in the reliability of capacitors when the ambient temperature drops. It is desirable therefore to take all practicable measures to keep the ambient temperature around capacitors as low as possible.

As the reliability of many other components used in electronic equipment, such as transistors, ICs, resistors, also depends similarly on the ambient temperature, it is prudent to house such equipment in air-conditioned rooms or cabinets. It is further desirable to ensure that the heat generated in devices is dissipated without causing local high temperature zones. Separate enclosures, ventilation ducts and fans may be provided as may be necessary.

40.3 Capacitor Failures due to Excessive Voltage

While it would be obvious that an increase in the operating voltage beyond the rated voltage of a capacitor would increase the failure rate, it is generally not realised that a reduction in the operating voltage below the rated voltage helps to reduce the failure rate. In fact this is so important that it is common practice, where high reliability is expected, to select capacitors of a rated voltage which is two or three times the maximum operating voltage. This is often expressed as a de-rating factor (0.5 and 0.33, respectively) to the rated voltage of a capacitor. The reliability of a capacitor can be shown to be a function of the de-rating factor (ratio of peak operating voltage to rated voltage) as in Figure 40.2.

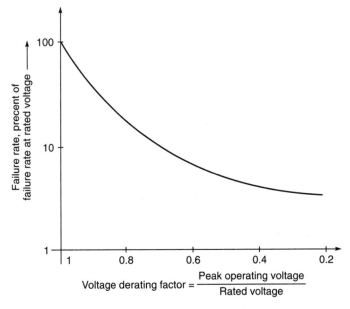

Figure 40.2 Variation of capacitor reliability with voltage derating factor

40.4 Capacitor Failures due to Over-age

Amongst various types of devices used in electronic equipment, capacitors are relatively more susceptible to the effect of ageing as far as their reliability is concerned.

The capacitance of capacitors is directly proportional to the area of the parallel conducting surfaces and inversely proportional to the distance between these surfaces. For reasons of economic design, the dielectric thickness in capacitors has to be as small as possible. This increases the electrical stress in the dielectric. Therefore the design of capacitors involves the choice of the dielectric thickness so as to obtain the most economical design consistent with reliability.

It is easy to decide on the dielectric thickness and the dielectric stress in order to ensure that the capacitor does not fail immediately at the specified peak voltage; but it is more difficult to ensure that the capacitor lasts a long time in service. There are certain dielectric degradation processes involving partial discharges, which can cause failures after a considerable time in service.

In this competitive world of capacitor manufacture, these devices are generally designed to perform with certain guaranteed maximum failure rates at the specified peak operating voltages. It is then up to the equipment designer to select capacitors of higher specified peak voltages if for his particular requirement a lower failure rate is expected.

The net result of the foregoing situation is that in general, capacitors have a limited life in service, which depends on the following factors:

  (a) The ratio of the operating voltage to the specified or rated voltage of the capacitor;

  (b) The temperature of the ambient in the immediate vicinity of the capacitor;

  (c) The age of the capacitor.

The age at which the failure rate of capacitors becomes unacceptably high is difficult to predict. Testing of capacitors in service to determine partial discharge rates at operating voltage can help in taking decisions whether or not to replace capacitors in order to minimise failures in service. Measurement of partial discharge rates requires special instrumentation. It is often possible to monitor the condition of the capacitor by measurement of the stabilised dc leakage current. It is best to consult the manufacturer for advice on the anticipated failure rate as a function of the age and for recommendations regarding monitoring the condition of old capacitors in service.

40.5   Failures due to Poor Intrinsic Quality

The most common failure mode of capacitors is 'short-circuit' between the electrodes through the dielectric. 'Open circuits' in capacitors do occur sometimes but they are relatively rare and are usually due to failures of internal connections between the electrodes and the terminals.

The short-circuit mode of failure is common to all the three causes mentioned above viz. over-voltage, over-temperature and over-age. What happens within the dielectric is that there are weak spots in the dielectric (which may be in the form of voids or inclusions) and partial discharges continue to occur within such voids or inclusions during normal operation of the capacitor. These partial discharges cause degradation of the material of the dielectric. The weak spots become weaker and the partial discharges become stronger. This establishes a vicious cycle of continuing degradation at an increasing rate until, finally, the weakened dielectric breaks down completely at normal operating voltage. In other words a short-circuit develops and the capacitor fails.

The frequency and the intensity of the partial discharges depend on the ratio of the operating voltage to the rated voltage, the size and number of the initial weak spots in the dielectric and the service age of the capacitor. The rate of degradation of the dielectric depends on its quality and operating temperature apart from the three factors just mentioned. In short the overall rate of degradation of the capacitor depends on both the internal quality of the capacitor and the operating conditions.

The internal quality of the capacitor depends on the following:

- The size and number of voids and inclusions in the dielectric.
- The purity of the dielectric material.
- The quality assurance measures taken by the manufacturer in regard to inspection and selection of the raw materials, the cleanliness and the control on the manufacturing processes.
- The screening and inspection of the finished products.

It is difficult for the ordinary buyer of capacitors to check on these factors at the manufacturer's works and it is equally difficult to ensure the same through tests or measurements on the finished products at the stage of purchase. One has to depend partly on the reputation of the make or brand of the product and partly on a vigilant watch on the performance of their products in service.

Reputed manufacturers usually publish reliability data about different grades of their products and these should invariably be asked for initially while purchasing the capacitors and verified if and when failures do take place.

40.6 Failures due to Open Circuits

In case of open circuits, which as stated earlier are relatively rare, the capacitor casings should be carefully opened out to determine the exact location of the open circuit and thereafter to determine the cause of the open circuit. In some cases there may be a short-circuit in the capacitor, which may cause the open circuit. In fact in some capacitors fusible connectors are provided to isolate shorted capacitors automatically from the supply because the consequences of short-circuits are likely to be more serious than those of open circuits.

In such cases the problem should be examined basically as a short-circuit problem on the lines discussed above.

If however there is no short-circuit in the capacitor itself and the open circuit is in the internal connector only, the problem is a true open circuit probably due to mechanical vibrations or manufacturing defects. These will have to be resolved through dialogue with the manufacturer or detailed examination of the fractured or melted connector.

40.7 Failures due to Over Temperature

CASE STUDY

In 25 kV ac electric locomotives, bridge rectifiers are used to rectify the single phase alternating supply to direct current for use by the DC traction motors. The diodes used in these rectifiers are susceptible to damage by high voltage spikes that may be received either in the incoming supply or generated in the load circuit. In order to absorb or suppress such spikes, networks comprising capacitors and resistors are connected across the input and output terminals of the rectifiers.

In the case of one such design of an electric locomotive, the capacitors in these RC networks had a poor reliability. Capacitors are known to have a limited life and replacement at least once during the lifetime of the locomotive is often required; but in this particular case capacitor failures were experienced even in the first year of service and the failure rate increased in the second year. Clearly there was something wrong.

Of the three principal causes of capacitor failures (viz.: overage, over-voltage and over-temperature), the first cause was ruled out. A simple calculation and check with the voltage rating marked on the capacitors ruled out cause number two. That left cause number three, i.e. over-temperature. Inspection of the capacitors at site revealed that the arrangement was as shown in the Figure 40.3.

The capacitors were mounted above the resistors on a panel. The hot air convecting from the resistors flowed over the capacitors and the ambient temperature around the capacitors was at least 30 °C higher than the general ambient. The capacitors and the resistors were interchanged physically. The resistors were mounted above the capacitors. The failure rate of capacitors dropped by an order of magnitude. The failure rate of replaced capacitors was even lower.

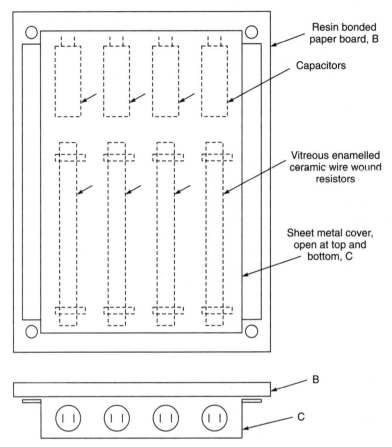

Resin bonded
paper board, B

Capacitors

Vitreous enamelled
ceramic wire wound
resistors

Sheet metal cover,
open at top and
bottom, C

B

C

Figure 40.3 Failures of capacitors due to high ambient temperature

40.8 Failures due to Over-voltage

CASE STUDY

A static voltage stabiliser comprising a special type of transformer and capacitors across a tertiary winding was prone to frequent failures due to failures of capacitors. The capacitor compartment was well ventilated and there was no rise in ambient temperature around the capacitors. The voltage developing across the capacitors was measured throughout the operating range of the input voltage at no-load to full-load conditions. It was observed that under certain conditions the voltage across the capacitors was slightly more than the rated voltage.

The capacitors were replaced by new ones of rated voltage 50 per cent higher than the maximum measured voltage across the capacitor. The failure rate on the modified regulators came down to negligible levels.

40.9 Failures due to Over-age

CASE STUDY

The smoothing capacitors used in the power-supply sections of certain electronic devices, which had been in service for nearly ten years started failing. There were three failures within one year in a population of 20 such devices. Since they had worked without failures for the first eight years it was concluded that these capacitor failures were due to over-age. The remaining 17 devices were also fitted with new capacitors to prevent any failures in service. There were no further failures for the next three years.

40.10 Specifying and Testing of Capacitors for Reliability

Operating tests such as measurement of capacitance and leakage current/power at the rated voltage and insulation resistance are some of the basic tests to be carried out for determining the operational suitability of the capacitors. Other tests have to be made to determine the reliability and durability of the capacitors. These are determination of the leakage current at different dc voltages somewhat higher than the rated voltage at different temperatures below and above the specified maximum ambient temperature. These are routine tests to be carried out on all capacitors. If any capacitors have leakage currents in excess of the average plus three times the standard deviation, such capacitors should be discarded.

Endurance tests with similar measurements at voltages and ambient temperatures significantly higher than the rated voltage and maximum specified ambient temperature may also be carried out on random samples. In these endurance tests the growth of the leakage current with time would be the crucial factor.

40.11 Do's and Don'ts for Preventing Capacitor Failures

(a) Measure the temperature of the air around the capacitor and also the case temperature of the capacitor. The measurement should be made after it has stabilised and when all covers and components that may affect the ventilation are in their appropriate places.

(b) Determine the Service age of the capacitor in hours of operation.

(c) Determine the derating factor, i.e. ratio of rated voltage of the capacitor to maximum operating voltage.

(d) Ascertain the Intrinsic reliability of the capacitor as declared by the manufacturer in terms of temperature, voltage derating factor and age.

(e) Check the possibilities of short time exposure to high voltage.

(f) On the basis of the above data ascertain whether the failure rates can be reduced by one of the following methods:

- Increase in the voltage rating of the capacitor.
- Change of manufacturer.
- Change of capacitor grade or type.
- Reduction, if possible, in ambient temperature around the capacitor.
- Replacement of over-aged capacitors.

(g) In the case of true open circuits (without a simultaneous short-circuit in the capacitor), examine closely the point or section of the fractured or melted internal connector in order to determine further course of action.

Chapter 41

Resistor Failures

In this chapter we learn

- *Types of resistors in common use and their applications.*
- *Principal factors which determine the reliability of resistors and the usual methods of improving reliability available to the user.*
- *Causes of resistor failures and steps to be taken for investigation of such failures.*
- *Case studies of a few typical failure modes.*
- *Selection and testing of resistors.* ❏❏

41.1 Introduction

There are several types of resistors in use in electrical equipment. Wire-wound resistors are probably the earliest types, but now in use only for high power applications. Other common types are composition resistors, solid type or film type and are used mostly in low power, low current applications in electronic circuits for telecommunications, computers and industrial electronics. We will first deal with wire wound resistors.

Wire wound resistors are used in electrical installations for a number of purposes such as starting resistors for motors, RC network filters or surge suppressers, economy resistors, voltage dropping resistors, etc. Wire wound resistors (as opposed to carbon or metal film resistors used in electronic equipment) are preferred for such applications as the currents, voltages, wattages and operating temperatures are generally higher.

Wire wound resistors usually consist of ceramic tubes wound with wire of high resistivity such as those made from alloys of nickel, chromium, manganese and iron. The wound tubes are often covered with a vitreous enamel coating to protect the wire and to facilitate the dissipation of heat. The composition of the ceramic tube as also of the vitreous enamel coating has to be compatible with the material of the resistance wire so that there is no chemical action which could damage the wire. This is important because the operating temperatures of these resistors can be high, i.e. of the order of 150 to 300 °C. High temperatures accelerate chemical reactions.

High power resistors may be made of flat edge-wound strips supported on ceramic insulators. They may be air cooled with fans and blowers. In such cases resistor reliability depends on an additional factor viz. reliability of the cooling system and its effectiveness in cooling the entire resistor element. If even one small part of the resistor element is not cooled properly while the rest of it is well cooled, failures of resistors are inevitable.

Composition resistors are made of specially prepared materials comprising carbon and other high resistivity materials, which are either moulded into solid rods or coated on insulating rods. The composition of the material and the dimensions of the resistor determine the ohmic value of the resistor.

All resistors, whether wire wound or composition, dissipate heat and they have to be proportioned in such a way that the temperature rise of the resistor does not damage the resistor itself. Temperature is the principal factor, which determines the rating of a resistor and also its reliability. The resistive element of the resistor is sensitive to rise in temperature and its rate of degradation increases exponentially with temperature. It is necessary to emphasise here that peak (point) temperature and not the (spatial) average temperature of the resistor element is the crucial factor for reliability.

Where the resistor elements are in strip form, the possibility of strip failures due to vibration, fatigue cracks and thermal stresses may have to be considered.

For reasons of economy, the operating temperature of the resistor element has to be high but for reliability the temperatures should be low. A proper balance has to be achieved by the designer and manufacturer between these conflicting requirements.

41.2 Causes of Failures of Resistors

The resistor may burn out through general overheating due to the passage of a current, which is much higher than the normal designed operating current. This may happen due to failure of some other component in series, such as a capacitor. In such a case it is not a failure of the resistor itself. Its burning out is the consequential effect of the failure of the other component, which should be investigated and corrective action taken accordingly. In this chapter we will discuss true failures of resistors which take place due to the occurrence of defects within the resistor itself or in the environment in comparison with that assumed in the design while the electrical loading of the resistor is within the designed limits.

The most common failure mode is the 'open circuit' where the wire parts or breaks at some point due to local overheating, melting or burning.

In wire wound resistors, the failure mechanism may be initiated either by mechanical damage to the wire during manufacture or due to corrosion of the wire during service. This type of damage can occur anywhere in the resistor but it is more likely at the points where the resistance wire is connected to the terminals for the external connections. The resistance wire at the ends is often doubled on itself to increase its cross-section and to reduce the temperature at the point of connection to the terminal. The wire should be brazed with special brazing compounds since soldered and bolted joints are very unreliable at temperatures above 90 °C. The terminal should also be liberally designed to keep its operating temperature low despite the high operating temperature of the resistance wire. The temperature of the resistance wire should normally be less at the ends than at the centre of the tube, as a result of increased spacing between turns.

If resistance wire parting takes place in the vitreous enamel protected zone, the operating temperature should be checked either on another resistor of the same type and make, under similar operating conditions or in the same equipment after replacing the defective resistor. If it is within the limit prescribed by the manufacturer, the failure can be considered to be due to an internal defect in the resistor. The reliability of the particular type of resistors can be determined by carrying out environmental tests involving humidity, thermal cycling, vibration and dust as specified in the relevant standards.

In composition resistors, overheating is again the cause of failure. The resistive material may melt, burn or fracture if the temperature exceeds the limiting temperature.

Mechanical fracture of the ceramic base of wire-wound resistors or the composition base of the composition resistors can also be a cause of resistor failure even when there is no abnormal rise in temperature. Such cases are rare but if they do occur the cause is found in factors such as vibration, mounting defects and defects in the composition of the base material.

The reliability of resistors depends on their operating temperature. This should be minimised by providing effective cooling or ventilation. The watt loading of the resistor under worst-case conditions should be 30 to 50 per cent of its watt rating.

When investigating resistor failures, the failure mode should be determined first. Is the failure due to electrical overloading or mechanical stresses? Careful examination of failed resistors, of resistors in normal service and of the environment is the starting point.

41.3 Failures of Bolted Connections

CASE STUDY

In one resistor, the resistance wire was connected to the terminal by merely looping through the terminal band as shown in Figure 41.1. The resistor failed after many years of service, by melting of the resistor wire at the point of entry.

Similar resistors in service were examined and bolted connections tightened. A few resistors, which showed signs of degradation on the wires at the terminations, were replaced.

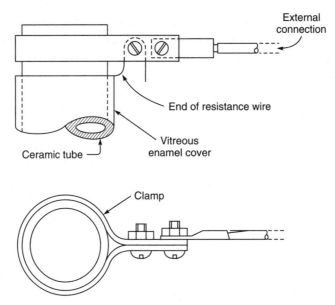

Figure 41.1 Wire-wound resistor failure at termination of resistor element

After completing this cycle of immediate action on all resistors of similar type, they were all modified by brazing the ends of the wires to the terminal straps. All the resistors were examined carefully to verify that there were no local damages to the wire.

The temperature of the resistor surface was measured after operation for several hours at maximum current. It was considerably less than the limit specified by the manufacturer.

41.4 Failures due to Over-Loading

CASE STUDY

In this case, the resistance wire got open circuited inside the vitreous enamel after 3 years service. Examination of the failed resistor as also similar resistors in other similar equipment showed cracks in the enamel. Measurements showed operating temperatures to be close to the limit and the rating/loading ratio was 1.05.

All similar resistors on other similar equipment, where there were cracks in the enamel, were replaced immediately. By way of long-term action, all the resistors were replaced by new ones of rating 50 per cent higher than the original design in order to reduce the operating temperature. There were no further failures of these resistors.

41.5 Failures due to Over-Heating

CASE STUDY

In yet another case, resistors mounted inside a sealed cabinet were getting overheated due to accumulation of heat dissipated by the resistors themselves. The heat dissipation by the cabinet to the air outside was obviously not adequate and the temperature of the air inside was rising gradually. This problem was not detected initially as the equipment was used only for short periods. However, later when the equipment was kept on almost continuously, there were failures of not only the resistors themselves but also of other temperature sensitive items like transformers and thyristors.

In order to minimise the failures, the cabinet was divided into two parts by a vertical partition. Items dissipating about 60 per cent of the total heat dissipated, i.e. high power resistors and one transformer were placed on one side which was provided with improved ventilation by drilling holes on the top and bottom covers. All PCBs, capacitors and semiconductor devices were retained in the dust tight sealed compartment.

These modifications eliminated the failures of resistors and also reduced the failure rate of other components by an order of magnitude. This episode showed again that high operating temperature of the environment is one of the chief causes of failures of electronic equipment.

41.6 Failures due to Poor Quality of Resistors

CASE STUDY

In an installation, comprising 150 speed regulators of a particular make, it was observed that in one position in the circuitry there were two makes of resistors. 95 were of make A and 55 of make B. However, the failures of these two makes were 14 of A and nil of B over a three year period.

Action taken: When this analysis was shown to the manufacturer of the equipment, he agreed to replace all the resistors of make A by those of make B. Cases of this type were experienced in both wire wound and composition resistors.

New samples of improved resistors supplied by manufacturer A were subjected to a series of endurance tests to determine their failure rates and operating temperatures before re-instating them on the list of approved makes. Any test program for the purpose of evaluating any new brand or manufacturer must include endurance tests at rating factors as stipulated in the specifications.

41.7 Failures of Leads due to Sharp Bends

CASE STUDY

There were several cases of fractures of resistor leads at the points where they were bent through 90° for mounting them on PCBs. The PCBs and the resistors were subject to vibration, which was unavoidable. It was observed that the leads had a sharp bend. On checking at the manufacturer's works it was observed that the resistor leads were being bent by hand on flat pliers.

The sharp bend was avoided by using round pliers instead of flat pliers as shown in Figure 41.2(b).

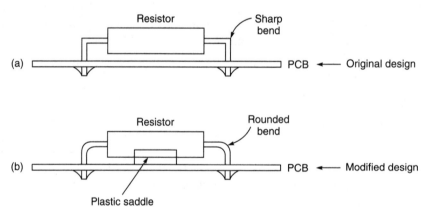

Figure 41.2 Resistor failures due to fracture at sharp bends in wires

Small plastic saddles (supports) were provided below the resistors as shown above to minimise the relative vibration between the resistors and the PCBs. There were no further cases of failures due to lead fracture after these modifications.

41.8 Failure due to Overheating

CASE STUDY

Certain resistors in a PCB used in an industrial control application were prone to failure. There was no vibration; the resistors were of a reputed make and the de-rating factor was about 2.5, which was considered satisfactory.

Examination of the location where the PCB was mounted showed that just close to the resistor in question, there was another wire-wound resistor operating at a high temperature.

A small heat resisting vertical baffle was provided between the wire-wound resistor and the PCB resistor. This stopped the radiation of heat to the PCB resistor. The baffle did not interfere with the vertical convection of air around the wire-wound resistor. The temperature of the PCB resistor dropped by about 30 °C. The failure rate dropped to zero.

41.9 Failures due to Local Overheating

CASE STUDY

In a large resistor used in the rheostatic braking circuit of an electric locomotive, the resistor element consisted of edge wound metal strips supported in ceramic insulators. The elements were cooled by forced air. Open circuits in these elements were a cause of failures of these resistors. As these failures usually occurred while operating on heavy gradients, initially it was suspected that perhaps the resistor rating was inadequate for the operating duty.

On, close examination several failed resistors showed that the open circuits were in similar locations. It was further observed that the ventilation of the resistor elements might be partially blocked by certain structural members of the enclosure in which the resistors were mounted. Temperature measurements showed that the element temperatures were of the order of 50° to 60 °C higher in these locations as compared to other parts of the resistor. Shifting the obstructing structural members solved the problem.

41.10 Selection and Testing of Resistors

It is desirable to select resistors with reliability rates guaranteed by the manufacturers. Based on these data suitable de-rating factors should be determined for the ratings to be selected. Thereafter, care should be taken to ensure that the operating temperature of the resistor does not increase due to the presence of other heat generating units in the vicinity or due to lack of ventilation around the resistor in question. Finally it should be ensured that no mechanical stresses are brought to bear on the resistor either during service or during manufacture. If vibration is unavoidable care should be taken in mounting the resistors to prevent relative vibration between the resistor and the panel or PCB on which the resistor is mounted.

All incoming supplies of resistors must be subjected to routine and acceptance tests as specified. Before placing the names of any suppliers on the list of approved vendors, endurance tests must be carried out on the resistors.

41.11 Do's and Don'ts for Preventing Resistor Failures

- Select resistors manufactured by reputed manufacturers.
- Carry out detailed prototype and routine tests before accepting them.
- Adopt de-rating factors, in fixing the ratings of the resistors, on the basis of reliability data from the manufacturer and conditions of service.
- When confronted with failures of resistors verify the above three pre-requisites.
- Examine the failed resistors to determine the exact location of the failure and the mode of failure.
- Check the worst-case operating temperature and compare with the manufacturer's guidelines.
- Check the surroundings of the resistor for sources of heat and obstructions in the dissipation of heat from the resistor.
- Consider extraneous causes of failures such as vibrations, terminal defects, etc.

POTENTIOMETER FAILURES

In this chapter we learn

■ *Potentiometers are built like fixed resistors but with the addition of a moving or sliding contact.*

■ *The causes and preventive measures for potentiometer failures are the same as those described in Chapter 41 about fixed resistors. In addition there are additional failure modes at the sliding contacts.*

■ *Failure modes, failure mechanisms and preventive measures for the sliding contacts.*

■ *Case studies of potentiometer failures.* ❏ ❏

42.1 Introduction

$\mathcal{P}$otentiometers are variable resistors. They are very much like the resistors discussed in Chapter 41, with the additional provision of the moving or sliding contact. They may be either straight or circular. In straight potentiometers there is a linear sliding contact; and in circular potentiometers there is a rotary sliding contact. See Figures 42.1 and 42.2.

The failure modes and mechanisms discussed in Chapter 41 on resistors apply potentiometers as well as—whether linear or circular. In addition, there are failure modes and mechanisms which apply to the sliding contact itself.

42.2 Sliding Contact Failures

In case of wire wound potentiometers the ceramic or vitreous enamel coating on the resistance wire is usually removed (or not applied in the first place) to provide a clear track for the sliding contact. In the case of composition film potentiometers, the surface is conducting as it is. The sliding contact presses down on the exposed, or bare, resistance wire or on the composition film surface to make electrical contact. A certain amount of minimum force is required to ensure that there is good electrical contact between the wire and the sliding contact. If this force is inadequate there can be additional contact resistance and

Contact force at A & B must be within specified limits, for reliability

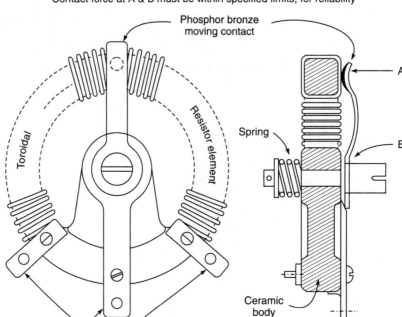

Figure 42.1 Circular or rotary potentiometer

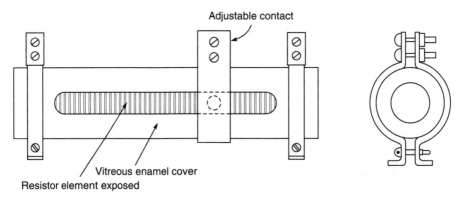

Figure 42.2 Linear potentiometer

additional heat generated at the point of contact. This can start a vicious cycle as shown in Figure 42.3.

This process leads eventually to arcing, melting, open circuits and failure of the potentiometer. This type of failure can be prevented by taking the following measures.

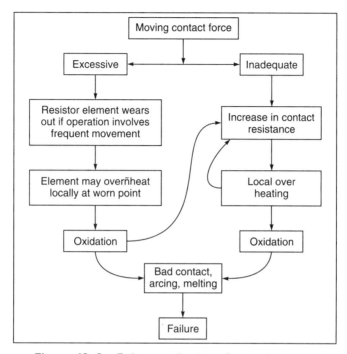

Figure 42. 3   Failure mechanism of potentiometers

(a) The initial contact pressure at the sliding contact should be high enough and the cross-section and material of the contact should be adequate to ensure that the temperature of the contact is well below the limit at which oxidation and other forms of degradation begin to act.

(b) If the potentiometer is in regular use in the sense that the sliding contact is moved frequently, the resistance wire or the composition film will wear and it is then necessary to monitor this wear and to replace the potentiometer in time to prevent any failures in service.

(c) The spring and the contact arrangement should be so designed as to prevent any reduction in contact force during service.

(d) Due to the presence of wearing parts potentiometers have a limited life and they have to be replaced at intervals that depend on the frequency with which they are operated.

Contact force should not be so high as to cause excessive wear or damage to the resistance wire. If the force is too high the potentiometer may fail prematurely. If the contact force is adequate but not excessive, the wear will be normal and some wear-out failures are to be expected after long and intensive usage of the sliding contact if they are not replaced in time. The failure mode will be open-circuit in the wire due to overheating and melting at some point in the worn portion of the winding.

Examination of the failed potentiometer on the worn track will show whether there is excessive wear. Other potentiometers if any in similar duty in the industrial unit should be examined carefully. If there is general wear and the potentiometers have been in service for a long time, they should all be replaced. The contact force should also be verified to make sure that there are no failures due to over-heating at contacts due to low contact force.

In some potentiometers, carbon blocks may be used for the sliding contact. In such cases blocks wear out faster and there is little wear on the resistance wire. If the blocks wear beyond the designed permissible limit, they must be replaced.

42.3 Poor Intrinsic Reliability of Potentiometers

The reliability and durability of potentiometers will depend on the design details such as the dimensions of the active parts, the composition of the resistive elements, the materials of the sliding contacts and the design of the terminations or connections. As far as the user is concerned this means it depends on the manufacturer of the potentiometers.

The only way on a long-term basis for large organisations to obtain potentiometers of high reliability and durability is:

(a) To maintain a list of approved suppliers or vendors. This list should be updated periodically on the basis of experience and test results.
(b) To specify and to carry out acceptance tests that include endurance tests on the potentiometers even when purchases are made from approved suppliers.
(c) To maintain records of all failures and all replacements and to investigate all failures.

The three measures mentioned above are applicable to all electrical and mechanical equipment but are mentioned specifically because in case studies of potentiometer failures, these were the solutions adapted for solving the problems. The numbers of such devices in use is very small and greater involvement with design details and failure investigations is generally not feasible.

The same procedures apply to many other small, mass produced, devices like resistors, capacitors, diodes, transistors, ICs, connectors, etc.

42.4 Failures due to Inadequate Contact Force

CASE STUDY

In one particular case a linear potentiometer was prone to failures due to burning out of the resistance wire, and it was observed that in all cases the failure occurred under the sliding contact.

Examination of a number of other components of the same type, make and application showed that the temperature of the sliding contact was excessive in many cases and that the contact force was too low. Replacement of the springs by stiffer ones that developed adequate contact force, solved the problem.

42.5 Failures due to Poor Quality of the Potentiometer

CASE STUDY

Rotary potentiometers used in a certain make of voltage regulators fitted in association with train lighting alternators, were prone to failures in service. These potentiometers were adjusted only once initially and rarely disturbed thereafter. The potentiometers that had failed and also some of those that had not failed were examined carefully. It was observed that although the manufacturer of the regulator was the same there were two makes of potentiometers in service, and all the failures were on the same brand. They were all replaced by more reliable make. (Further examination revealed that the wire was subject to corrosion). On carrying out the specified environmental tests, they failed. (Obviously, the resistance wire was not of the quality suitable for the environment.) This case showed that if the manufacturer of the regulators had carried out environmental tests on the potentiometers, he would not have selected the defective brand.

42.6 Do's and Don'ts for Preventing Potentiometer Failures

- Refer to the Do's and Don'ts given in Section 41.10. They are all applicable to potentiometers also.
- Before acceptance or installation, check whether the contact force is within limits as specified by the manufacturer.
- Carry out endurance tests, before acceptance of bulk supplies.
- Check the condition of the sliding track on the element, to see whether there is excessive wear.
- In the case of potentiometers that are used frequently, fix suitable intervals for their replacement based either on experience or on the manufacturers' recommendations. Mark these dates on the potentiometers while installing them.
- If carbon blocks are used for the moving contacts, carry out periodical inspections and replace them when they reach the limiting size indicated by the manufacturer.

Chapter 43

CONTACT FAILURES

In this chapter we learn

- *Classification of contacts based on conditions relating to interruption of electric current.*
- *Common failure modes and factors that influence reliability of contacts.*
- *Importance of contact force and contact temperature.*
- *Case study relating to tap-changer failure in a power transformer.*
- *Case study of contact failure due to inadequate clearance.*
- *Summary of 11 case studies discussed in other chapters.*
- *Comments about cleanliness of contacts.*
- *Comments about alignment and matching of contact surfaces.*
- *Dangers of electrical fires inherent in pressure contact failures.*   ❏ ❏

43.1 Introduction

$\mathcal{T}$here are several different types of contacts. They can be classified in many different ways. One method of classification based on the operating conditions, being most relevant to reliability questions, is given below:

(a) Contacts which generally open while carrying a current or close a circuit and cause a current to start flowing immediately.

  (i) Main and auxiliary contacts of circuit breakers. The main contacts can interrupt very high fault currents.

  (ii) Main and auxiliary contacts of contactors, and on-load tap-changers. The contacts can interrupt normal operating currents.

  (iii) Main contacts of some types of relays.

(b) Contacts, which never open while carrying current, and which never complete the circuit thereby starting current flow.

  (i) Main and auxiliary contacts of isolators.

(ii) Contacts in multiple line plugs/sockets as in multi-core cable couplers, PCB connectors.

(iii) Contacts in some types of relays.

(iv) Main and auxiliary contacts of off-load tap-changers.

(v) Contacts in bolted connections.

(vi) Contacts in off-load rotary switches or cam switches.

(c) Contacts, which are never opened after installation and which remain closed permanently.

(i) Contacts between crimped sockets and cables.

(ii) Contacts between cables in twisted joints.

(iii) Contacts between wires and pillars in wrapped connections.

Type (a) contacts can be and are sometimes used for functions where type (b) contacts are sufficient but not vice versa.

43.2 Failure Mode and Mechanism of Failure

The usual failure mode for contacts and the mechanism of failure develops as follows:

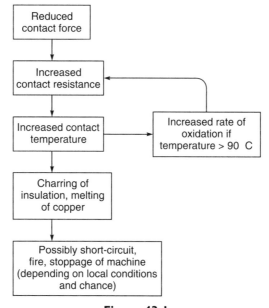

Figure 43.1

43.3 How the Failure Mechanism Gets Started

In all the above types of contacts there are two common factors, which are of the utmost importance for obtaining reliable service, since deficiencies therein are the starting points for all failures. These are:

  (a) The normal (i.e. the perpendicular) force between the two elements of the contact.

  (b) The resistance and the temperature of the contact.

The magnitude of the force determines the contact resistance, the voltage drop across the contact, the heat energy developed, and the temperature of the contact. The initiation of the vicious cycle shown in Figure 43.1, which leads to the failure of the contact depends on the last parameter, i.e. the temperature of the contact. The oxidation rate of the contacting surfaces and the consequent degradation of the contact becomes significant when the temperature is higher than 90 °C. The greater the excess over this temperature, the faster is the degradation.

If the temperature of the contact rises above 90 °C either because the contact force is inadequate or because the ambient temperature is too high, the failure mechanism gets into action and failure is certain.

43.4 Other Relevant Factors in Failure Mechanisms

Other factors such as cleanliness of the contacting surfaces at the time of installation, cross section of the conducting parts, contact area, material of the conducting parts, etc. have also some influence on the temperature rise but all such factors do not change with time. If the original installation is satisfactory these factors will not undergo any change. But the perpendicular force between the contacts can and often does get reduced; and when this happens the change is usually downwards as regards the force and upwards as regards the contact resistance, the voltage drop and the temperature. This initiates the failure mechanism.

The limit of 90 °C mentioned above refers to copper components. It is prudent to design contacts in equipment to operate at temperatures of the order of 75 °C. Copper contacts are used when current interruption arcs have also to be dealt with as in circuit breakers and contactors.

The limit for silver contacts is 115 °C, but the recommended operating temperature is 95 °C. Silver contacts are not used for arcing contacts.

Gold contacts are used mainly where high reliability is required in very light, low force, and low voltage contacts as in PCB connectors and multi-core couplers in electronic equipment. Gold contacts are not used for arcing contacts.

From the case studies which follow, it is evident that failures are due to the loss of the contact force for one reason or another or to high ambient temperatures. The first thing to check in the event of any case involving overheating of contacts, is the perpendicular or normal force between the contacts. In 90 per cent of the cases, this will be the source of the problem. Only when this probable cause is eliminated the other possibilities, such as high ambient temperature, lack of cleanliness of the contacts, defective material of the contacts, adverse environmental conditions, etc. may be considered.

43.5 Failures of Contacts due to Inadequate Contact Force

CASE STUDY

After a power transformer explosion, investigation revealed that the source of the arcing and the heat was at the off-load tap changer inside the transformer tank. The fixed and the moving contacts and some of the surrounding structural members showed signs of arcing and melting. It was a three phase transformer with 5 taps on each phase. There were thus 15 fixed contacts and three moving contacts. Of these, three fixed contacts and one mobile contact were very severely damaged. The remaining part of the tap changer was in reasonably good condition.

Measurement of the contact force on the remaining two moving contacts showed that the contact forces were barely 15 and 20 per cent of the minimum force indicated by the manufacturer. Similar measurements on another (undamaged) transformer of the same make and type showed similar defects. Temperature rise of the moving and fixed contacts were then measured in a test rig simulating actual conditions. It was found that the temperature rise was not only too high (nearly 70 °C) but also rising gradually.

The low contact force was found to be due to a defect in the spring, which caused it to take a permanent set at any attempt to increase the contact force.

It was concluded that the failures were due to a low contact force in the tap-changer. The springs were replaced by new stiffer ones and the temperature rise was verified to be now less than 30 °C. An endurance test of 30,000 operations was carried out on the tap-changer. There were no further failures of this type on the transformers.

43.6 Contact Failure due to Inadequate Clearance

CASE STUDY

Figure 43.2 shows an auxiliary contact used in a locomotive control circuit. There were cases of contact failures due to an unusual effect, which had nothing to do with normal contact operation.

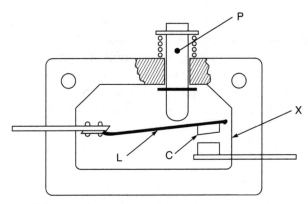

Figure 43.2 Auxiliary contact failure due to thermal expansion

There were locomotive failures when the auxiliary contact failed to open even when the plunger P was retracted by the main mechanism which was to be monitored by the contact. The contact leaf L got stuck in the closed position leading to malfunctions in the equipment. The locomotive had to be switched off and returned to the shed for repairs. However, when the locomotive reached the shed all the equipment operated satisfactorily.

This type of incident happened several times before the true cause of this intermittent fault was discovered to be as follows:

(a) When the contact plunger P was pushed in during normal operation by the equipment being monitored by the contact, the contact C closed and started carrying the normal current through the contact leaf L. The length of the leaf increased due to its rise in temperature and thermal expansion. The end of the leaf touched the casing at X and remained stuck as long as the current flowed through it.

(b) When the locomotive was switched off because of malfunctioning of the equipment, the current stopped flowing through the leaf L and it cooled and returned to its original length. The leaf L was no longer in contact at X. Hence when the locomotive was tested in the shed it worked normally. It was only after the locomotive was in service for about one hour that the leaf temperature was high enough to cause the trouble.

(c) The root cause of the trouble was thus seen to be inadequate clearance between the end of the contact leaf L and the casing. This problem was easily corrected by filing away a small length of the leaf at its end. There were no further failures of this type.

(d) Checking of the relevant working drawings, in the manufacturer's works revealed that the actual leaf lengths were longer than the specified dimensions.

Intermittent faults of this type are difficult to detect. Every designer must consider the effects of phenomena like thermal expansion, metal creep, metal fatigue, eddy currents, etc., referred to in Section, 1.4, while deciding the component dimensions and tolerances. Failures can be prevented only by careful design of equipment, and fanatical adherence to manufacturing tolerances.

43.7 Summary

CASE STUDY

The same failure mechanism can be seen in many different types of equipment. They have been described in the relevant chapters about equipment or components and some of them are summarised here.

EQUIPMENT OR COMPONENTS	REFERENCE	FAILURE MECHANISM WHICH CAUSED LOSS OF CONTACT FORCE AND LED TO OVERHEATING, BURNING ETC.
Terminal board Bus-bar joint	CASE STUDY 39.3 CASE STUDY 46.4	The force circuit of the threaded fastener included insulating material, which shrank under heat and pressure, resulting in relaxation of contact force.
Terminal board	CASE STUDY 37.4	The force circuit of the threaded fastener included aluminium bars, which are subject to metal creep and relaxation of force at normal operating temperatures.
Non-ferrous threaded fasteners	CASE STUDY 37.6b	Due to the danger of breaking during tightening, non-ferrous fasteners were not tightened sufficiently. The contact force was inadequate to start with.
Bus-bar joints	CASE STUDY 46.3	The initial design of the fastener size was inadequate.
Knife switches	CASE STUDY 37.3	Due to defects in design, the force between the knife blade and the fixed contact was inadequate.
Crimped sockets	CASE STUDY 14.4	Due to inadequate thickness of the barrel wall of the crimped socket, the elastic forces exerted by the barrel on the wire after crimping were inadequate.
Crimped sockets	CASE STUDY 32.5e CASE STUDY 32.5f	Due to incorrect selection of socket size or due to inadequate crimping, the contact force between the barrel of the socket and the cable was inadequate.
Parallel clamps	CASE STUDY 38.2 CASE STUDY 38.3	The majority of parallel clamp failures were due to the loss of contact force, for one reason or another.

43.8 Cleanliness and Matching of Contacts

It is certainly desirable to ensure general cleanliness of contacts, as also to avoid excessive effort in this direction. Adequate contact force takes care of any minor deficiencies in cleanliness.

Contact cleanliness becomes very important only in the case of very small or light contacts such as those in relays and micro-switches in which the contact forces are necessarily small because of the small size. In such cases the contacts should be housed in air tight or dust tight enclosures. During manufacture or maintenance, special care should be taken to do the work in air-conditioned and clean surroundings.

Clean lint-free cloth should be used for cleaning the contact faces. Solvents used for cleaning must not leave residues. Emery papers, glass papers or sandpapers should not be used for cleaning carbonised or pitted contacts. They can leave behind fine grit. Contact cleaning files, which are similar to manicurist's nail files, may be used instead.

Light dust and oil films are not harmful in contacts where the contact force is of the order of 100 grams or more and the operating voltage is 100 volts or more.

A very important feature of contacts is the wiping-action between the mobile and the fixed contacts, after the contacts touch. This is a very small but significant lateral movement, which causes the mobile contact to slide or wipe across the surface of the fixed contact, while being pressed down. This action ensures good contact despite small quantities of dust or oil film on the contact faces. Contacts with wiping action are more reliable than those without wiping action. See Figure 43.3.

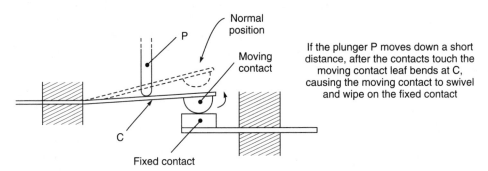

If the plunger P moves down a short distance, after the contacts touch the moving contact leaf bends at C, causing the moving contact to swivel and wipe on the fixed contact

Figure 43.3   Principle of wipe-action on contacts

43.9 Matching of Contacts is Unnecessary

It has been observed that conscientious workers spend a great deal of time matching electrical contact surfaces until contact over the full surface is obtained in the belief that this improves the electrical contact. This matching is not necessary and leads to unnecessary wear of the contacts. The true contact area is in any case less than 10 per cent of the apparent contact area. All that is necessary is to ensure proper alignment of contacts and no visible air gap between the contact surfaces.

If the operating temperature rise of an electrical contact is low (less than 30 °C) and if the voltage drop at rated current is within the specified limits it is best to ignore any minor irregularities in the contact faces.

43.10 Dangers Inherent in Pressure Contact Failures

Pressure contact failures of the type described in Sections 43.5 and 43.7 are more dangerous than short-circuits caused by insulation failures. There is no protection system against defects of the former type whereas defects of the latter type are usually detected by the protection systems always provided in any electrical installation.

There are thousands of pressure contacts in most electrical installations and most of these pressure contacts are made manually, in the final stages of assembly and installation.

A large number of electrical fires that take place every year are due to this problem. There is a very great need for special efforts to make all electrical artisans, supervisors and engineers aware of these facts if we are to make a significant reduction in the number of electrical fires. Most failures occur, neither because there is any technical problem in doing the job correctly nor because there is any increase in costs of equipment on this account, but merely because the staff concerned are simply not aware of these dangers.

43.11 Do's and Don'ts for Preventing Contact Failures

- Ensure that the force between the two parts of the contact system is adequate and as specified in the design.
  - If threaded fasteners are used for this force, ensure that they are tightened fully.
  - If springs are used for this force, ensure that their free length and deflection are correct.
  - If crimping is used for this force, ensure that the crimping is done correctly (Refer Chapter 32).
- Ensure that the contact temperature is less than 75 °C under the worst case conditions of current and ambient temperature.
- If the contact force provided originally is getting reduced in service due to any reason, identify that process and ensure that the contact force is not allowed to drop in service.
- If any particular contact is failing frequently, monitor periodically, its temperature, contact force, appearance, milli-voltage drop across the contact when carrying full current and contact resistance. These parameters will point to the probable cause of failure.

Chapter 44

RELAYS FAILURES

In this chapter we learn

■ *Defects due to pick-up and drop-out parameters of relays going outside the specified limits.*

■ *Significance of the terms pick-up and drop-out parameters of relays.*

■ *Causes of differences between the above two parameters.*

■ *Case studies relating to defects in these two parameters:*

   • *Due to sharp corners of metal parts cutting rubber diaphragms.*

   • *Due to differences in the service conditions and calibration conditions.*

   • *Effects of resistance of long leads on relay calibration due to failure of a spring.*

   • *Due to calibration errors caused by defects in calibration instruments.* ❏ ❏

44.1 Introduction

$\mathcal{V}$arious kinds of relays are used in electrical engineering practice. Some of these are: over-current relays, over-voltage relays, earth-leakage relays, interposing relays, etc. These are purely electrical relays in the sense that the inputs and the outputs are both electrical. Other types of relays sense physical or mechanical inputs and give electrical outputs. For instance, airflow relays, air pressure relays, oil pressure relays, temperature relays, etc.

For each kind or type of relay there are hundreds of makes and ratings. Their shapes, appearance, operating principles and sizes would be different but they all have very few and common failure modes.

(a) Short or open circuit in solenoid.

(b) Overheating, burning, arcing, melting and welding, of contacts.

(c) Overheating, burning, arcing, melting and welding, of leads and terminals.

(d) Pick-up or drop-out parameters outside specified limits.

Items (a), (b) and (c) listed above are discussed in detail in Chapters 50 (Solenoid failures), 43 (Contact failures) and 39 (Terminal failures) respectively. In this chapter we will discuss only item (d) viz. variations in pick-up and drop-out parameters.

44.2 Variations in Pick-up and Drop-out Parameters

If the applied voltage to an over-voltage relay set to operate at 260 volts, is increased gradually from a low voltage, the relay remains unaffected until the voltage reaches 260 volts. At that point the relay will operate, i.e. normally open (N/O) contacts will close and normally closed (N/C) contacts will open. There will be no change if the voltage continues to rise. If now the voltage begins to fall, the relay will remain un-affected until the voltage falls to, say 244 volts and then the N/O contacts (which had closed on the upward swing of voltage) will re-open. Similarly the N/C contacts will re-close.

For such a relay, 260 volts is the pick-up voltage and 244 volts is the drop-out voltage. Similarly, relays operating on other parameters will have pick-up current and drop-out current, or pick-up pressure and drop-out pressure, or again, pick-up temperature and drop-out temperature.

These differences in the pick-up and the drop-out parameters are unavoidable due to practical limitations in devices such as mechanical backlash and magnetic hysteresis. These differences are generally a blessing in disguise. They prevent 'chattering' of relays and, sometimes, special steps are taken to increase the difference between these two parameters beyond the minimum levels due to the physical limitations. The exact values needed depend on the system design and these are indeed considered and specified by the designer of the system. Once these are so determined, they must be respected. If the relays develop changes in these values or replacement relays have different settings, failures of main equipment could possibly occur due to these changes. Therefore, it is important to verify these settings on all relays periodically and specially after overhauls, repairs, or replacements.

The pick-up and drop out settings of relays can be regulated by adjusting the screws or nuts provided for this purpose. The adjustment has to be made not only initially but also periodically because wear and tear of the contacts or stops and wear in the bearings or pivots of moving parts have the effect of causing the adjustments to drift. Steps must also be taken to prevent adjusting screws or nuts from turning due to vibration. The lock nuts must be tightened fully or locked in some other way after verification of the settings.

44.3 Case Studies

There are two main types of relays viz. operating relays and protective relays. Operating relays come into play all the time under normal operating conditions but protective relays remain dormant most of the time. They operate rarely, when there is some short-circuit, earth fault or some other abnormal condition. Defects in operational relays get detected soon enough but those in protective relays may not get detected under normal conditions. The latter case is dangerous because when these defective protective relays fail under abnormal conditions there may be serious consequential losses. Therefore protective relays should receive special attention and they should be tested periodically in situ.

The failure modes and failure mechanisms of both these types of relays are however the same. A relay may mal-function due to defects like bad contact, broken spring, cracked diaphragm, open circuit or short circuit in a solenoid, incorrect adjustment, etc.

As failures due to incorrect calibration or adjustment are being discussed, a few such cases are summarised in the following table:

CASE STUDY NUMBER	FAILURE MODE AND FAILURE MECHANISM	REMEDIAL MEASURES
44.3.1	Current limiting relays fitted in Electric Multiple Unit Stock in one Depot were not operating at the desired currents. Investigation showed that the relays were calibrated correctly in the depot and the variations were due to the resistance of the long leads on the EMUs between the relays and the shunts.	The shunt resistance and the lead resistances were used for calculating the calibration values in the depot so as to give correct operation in actual service. The nuts used for adjusting and locking the calibration were sealed with Loctite to prevent possible shift.
44.3.2	Vane type airflow relays failed due to fracture of a spring provided for keeping the relay in the open position in the absence of airflow. The failures were due to sharp bends in the spring termination.	The problem was solved by redesigning the shape of the spring termination as discussed in Chapter 18.
44.3.3	The rubber diaphragm of compressed air pressure regulating relays developed circular cracks as shown in Figure 23.4, Chapter 23. These cracks were due to sharp edges of the flanges.	The edges of the flanges were rounded off (Refer to Chapter 23). The diaphragm thickness was increased slightly and the material changed to a polychloroprene with a higher oil resistance.
44.3.4	Certain electromagnetic relays were not picking up at the correct currents when mounted on the locomotives; but checking the relays in the laboratory showed no problems.	The relays were mounted on vertical panels in the locomotive, but the calibration had been done with the relays horizontal. As the weight of the armature was significant in relation to the force exerted by the spring, the position of the relay—whether vertical or horizontal—was important. Calibration of the relays in the same position as in actual use solved the problem.

CASE STUDY NUMBER	FAILURE MODE AND FAILURE MECHANISM	REMEDIAL MEASURES
44.3.5	The calibration of a particular type of relay was found to be defective on several locomotives within one week. The error in all the relays was the same. No satisfactory explanation could be found for such defects.	Checking of the instruments used on the test bench exposed a defect that had developed in one of the instruments. The same was replaced. This episode highlighted the fact that efforts must always be made to determine the possible physical causes of defects, particularly when new types of defects suddenly make their appearance.

44.4 Do's and Don'ts to Prevent Relay Failures

• Refer to the Do's and Don'ts in Chapters 50, 43 and 39 relating to failures of relays due to failures of solenoids, contacts and terminals. If there are resistors or capacitors in the relays, refer to the appropriate chapters in regard to their failures.

• Do the calibration of relays in the laboratory, but ensure that the conditions (such as verticality or otherwise, lengths of leads, etc.) are the same as in actual use.

• Ensure that the instruments used for calibration are of a higher accuracy and that they are calibrated periodically and also when inexplicable defects in many relays are noticed.

References

(a) Chapter 50 ibid Solenoid Failures.
(b) Chapter 43 ibid Contact Failures.
(c) Chapter 39 ibid Terminal Failures.

Chapter 45

FAILURES OF FUSES AND
MINIATURE CIRCUIT BREAKERS

In this chapter we learn

- *Why all electrical equipment and installations (except a few low voltage, low current, and hand held devices) are invariably provided with protective devices.*
- *Three main categories of protective devices of which one is the category 'Fuses and miniature circuit breakers'.*
- *The applications where fuses and miniature circuit breakers are generally used.*
- *Definitions of two types of failures of fuses and miniature circuit breakers.*
- *Possible modes of failures and their causes. Measures to be taken for prevention of failures.*
- *Specifications and testing of fuses and miniature circuit breakers.*
- *Case studies of fuse and miniature circuit breaker failures.* ◻ ◻

45.1 Introduction

*A*lmost all electrical equipment include some protective gear which disconnects the source of power from the circuits and devices, *automatically*, whenever a short-circuit or earth fault develops therein. The exceptions to this rule are very small, low power, low voltage devices such as hand held electric torches.

Protective devices are essential because defects like short-circuits can cause equipment damage if the power source is not switched off at once. Even fires can start and these may spread into major conflagrations if flammable and combustible materials are around the location of the defect. Only very small, low power, low voltage devices comprising one or two primary cells in series, are free from this danger of fire upon being shorted. A short-circuit will discharge the cells but a fire is unlikely. It is not necessary to provide a protection system in such cases.

It is impossible to prevent short-circuits because insulation can fail without notice and for reasons beyond control. However, it is possible to prevent expensive and irreparable damage resulting from insulation failure and short-circuits.

The supply circuits are capable of pumping in vast quantities of energy into very small fault zones. When this happens, very high temperatures of the order of several thousand degrees Celsius are developed; and any combustible material within the radiant zone, starts a fire if the electrical arc at the short-circuit persists long enough. Since such incidents are highly probable in any electrical system, protective devices that detect the abnormal condition, and disconnect the source of power from the short-circuit zone within a few milliseconds, are essential.

If the environment is highly flammable and explosive as, in a coal mine, with a mixture of methane and air in the atmosphere, even a momentary spark can cause an explosion. In such cases special protection devices have to be provided to contain the entire electrical system in sealed pipes and containers. These are exceptional situations and the majority of electrical installations do not call for such extreme measures.

Protective devices in electrical equipment can be divided into three main categories as follows:

(a) Electric circuit breakers. These are generally used in medium/high current, medium/high voltage applications. They may be of different types such as electromagnetic, compressed air, compressed gas, vacuum, oil, etc.

(b) Miniature circuit breakers and fuses. The former are usually electromagnetic and thermal tripping devices and the latter consist of fusible elements. They are generally used in low voltage, low current applications such as household electrical installations, control and indication circuits of high power installations, and in power supply circuits of electronic devices.

(c) Electronic, semi-conductor devices such as thyristors, which can interrupt currents by the suppression of control pulses.

In this chapter we will discuss the failures of only the second category viz. Miniature Circuit Breakers (MCBs) and fuses.

45.2 Definition of Failure of Fuse or MCB

It is necessary to define what we mean by failure of a fuse or a miniature circuit breaker. If a fuse melts or an MCB opens, interrupting the service, due to an electrical fault in the controlled circuit, it is *not* a failure of the fuse or of the MCB. It is a *successful operation* of the device. In such cases, the causes of the electrical fault in the controlled circuit need to be investigated.

A fuse or an MCB can be considered to have failed only if one of the following two things happen:

(a) The MCB or the fuse interrupts the circuit even when there is no fault (as defined in its specification) in the controlled circuit, or

(b) The MCB/fuse fails to interrupt the current when there is a fault (as defined in its specification) in its controlled circuit. The MCB or the fuse may itself suffer damage. There may be damages or fires in other equipment controlled by the MCB/fuse.

Obviously, failures of type (b) are more serious than those of type (a). Type (b) failures can lead to expensive damages apart from interruption in the service. In type (a) failures there is only interruption in service and the repair costs are negligible. The root causes of the two types of failures are different.

Whenever a fuse or an MCB is involved in a failure it is desirable first to determine whether there is indeed a failure of the fuse/MCB and if so whether the failure is of type (a) or type (b).

45.3   MCB/Fuse Failures of Type (a)

MCB/fuse failures of type (a), i.e. uncalled for interruptions caused by these devices in the absence of any fault (as defined in the specification) in the controlled circuit are often neglected by the user. The MCB is re-closed or the fuse is replaced and the incident is overlooked because the service is restored.

If such incidents take place repeatedly or if these unnecessary interruptions in service are not acceptable, further investigation becomes necessary. Possible causes of such failures are described below, generally in order of probability of occurrence:

(a) The fusing characteristics of the fuses or the tripping characteristics of the MCBs may be unsuitable for the operating conditions. These characteristics are generally of the shape shown in Figure 45.1.

(b) If the current/time combination ever falls to the right of the characteristic, either during starting or after some temporary overloading even during normal operation, the fuse will blow or the MCB will trip. The cause of failure is, incorrect selection of the fuse or MCB. Such failures will continue to occur whenever the current/time combination point falls to the right of the characteristic.

(c) In the case of fuses, even though the initial characteristic of the fuse is correct, it is possible that the characteristic may shift to the left as a result of degradation of the fuse caused by oxidation of the fusible element. This possibility exists because fuse elements have necessarily to operate at high temperatures in order to possess the required fusing characteristic for the protection of the wiring and equipment.

If either tinned or bare copper wires are used for the fusible elements this kind of problem is sure to arise sooner or later depending on the ratio of the operating current

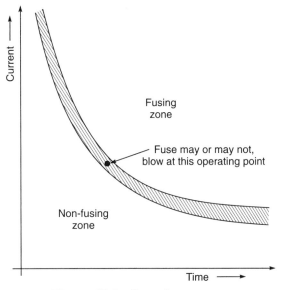

Figure 45.1   Fuse characteristic

to the rated current of the fuse. When this ratio is unity the temperature of the fusible element is of the order of 200 °C, and oxidation is rapid at this temperature. Such fuses are bound to blow very soon at normal currents since the temperature of the fusible element will continue to rise in a vicious cycle of rising temperature and increasing rate of oxidation. For this reason fusible elements should be made of either pure silver or of silver-plated copper.

This type of failure may be considered to be a 'safe' failure since the consequences are not serious. There is a discontinuity of service but that is all. There are no damages to equipment. Fuses are therefore considered to be fail-safe since they can fail only in this manner.

(d) MCBs can sometimes fail in mode (a) described above, i.e. trip even when the current-time combination does not fall to the right of its characteristic. Such failures may be due to weakening of the internal springs due to corrosion or wear and tear on the latching mechanisms.

45.4   Fuse or MCB Failures of Type (b)

MCBs can however fail also in the more dangerous mode (b), i.e. where the MCB fails to open the circuit even when the current-time combination falls to the right of the tripping characteristic. An extreme and dangerous case would be when the MCB fails to open despite an excessive current through it. This could damage equipment in the protected circuit or even start a fire. Clogging or jamming of the mechanism causes such failures either due to welding of the contacts or collection of dust and grit in the MCB enclosure or corrosion of components.

MCB failures can be minimised by:

(i) Using MCBs, which have successfully undergone environmental and endurance tests.

(ii) Selecting MCBs of characteristics, which are properly coordinated with the loading characteristics.

(iii) Installing the MCBs in dust-proof and roomy enclosures.

(iv) Annual checks on the tripping characteristics to ensure that there is no drifting of the same. Defective MCBs should be replaced.

45.5 | Specifications and Testing of Fuses and MCBs

Whenever electrical protective devices such as circuit breakers or fuses are to be selected one has to consider two major requirements. These are:

(a) The device must under specified abnormal conditions open the circuit automatically without getting damaged or causing any damage to other equipment. This is particularly important when short-circuit currents and voltages are high.

This requirement is of the greatest importance. If it is not met the entire purpose of providing these devices is defeated. Every such case must be fully investigated and steps taken to prevent recurrence. When these devices are first selected and installed it is desirable to check this aspect fully by simulated faults under controlled conditions.

(b) The device must not open automatically under any normal operating conditions thereby causing an inconvenient or even harmful interruption in power supply.

It is necessary to specify fully not only the limiting values but also the full range of normal operating values of the currents, voltages and durations.

The simplest but most effective test on an MCBs consists of:

(i) **The one-hour non-tripping current test.** In this test, the specified test current of about 140 per cent of the rated current, is passed through the MCB. It **must not** trip in less than 1 hour.

(ii) **The one-hour tripping current test.** In this test. the specified test current of about 160 per cent of the rated current is passed through the MCB. In this test the MCB **must** trip in less than one-hour.

It may be noted that the same test rig can be used for these two tests and the MCB must pass both these tests. This test can be a routine test and the tested product can be used.

In the case of fuses, similar one-hour fusing current test and one-hour non-fusing current test can be carried out. They can be made only on random samples drawn from the offered lots because the second test is destructive. For routine testing, It is customary to measure the milli-volt drop across the fuse at a current, which is approximately equal to the rated current. The resistance of the fuse is calculated from these two measurements and compared with the permissible limits as determined on the samples used for the fusing and non-fusing tests.

When there are failures of fuses or MCBs it is desirable to carry out the fusing and non-fusing (or the tripping and non-tripping) tests to verify the suitability of the design for the application. These tests may be made on fuses or circuit breakers of identical design. If these are satisfactory and the failures have occurred on fuses/MCBs after much service, endurance tests may be carried out to verify the effects of ageing on the fuses or MCBs. The used or failed fuses/MCBs may be opened out, after first checking their electrical characteristics such as resistance, and non-fusing test if possible, and examined carefully.

45.6 Probable Causes of Failures of Fuses and MCBs

Here are some of the probable defects that may be responsible for the failures:

(a) **Fuses**

- Fusible element of incorrect size.
- Fusible element oxidised or nicked.
- Poor connection or joint between the fusible element and the terminal.
- Lack of ventilation around the fuse carrier.
- Fusible element made of lead, aluminium, or copper.
- Bad contact between the fuse and the fuse carrier.
- Overheating of fuses or MCBs by heat conducted from bad contacts in fuse carriers.

(b) **MCBs**

- Internal mechanism jammed due to dust, grit, etc.
- Internal mechanism jammed due to corrosion.
- Internal mechanism worn at pivots and latches.
- Contacts welded due to excessive current/voltage or reduction in contact force.
- Excessive sparking/arcing due to inductive loads and unsuitability of the design for such loads.
- Bad contact between the connecting wires and the MCB terminals.

45.7 MCBs Failed to Clear Faults

CASE STUDY

MCBs were fitted on certain classes of locomotives. There were a few cases of fires in the control circuits the causes of which were traced to the failure of the concerned MCB to disconnect the faulty circuit from the battery. On checking the characteristics of the MCBs (which were several years old) it was found that they did not pass the 1-hour tripping current test. New MCBs of the same make and type did pass the same test. It was concluded that the MCBs were deteriorating in service.

Opening out some of the MCBs revealed signs of corrosion and accumulation of dust.

The action taken immediately was to replace all the MCBs by proven non-deteriorating fuses. A programme of testing different brands and types of MCBs to determine their deterioration qualities was undertaken. The test programme included comprehensive environmental tests.

45.8 Fuses Blowing in the Absence of Faults

CASE STUDY

In one railway division, hundreds of 5-ampere fuses were used for the protection of circuits feeding power to the signal lamps. There were signal failures from time to time, which were traced to melting of these fuses even when there were no faults in the circuit. Each failure of this type led to detention of many trains.

Investigation showed that all the fuses were wired with fine copper wire soldered at the end caps. The failure mechanism was as follows: The operating temperature of the fusible element viz. the bare copper wire, was around 200 °C. At this temperature the wire oxidises rapidly and this starts a vicious cycle as shown in Figure 45.2.

The remedy in such cases is to use silver or silver plated copper for the fusible elements. Silver does not oxidise as easily as copper at the fuse operating temperatures. Fuses with silver or silver plated copper fusible elements do not, deteriorate in service and their characteristics remain stable.

Replacement of copper-wire fuses by silver-wire fuses solved the problem of fuses melting at normal operating currents. Although these fuses are more expensive the savings due to prevention of signal failures and the longer life of the fuses justify the slight increase in initial cost.

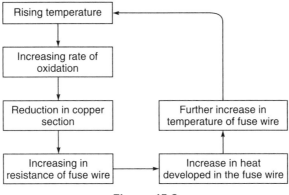

Figure 45.2

45.9 Fuse Failures due to Bad Contact between the Fuse Cap and Fuse Carrier Clips

CASE STUDY

In one installation, there were instances of even non-deteriorating, silver-wire fuses failing in service in failure mode (a) referred to in Section 45.2. Tests on cartridge fuses of the same age group did not reveal any degradation in the fuse characteristics. It was, however, noticed that there was repetition of the fuse carrier location in the failure statistics. Careful checking of those fuse carriers showed that the fuse cartridges were rather loose in the fuse carrier clips. Further checking of the dimensions and contact forces revealed deficiencies in the same. The voltage drop at these contacts was excessive and the temperature rise was also excessive. The high temperature of the cartridge fuse caps led to the shifting of the fuse characteristics to the left and down causing them to melt at normal currents. See Figure 45.3.

The pull out force for the test plug was less than the required level. Note that the test plug is to be pulled out in the axial direction to assess the true contact force indirectly. The pull out force in a direction at right angles to the axis—which is the force required to be applied to remove the fuse from the carrier—is not relevant to the question of contact force between the cartridge and the carrier.

Replacement of the defective fuse carriers solved the problem.

There was a similar problem of fuse failures in mode (a) of Section 45 .2, where the over-heating of the fuse was due to heat conducted from a bad terminal joint discussed in greater detail in Chapter 39. The solution in this case was similar: i.e., to solve the problem of overheating at the source of heat viz. the terminal defect.

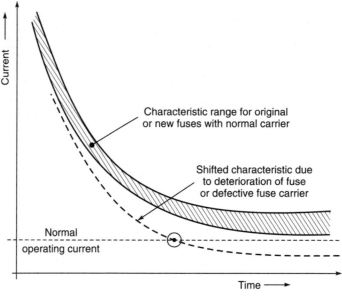

Figure 45.3 Shifting of fuse characteristics

45.10 Failure of MCB to Trip on Overload

CASE STUDY

There was an incident in one industry, where there was a fire in the control circuit wiring. It was found that the MCB did not trip despite a heavy current through it due to welding of the internal contacts that were meant to open the circuit. The MCBs had complied with all the specified routine tests. The manufacturer of the MCB claimed that the welding of the main contacts was due to current exceeding the rating of the MCB.

Twenty new MCBs of the same type and make as the defective one were connected in series and the rated current passed through them for about ten days continuously, and the mV drop across each MCB was monitored every eight hours. While the mV drop increased initially in all MCBs, it stabilised after a few days in most of them but in two MCBs it continued to rise day after day and was still rising when the test was terminated after ten days. The plastic casings of these MCBs in question were cut open carefully to expose the main contacts from one side without affecting the MCB mechanism. The test was repeated and test results were observed. The test was continued for ten more days and the fixed contact temperature was measured with the help of an infrared radiation thermometer. It was confirmed that the main contact temperatures of these two MCBs were not only higher than 90 °C but also continuing to rise after 20 days of operation at rated current. The temperatures of the contacts in the18 MCBs that had not registered a rise in the mV drop, was less than 45 °C.

From the above series of tests it was concluded that the contact force between the main contacts was inadequate in *some* of the MCBs and this was the cause of rising mV drop and rising temperature of the main contacts. This process must have led to the welding of contacts on one MCB that was involved in the case of the fire in the control wiring.

The particular make of MCB was removed from the list of approved suppliers. All MCBs of that make were discarded and replaced. Similar tests were conducted on other makes on a number of samples according to a sampling plan. Only those, which showed no degradation, were retained in the approved list. It may be noted that mV drop has to be measured accurately across each MCB in the series.

Most of the recommendations made above apply also to MCCBs and to HRC fuses. Their common failure modes and failure mechanisms are very similar to those discussed here. As regards HRC fuses, their special application is where the required rupturing capacity is very high due to proximity to power stations. If such fuses fail to clear the fault, i.e. are destroyed through explosion, the only remedy is to use circuit breakers of the required rupturing capacity.

45.11   Do's and Don'ts for Preventing Failures of MCBs and Fuses

- Select fuse and MCB ratings carefully to ensure that they are lower than the carrying capacity of the wiring system to be protected and higher than the normal loads to be carried.
- Select manufacturers whose MCBs or fuses have passed all the specified tests including endurance tests.
- Carry out acceptance tests as per relevant specification on random samples in accordance with approved sampling plan. In critical applications carry out routine tests also on bulk supplies.
- As far as possible use fuses with silver or silver-plated elements. If fuses with copper elements are unavoidable, replace them periodically before they become unreliable.
- If MCBs are used, ensure that they are mounted in dust-proof but roomy enclosures. Carry out 1-hour tripping and 1-hour non-tripping tests periodically at the appropriate currents.
- When correctly rated and tested fuses blow or MCBs trip, consider the possibility of overheating due to bad contacts in fuse carriers and external connections.

Chapter 46

Bus-bar Failures

In this chapter we learn

- *Applications where bus-bars are used for electrical connections.*
- *Four common failure modes, and the mechanisms of failures for each failure mode.*
- *Case studies of bus-bar failures.*
- *Measures for preventing busbar failures.*

❑ ❑

46.1 Introduction

Bus-bars are usually rectangular, sometimes tubular, bare copper bars, supported on insulators, in enclosed chambers and connected with threaded fasteners to other bus-bars or cable sockets. Tubular bus-bars are used where greater mechanical stiffness or corona resistance is needed. Where a number of machines or equipment are located side-by-side, this method of making electrical connections is often the neatest and also the most economical.

Bus-bars are also the preferred method of making electrical connections when the currents are very high as in electro-chemical equipment, substations, large or high power machines, etc.

46.2 Failure Modes

The common failure modes, generally in order of importance, are as follows:

(a) Overheating at bolted joints between busbars or between cable sockets and busbars.
(b) Fracture of bus-bars.
(c) Failure of support insulators.
(d) Shorts between bus-bars.

46.3 | Failures of Bolted Joints on Bus-bars

Failures of bus-bars are relatively rare in comparison with other electrical equipment failures but amongst the few failures of bus-bar systems, failures of bolted joints are most common.

The contact resistance between two bus-bars that are bolted together depends on the force exerted by the fasteners and not on the contact area. There is an inverse relationship between the contact resistance and the force as shown in Figure 46.1.

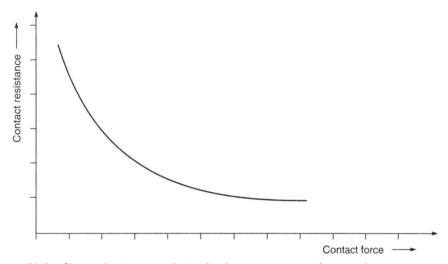

Figure 46.1   Shows the inverse relationship between contact force and contact resistance

The heat developed in and the temperature rise of the joint is directly proportional to the contact resistance. If the joint temperature rises to 90 °C due to either defective design or inadequate tightening of fasteners, the oxidation rate of the contact surface becomes significant and this then starts off the double vicious cycle shown in Figure 46.2. When the temperature rises beyond 140 °C, increased creep adds to the problem of increasing contact resistance.

This process is slow but inexorable. It can be stopped by tightening of the fasteners if the design itself is not defective. In the latter case the design would have to be modified. If the bus-bars are allowed to continue in service, it can only lead to overheating, melting of the joint, perhaps even fire.

Even when the fasteners are fully tightened initially during installation, design defects such as the interposition of materials that can creep or shrink in the joint can lead to relaxation of joint force and eventual failure. Case study 46.4 will make this clear.

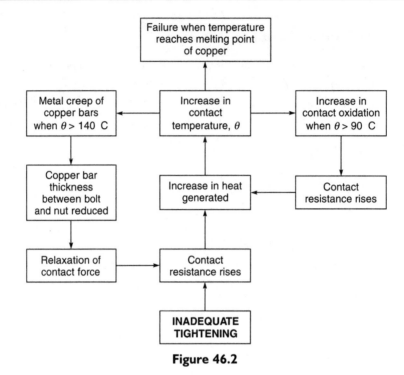

Figure 46.2

46.4 Failures due to Shrinkage of Laminates in Fastener Force Circuit

CASE STUDY

In one particular installation, laminated insulated bars as shown in Figure 46.3 were used as supports for the busbars.

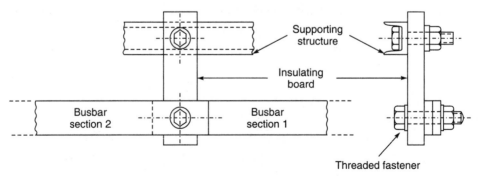

Figure 46.3 Shows a defective design of busbar joint including an insulating board

A cost saving design turned out to be the root cause of an expensive failure. The laminated bar developed shrinkage under the effect of the force exerted by the fastener and this led to relaxation of the contact force between the bus-bars. The failure mechanism described in Section 46.3 became active and finally (more than two years after installation), the bus-bar joint failed due to over-heating and melting. The remedy was simple. The replaced bus-bars were supported through a separate hole adjacent to the joint, with another bolt and nut. The fastener used for the bus-bar would not then be affected by the shrinkage of the laminate. All other similar locations were also modified. See Figure 46.4.

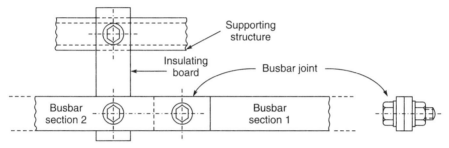

Figure 46.4 Shows modified arrangement of the joint shown in Figure 11.3. There is no insulation in bus-bar joint

46.5 Failure due to Inadequate Bolt Size

CASE STUDY

In this case the failure mode and the failure mechanism were the same as in case study 46.4 but the root cause was different. It became clear after investigation that the bolt size used was too small and the force exerted by the bolt was not adequate. As a result, the contact resistance was too high; but nothing untoward happened at first because the current through the joint was initially much less than the rated capacity. However, in the course of several years, the loading increased due to the connection of additional loads. The temperature of the joint must then have exceeded 100 °C when the ambient temperature was about 45 °C. This then started the degradation process described in case study 46.3 and culminated in failure of the joint. The action taken was to replace all the bolts by larger bolts, to tighten them fully, and to verify that the temperature rise at the joint under full load conditions was less than 25 °C. See Figure 46.5.

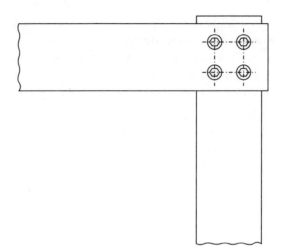

Figure 46.5 Busbar joint with fasteners of inadequate size. The force exerted is too small, contact resistance is too high

46.6 Failure due to High Ambient Temperature

CASE STUDY

In yet another case the basic failure mechanism was the same as above but it was initiated by a different phenomenon. The busbar joint was inside an oven where the operating ambient temperature was nearly 150 °C. At this temperature copper is subject to metal creep. See Figure 46.6.

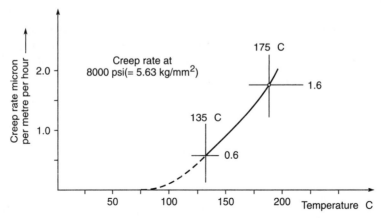

Figure 46.6 Measured creep rates of tough pitch, high conductivity copper above 135 °C

As a result of metal creep, under the effect of stress and temperature above the creep temperature limit of 140 °C, the copper bus-bars and sockets took a permanent compression set. This caused loss of the tension in the bolt and the compressive force between the socket and the busbars. The vicious cycle shown in Figure 46.7 was set into action and the joint eventually failed due to over-heating, arcing and melting. In this process, the high operating temperature (above 200 °C) caused the oxidation rates to be very high and this speeded up the failure mechanism. A double vicious cycle was set up as shown in Figure 46.7.

The vicious cycle shown in Figure 46.2 is slightly different from that in Figure 46.7. In the former the starting point is the inadequate tightening of the fasteners. In the latter the starting point is the high ambient temperature and creep starts at once along with high oxidation rate.

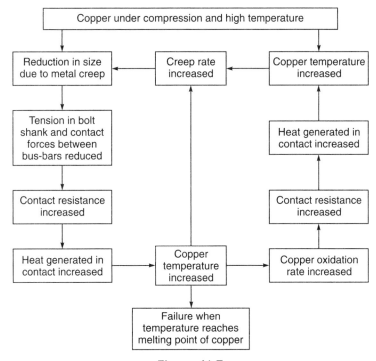

Figure 46.7

The action taken was:

(a) Immediately to replace the bolts by longer bolts, to insert disc springs as shown in Figure 46.8, to clean the joint surfaces and tighten the bolts/nuts every six months, and

(b) As a long-term measure, to replace the threaded fastener joints by brazed joints at all locations where the ambient temperature was higher than 50 °C.

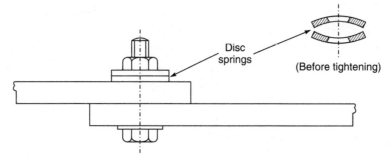

Figure 46.8 Use of disc springs to compensate for creep in busbars operating at 150 °C

46.6 Bolted Joint Failures due to High Temperature

CASE STUDY

In the 1940s it was the normal practice to use threaded fasteners for connections between the field coils of dc motors and generators. The field coil insulation was Class B and the coil temperature rise was generally less than 60 °C and the joint temperatures were less than 80 °C. There was no problem with the bolted joints between the field coils.

Thirty years later, when class H insulating systems were introduced and traction motor power/weight ratios were rising, the conductor temperature rise in the field coils rose to about 120 °C and the joint temperatures rose to about 140 °C. Occasional failures of these joints ascribed to metal creep and high oxidation rates of copper, led to the replacement of bolted joints by brazed joints for the connectors between field coils.

46.7 Fractures of Bus-bars

Copper is a ductile and reasonably strong material. Fractures of busbars in the straight portions are very rare. There have been, however, some cases of busbar fractures at bends.

When a copper rod or bar is bent, the metal on the outer radius is stretched while that on the inner side is compressed as shown in Figure 46.7. The central part of the bar is neither stretched nor compressed. The percentage stretch is given by the expression:

Per cent stretch on outer radius = $100/(2m+1)$, where m is the ratio of the inner radius to the thickness of the bar.

The values of per cent stretch for different values of m are given in the Table 46.1.

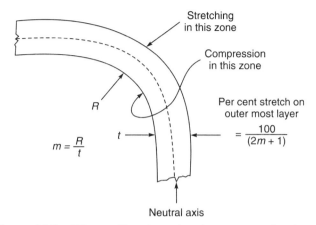

Figure 46.9 Effects of bending a bus-bar on its surface layers

Table 46.1

RATIO OF INNER RADIUS TO THICKNESS OF BAR	PER CENT ELONGATION OF THE BAR ON THE OUTER SURFACE
0.5	50
1	33
2	20
3	14
4	11

There is a similar compressive effect on the inner radius. The deformation of the metal in these proportions has very adverse effects on the structure of the material. In particular the stretching is very deleterious. Cracks develop on the surface. Usually they are micro-cracks not visible to the naked eye. There is stress concentration at such cracks if there are alternating stresses, and fatigue cracks can develop and finally cause fractures.

46.8 Bus-bar Failures in Transformers due to Sharp Bends

CASE STUDY

In a 4 MVA power transformer on an electric locomotive the secondary winding had many sections, which were paralleled on heavy copper busbars. Investigation of a transformer failure showed fracture and melting of a busbar at a bend. Examination of other transformers of the same make and same type showed a very sharp bend with an inner radius of about 0.5 the thickness of the busbar. See Figure 46.10.

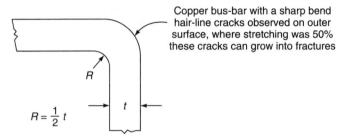

Figure 46.10 Cause of fracture of bus-bar at bend

In one case there was a visible crack at the bend. The bus-bar was well secured and mounted on insulators. Due to the alternate heating and cooling with varying currents through the bus-bars, there would have been alternating tensile and compressive stresses. These must have led to propagation of fatigue cracks and eventual fracture of the bus-bars.

The action taken was to replace all such bus-bars, giving priority to those which had visible cracks, by new ones with a bend radius = three times the thickness of the busbar.

46.9 Bus-bar Failures due to Vibration

CASE STUDY

In an unusual case, copper bus-bars in parallel were mounted on insulators as shown in Figure 46.11. The currents were very large in magnitude and alternating at 50 Hz. The busbars fractured through the hole where they were mounted on the insulators. The fracture occurred while the busbars were carrying high currents. There was arcing and melting at the fracture.

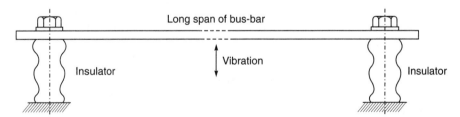

Figure 46.11 Busbar failure due to resonant vibration

On investigation it was found that the current density in the copper bars, even considering the reduced cross-section at the holes, was within permissible limits and the maximum temperature rise at the support insulators was also less than 30 °C. However the particular span between the supporting insulators seemed to be significantly larger than in other parts of the system.

Careful observation of the bus-bars at a similar location in another part of the installation showed that the bus-bars were vibrating and that the amplitude of vibration was changing regularly. Calculation showed that the resonant frequency of the bus-bar as an elastic beam was very nearly equal to 100 Hz. This led to the conclusion that the bus-bars were vibrating due to resonance between the forces of attraction between the two parallel bus-bars carrying heavy alternating currents and the natural frequency of the bus-bars. The action taken was to provide additional insulating supports in between the long spans that were prone to vibration.

46.10 Bus-bar Failures due to Thermal Stresses and Fatigue

CASE STUDY

In a rolling mill motor the interconnections between field coils were made with thick copper bars brazed between the coil terminations and going around the stator. These bars were covered with insulating tape and clamped to the stator through insulated clamps. There were a few cases of insulation failures and melting of the bars under the clamps and it initially seemed that these failures were caused by failure of insulation. However, calculations showed that the strength of the insulation was very high and tests made on some old machines showed little degradation of the breakdown voltage at the clamps.

The temperature rise of these connectors was measured in some of the old machines that had collected over the years thick deposits around these connectors. It was seen that the connector temperature was not only very high (about 70 °C) but also fluctuating under the intermittent duty of the mill operation. Opening out these connectors of a few old machines and carrying out dye penetrant tests revealed fatigue cracks under some of the clamps.

The failure mechanism was concluded as follows: The constant variation in the temperature of the connectors which were physically restrained from free expansion and contraction was producing alternate tension and compression in the copper bars, of a magnitude close to the yield point of annealed copper.

The Young's Modulus of copper is approximately 12500 kg/mm^2 and its coefficient of thermal expansion/contraction is 16.5×10^{-6} per °C. Therefore, if copper is restrained from expansion or contraction, the thermal stress is approximately 0.2 kg/mm^2/°C (The product of the two parameters just mentioned.) This shows that if the temperature variation is about 50 °C, the thermal stress resulting from physical restraint on expansion or contraction could be 10 kg/mm^2 which is close to the yield point of copper and higher than its conventional endurance limit corresponding to 5×10^7 cycles.

It was therefore concluded that this process or failure mechanism was leading to fatigue failures of the copper bars. When the final break occurred, the arc caused the insulation to be damaged and the failure appeared to be an insulation failure. The action taken was:

(a) To remove all the deposits around the connectors by high pressure, hot water jets followed by thorough drying out, and

(b) To replace all the connectors between the field coils of similar machines by connectors with an increase in cross-section by 20 per cent. Priority was given to those machines, which had a higher loading, and were in service for a longer period.

46.11   Failure of Support Insulators

CASE STUDY

Failures of bus-bar support insulators can be like any other insulators and these are discussed in Chapter 49. A reference may be made here to one particular type of insulator fracture caused by stresses produced by the busbars.

As already explained in Section 46.10, temperature variations in busbars generate very high stresses in the busbars if they are prevented from expanding/contracting freely. The force produced by a constrained bus-bar of size 50 mm × 10 mm, operating with a temperature rise of 50 °C, can be nearly 5 tonnes on a supporting insulator and this can cause the insulator to fracture. It is desirable therefore to provide expansion loops or flexibility in supports to allow the bus-bars to expand and contract freely.

46.12   Short-Circuits between Bus-bars

Bus-bars are usually made of bare copper and the insulation between bus-bars at different potentials is provided by air. Air insulation failures are very rare but shorts between bus-bars are not. Shorts are usually caused by birds or pieces of wire or moist twigs carried by birds, in out-door bus-bars.

Snakes, lizards and other vermin have caused failures in indoor installations. Provision of vermin proof enclosures at the design stage is the starting point of efforts to prevent such failures. Maintenance of such enclosures is equally important. Special attention should be paid to such matters in regard to unattended substations, particularly when they are situated in forests and uninhabited areas.

46.13 Short-Circuit between Bus-bars due to Vermin

CASE STUDY

In one railway traction switching station, situated in a forest, every possible precaution was taken to prevent entry of vermin. The floor was concreted, the entire area was fully enclosed, and all ventilation openings were sealed with fine wire mesh. Despite all these precautions, a snake caused a short-circuit. The incident occurred a day after maintenance staff had visited the switching station for monthly inspection. It was concluded that the snake must have entered through the main door left open for ventilation while the maintenance staff were working. Action taken was to install wire mesh doors inside the main doors, so that they could be kept closed during the maintenance visits.

46.14 Do's and Don'ts for Preventing Bus-bar Failures

- Ensure that the operating temperatures of all bus-bar joints with threaded fasteners are below 75 °C, when carrying the maximum current.
- Ensure that all joints with threaded fasteners are made with steel fasteners, with direct contact between the conductors being joined and all fasteners fully tightened. Provide adequate locking arrangements if the bus-bars are subject to vibration.
- Provide expansion joints or resilient mountings on heavy busbar systems to prevent excessive thermal stresses on insulating supports.
- Do not use bolted joints on bus-bars operating at high temperatures (>90 °C). Use brazed or welded joints.
- Do not bend copper bars with bend radius less than 3 times the thickness of the busbar.
- Ensure that the bus-bars are not allowed to vibrate due to electromagnetic forces by providing adequate number of insulating supports.
- If there are possibilities of vermin causing short-circuits between bus-bars, mount the busbars in well ventilated enclosures with wire mesh covered openings.

Chapter 47

FAILURES OF LEAD ACID BATTERIES

In this chapter we learn

- *Reliability with reference to durability of batteries.*
- *Routine checks when batteries are replaced.*
- *General measures for prolonging the life of batteries.*
- *Operating practices for prolonging the life of batteries.*
- *Maintenance practices for prolonging the life of batteries.*
- *Case studies involving other causes of battery failures.* ❑ ❑

47.1 Introduction

ƒailures of lead acid batteries are often treated as durability problems since these batteries have a limited life of three to five years. If the battery requires replacement after two years, the action taken is, merely to replace the battery. It is not easy to determine the true causes of such premature failures and the maintenance staff have many other problems to deal with.

Lead acid batteries are expensive items and it is possible to get improved life by taking a few simple precautions. While those directly connected with maintenance may not be motivated to investigate each premature replacement, it is desirable for higher management in large organisations to monitor the ages of batteries being replaced. The management should establish norms based on service conditions and organise routine studies on all batteries, which have to be replaced prematurely.

In this chapter we will discuss the practical measures taken to prolong the life of lead acid batteries and the checks made as a routine for replacements.

47.2 Routine Checks When Lead Acid Batteries are Replaced

(a) The number of years and months service rendered by the battery before replacement should be determined from the records. It is better also to mark indelibly on the battery container itself, the date of its commissioning, so that the ages of batteries being replaced can be seen instantly. This is an obvious and essential step but it is often neglected.

(b) If the age is less than the established norm, the following checks should be made immediately.

- The electrolyte level in each cell should be measured.
- The cells should be topped up to the correct level.
- 24 hours later, the battery should be charged fully and specific gravity in each cell should be carefully measured.
- The voltage of each cell should be checked first on open circuit and then again while delivering amperes equal to one tenth of the Ampere-hour capacity.
- Samples of electrolyte should be analysed for the impurity levels.
- If there are any dead cells or open circuit cells they should be opened out carefully to determine by visual examination the causes of such defects.
- The quantity of sludge collected at the bottom of each cell should be measured.

(c) It is usually possible to determine the cause of premature replacement of the battery on the basis of the above preliminary checks. Batteries that require premature replacement should be sent back from time to time to the manufacturers for their analysis and comments.

The effort involved in such analysis is very small when compared with the possible savings in battery replacement costs.

47.3 Measures for Prolonging the Life of Lead Acid Batteries

The life of lead acid batteries depends on the quality of the design and manufacture, i.e. the make or brand of the battery. It is desirable always to compare the life obtained with different makes under similar conditions of service and quality of maintenance. For this purpose, a register should be maintained indicating the make and type of battery, date of commissioning, dates of topping up with distilled water, quantity of distilled water added, dates of equalising charge, and finally date of replacement. It is even more desirable to continue the same register to indicate the routine measurements and observations on prematurely replaced batteries as suggested in Section 47.2 above.

The above details may be maintained in cards. These data are very useful, if reviewed periodically, for:

(a) comparing the durability of different makes.
(b) detecting any defects in maintenance practices.
(c) taking steps for prolonging the life of the batteries.

The durability data of different brands should be taken into account while determining the distribution of orders to the suppliers. The order should be divided between the top two or three makes classified according to average life obtained. The effects of this policy on quality of supplies received are very slow but they have long-term benefits.

The life of lead acid batteries is greatly influenced by the temperature of the battery electrolyte. The higher this temperature, the lower is the life. It has been estimated that the life of the battery is reduced by half for every 8 °C increase in the operating temperature of the battery. This is a manifestation of Arrhenius' Law applicable to all chemical reactions. After all, the operation of the battery, either in the charge mode or in the discharge mode, involves a chemical reaction between the electrolyte and the materials used in the construction of the positive and the negative plates. Even the 'local action' that takes place, when the battery is neither under charge nor under discharge and which is responsible for the degradation of the battery involves chemical reactions. Therefore it is advisable to reduce the operating temperature of the battery to as low a level as practicable.

In any given installation, the electrolyte temperature depends on (a) the ambient temperature; (b) the dissipation of heat from the battery to the surroundings; (c) the charging and discharging rates.

The ambient temperature here is the temperature of the air surrounding the battery. This may be a little higher than the general atmospheric temperature if the batteries are housed in closed boxes. The battery boxes should provide for the free convection of air. This factor is connected with the design and installation of the battery.

The fastest returns on action taken are in regard to operation and maintenance practices. These are discussed in the next section.

47.4 Operating Practices for Prolonging the Life of Lead Acid Batteries

As far as operation is concerned the only control which the maintenance organisation has is on the charging current. The charging equipment should be suitable for automatic regulation of the charging current. It should be so regulated that it is never too high but every cell gets fully charged. Excessive charging not only increases the average temperature of the battery but also increases the consumption of distilled water and plate degradation. If the battery is on float charge the best way is to provide regulated constant voltage at the battery charger terminals.

The discharge current depends on the load requirements and is generally beyond the control of the maintenance organisation.

47.5 Maintenance Practices for Prolonging the Life of Lead Acid Batteries

The most important part of battery maintenance is the electrolyte level and its quality.

The electrolyte level should be maintained within the prescribed limits as far as practicable and never allowed to go below the top level of the plates. The level should be constantly monitored and distilled water added as needed to bring the level back to the maximum limit. The design of the battery box, accessibility of the cell tops, and the equipment for checking electrolyte levels and refilling of distilled water are very important. If these are unsatisfactory, maintenance is likely to deteriorate.

As far as quality is concerned, it is necessary to verify by periodic random checks at the point where batteries are topped up, that the samples of water taken have the required purity. This should be verified by actual chemical analysis. The impurity levels should not exceed the limits specified in the standards for battery topping-up water. Such checks must also be made according to approved sampling procedures at the point where supplies are received.

Copper, iron and chlorides are some of the impurities that are very harmful and these are also around in different forms in any industrial environment. Care must be taken to ensure that they do not find their way into the battery electrolytes. In railway batteries, brake block dust can be a source for iron entry into the cells. Analysis of the electrolyte from the cells of batteries that are prematurely replaced is important for checking on the possibility of such impurities being the cause of premature failure.

47.6 Lead Acid Battery Failures due to Other Causes

CASE STUDIES

The following list of battery failures due to causes other than impure top-up water and high operating temperature are taken from many case studies. Many of these causes are trivial but true. It is necessary to take special steps to train the staff to avoid them.

- A locomotive battery, provided for starting the main diesel engine, was found to be totally discharged when required. It was not a true battery failure as it was caused by a defect in the battery charger. There was a delay in starting the locomotive.

  The battery charging circuit must include a charge/discharge ammeter which should be verified by the driver from time to time to ensure that the battery is being charged when the locomotive is in service.

- A similar case occurred in an electric locomotive where the battery is required for raising the pantograph and for the starting contactor of the 1-phase to 3-phase Arno convertor.

The battery got discharged because certain lights were not switched off before the locomotive was stabled. This case is also not a true battery failure. There was a delay in starting the locomotive.

- The battery seemed to be 'dead' when trying to start the locomotive. This was due to a bad contact between the positive terminal post and the connector clamp. The clamp faces were touching each other instead of having a clearance. See Figure 47.1. The contact pressure between the terminal post and the clamp was totally inadequate. Bad contact is also possible due to 'creep' of the lead pillar.

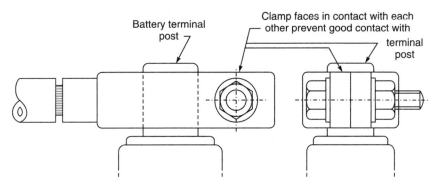

Figure 47.1   Shows how trivial defect in clamp causes failure of main contact

- In another case, the battery seemed to be 'dead' and investigation showed that the entire electrolyte had drained away from a leaky cell. The cause of this leakage was a crack in the cell container and obviously a manufacturing defect.
- In yet another similar case, the cause of a 'dead' cell was a fracture inside the cell between the terminal post and the horizontal paralleling strap supporting the positive plates. This was also a manufacturing defect—a defect in the lead burning process. The only way to prevent such failures due to manufacturing defects of this type is to involve the manufacturers in the effort by imposing penalties for such failures. It is not practicable to detect such defects at the inspection stage during purchase.

Many battery failures are caused by battery charger defects that remain undetected until it is too late. In un-attended substations and switching stations, protective relays should be provided to detect drop in battery specific gravity of a pilot cell and to send an alarm signal to the control centre.

As regards battery charger failures are concerned, it may be recalled that these are generally due to failures of components such as capacitors, resistors, terminals, fuses, solenoids, relays, etc. These failures have been discussed in detail in Chapters 32 to 65.

47.8   Do's and Don'ts for Preventing Battery Failures

- Buy batteries from approved suppliers and maintain records of life obtained from each battery.
- Open out all dead cells to determine the true cause of the defect. Where manufacturing defects are found return the cells to the manufacturer for free replacement.
- Take special steps to ensure quality of distilled water used for topping up. Ensure that adequate supplies are stocked at the maintenance depots. Ensure that electrolyte levels in all cells are kept within limits.
- Monitor the specific gravity of electrolyte of pilot cells and cell voltage on all cells during periodical maintenance.
- Investigate quality of water drawn from cells that have had to be replaced prematurely, on the lines indicated in Section 47.2 (b).
- Take steps to keep the battery temperature as low as possible while undergoing charge/discharge cycles in service by controlling the charging rates and providing adequate ventilation.
- Ensure that the battery chargers are reliable and that they are designed to regulate the charging current suitably.

FAILURES OF INCANDESCENT LAMPS

In this chapter we learn

- *Applications where incandescent lamps are in use at present.*
- *Construction and operating parameters of incandescent lamps.*
- *Failure modes and failure mechanisms in incandescent lamps.*
- *Measures that can be taken to minimise incandescent lamp failures.*
- *Condition monitoring possibilities in regard to incandescent lamps.*

$\square\ \square$

48.1 Introduction

*I*ncandescent lamps are being used for applications where the source of light has to be very small in size to enable the use of parabolic reflectors to produce long and parallel beams of light. Some examples of these are: automobile and locomotive headlights, search lights, portable or hand held flashlights, projector lamps, etc. For general lighting applications, incandescent lights are no longer the preferred option but they are still being used, because of their low initial cost for domestic and commercial lighting, despite their lower energy efficiency and lower life. Fluorescent lights are replacing these but not as fast as may be expected.

Incandescent lamps are of many different sizes and shapes. They all have one common feature of construction. The source of light is always a small filament, i.e. a length of fine tungsten wire, which is heated to white heat by the passage of electric current through it. The filament is housed in a transparent or translucent glass envelope, which is either evacuated or filled with an inert gas which is usually nitrogen.

Tungsten is used for the filament because of its very high melting point, which is 3410 °C. (Compare MPs of Aluminium 660 °C, Copper 1083 °C, Iron 1535 °C).

The operating temperature of the filament is usually in the range 2500 °C (for a 60 W lamp) to 2700 °C (for a 1000 W lamp). At these temperatures about 7 to 12 per cent of the input energy is converted into visible light energy.

At the operating temperature of the tungsten filament, a small but significant amount of metal gets continuously sublimated, i.e. converted from the solid state directly into the vapour phase. This then gets deposited on the inside of the glass envelope or on other internal parts of the lamp. All incandescent lamps have a limited life which is, on an average about 1000 hours with a standard deviation of about 200 hours.

The designers and manufacturers of the lamps try to maximise the average life and minimise the variance in the life by making various apparently minor improvements in the constructional details and manufacturing processes.

The life of an incandescent lamp is very sensitive to voltage fluctuations. Thus, a 5 per cent increase in voltage reduces the life by about 50 per cent and a 15 per cent increase in voltage reduces it by about 85 per cent. Due to the variations in the supply voltage and also to variations in the lamp construction, the actual life obtained from different lamps in an installation varies within very wide limits.

48.2 Failure Mechanism of an Incandescent Lamp

While the average temperature of the tungsten filament is usually in the range 2500 to 2700 °C, the actual temperature of the filament varies from point to point. The cross-section of the wire is not quite the same at every point due to normal manufacturing variations. Where the cross-section is less than the average cross-section, the operating temperature is more than the average temperature and at such locations, the rate of sublimation of the wire is higher. This

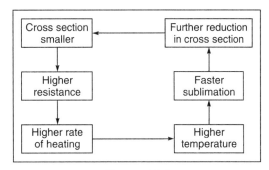

Figure 48.1

reduces the cross-section and thereby starts a vicious cycle that ends in fusing of the lamp as shown in Figure 48.1.

Eventually, the temperature of the filament at certain points becomes so high as to cause the filament to melt at such points. This usually occurs while switching the lamp on.

The cold resistance of tungsten filament lamp is in the range 8 to 15 per cent of the resistance at the operating temperature. Hence the power input to the lamp at the instant of switching on is 12 to 7 times the rated power of the lamp. The resistance rises and the power falls to the rated operating value within less than a tenth of a second. The temperature differentials between different parts of the filament are higher during these momentary starting current surges than during steady state conditions because conduction along the filament takes some time to have a significant effect. As a result, the conditions at which the filament actually breaks up are usually reached during such current surges. Most cases of incandescent lamp failures occur when the lamp is being switched on.

The statistical distribution of failures in a large lot of incandescent lamps, when operated at the rated voltage is somewhat like the following. If the operating voltage is different from the rated voltage, the distribution would also be very different but the general pattern would be similar. This table helps us to determine a suitable schedule for premature replacements of incandescent lamps, in order to minimise failures in normal service.

Life obtained, range in hours	Percentage of lamps that burn out in the range
less than 200	0.2
201 to 400	0.5
401 to 600	2.2
601 to 800	15.1
801 to 1000	41.3
1001 to 1200	27.5
1201 to 1400	11.4
1401 to 1600	1.8

From the above distribution of life as obtained with regulated voltage, it is seen that nearly 60 per cent of the lamps fail in less than the average life of about 1000 hours. However, if all lamps are replaced after 600 hours of use, the failure rate can be reduced to 5 per cent. This 'premature' replacement of lamps is not economical when the lamps are used for illumination and lamp failure has no serious repercussions due to the availability of other lamps in parallel.

The lamp replacement strategy has to be adjusted to suit the needs. In a domestic situation, or even in a small commercial establishment, the obvious method is to maintain ready stocks of spare bulbs, provide extra lamps in each room and to replace bulbs only when they actually burn out. In large workshops, where replacement of bulbs involves labour costs, it is economical to prepare a suitable group replacement plan to minimise the overall cost. In a 'Group Replacement Plan', lamps are replaced in groups when any one lamp in the group burns out after a specified time limit. This results in premature replacement of some lamps but there are savings in labour costs.

48.3 Measures to Minimise Incandescent Lamp Failures

Since incandescent lamps may burn out after a life that may vary within wide limits between 200 hours and 1600 hours, such burn-outs are not treated in the way other component failures are treated. Where illumination is essential several lamps are usually installed so that the burning out of one lamp has negligible effect on the overall illumination. This allows the maintenance organisation to replace the burnt lamps within a reasonable period. In applications like headlights or projector lamps, the user maintains a stock of spare lamp bulbs for instant replacement.

In certain applications such as railway signal lamps, burning out of lamps while in service constitutes a failure, which causes delays to trains and possibly some measure of partial responsibility for accidents. In such applications the method adopted is to provide a redundant filament inside the lamp itself which gets automatically switched on when the main filament burns out.

There are several measures that can be taken to minimise burning out of incandescent lamps in service. They all entail additional costs and should be adopted only when lamp failures in service are unacceptable. These are as follows:

(a) Carry out life tests on random samples drawn from each lot of supplies received to ensure that the relevant clause of the specification is fully complied with. Control and stabilisation of the supply voltage during the test is very important.

(b) Screen all lamps at 110 per cent of the nominal voltage until 5 per cent of the lamps tested burn out, or for 10 hours whichever comes first. This will prevent 'weak' lamps from going into service.

(c) If the supply system voltage is prone to excessive rise during certain periods, provide voltage regulators in the supply system to ensure that the lamp voltage does not rise beyond 102 per cent of the nominal voltage.

(d) Replace all lamps that have been in use for more than 600 hours, as a routine. Although this means premature replacement of some lamps, the reliability of the whole group of lamps increases.

(e) Where the lamp utilisation varies from location to location, maintain location-wise records of lamp life, and refine the replacement interval from 600 hours either upwards or downwards as may be considered necessary on the basis of statistical analysis of the life obtained in actual practice.

The additional effort involved in measures such as those mentioned and the cost of replacing lamps at 40 per cent below average life as above are financially justifiable in important applications like railway signal lamps if automatically switched redundant filaments are not provided in the system.

In the case of railway signal lamps, the lumen per watt efficiency of the lamp is not very important. It is therefore desirable to design the lamps with a higher wattage and reduced operating temperature of the filament in order to get the desired light output at a higher reliability and durability of the lamp.

48.4 Condition Monitoring of Incandescent Lamps

The process of filament degradation is, as explained in Section 48.3, one of progressive reduction in the diameter of the filament. Unfortunately, the degradation is not uniform and failures are due to localised degradation at higher rates at certain points only. Measurement of resistance of the filament cannot help in detecting the presence of such points on the filament.

A storage type video camera may be used to photograph the filament while a capacitor is discharged through the lamp. The size of the capacitor and the voltage to which it is charged can be determined by trial until only the hot spots appear on the screen. Comparison of new lamps with used lamps will then help to determine the limiting conditions. Such methods are of academic interest only at this time.

When many incandescent lamps are put into service, a small percentage of lamps fuse within the first ten hours. These are due to defects in the manufacture and assembly of the filaments. The only way to eliminate them is to screen all the lamps at 110 per cent of the rated voltage for about 5 hours.

Where incandescent lamps are used and lamp failure has serious consequences as in railway signal lamps, the best course is to use lamps with double filaments. Automatic relays should be provided for the changeover to the spare filament and for sounding an alarm to call for immediate replacement of the lamp. Service will be maintained in the meanwhile by the spare filament.

If the lamps are replaced after 600 hours use, the failure rate will be about 5 per cent instead of 60 per cent when the lamps are replaced after 1000 hours. This system can be used, as the second best system, where the lamps are not of the double filament type. The replaced lamps should be destroyed to prevent inadvertent reuse. The lamp replacement cost increases by nearly 67 per cent but the extra cost is still far less than the cost of signal failures in terms of delays to trains, and additional energy costs in stopping trains at signals.

48.5 Do's and Don'ts for Minimising Incandescent Lamp Failures

- Incandescent lamps have a limited life. Failures cannot be prevented. As far as possible, additional or redundant lamps should be provided to avoid total darkness when lamps fail.
- In order to increase durability of incandescent lamps, carry out acceptance life tests on samples drawn from lots. This will effectively control the manufacturer's contribution to the effort for improvement in durability.

- Ensure that the supply voltage is not allowed to increase beyond 5 per cent over the nominal voltage. If high voltage periods are common and large numbers of incandescent lamps are in use, automatic voltage regulators may become financially viable.
- In signalling applications, use double filament lamps with automatic changeover relays. If these are not provided screen all new lamps at 110 per cent of nominal voltage for 5 hours and replace all lamps at 600 hours.

Chapter 49

INSULATOR FAILURES

In this chapter we learn

- *Three failure modes of insulators viz. flash-over, puncture and fracture.*
- *Characteristics of the three failure modes.*
- *Mechanisms of failures for each of the three failure modes.*
- *Degradation of flashover strength.*
- *Six case studies with details of preventive action.*

49.1 Introduction

A few types of insulators in use on overhead lines, in sub-station structures, in electrical rotating machines, in transformers, and in any electrical equipment are illustrated in Figure 49.1. The main function of an insulator of any type would be to provide mechanical support to some conductor or equipment at a high voltage with reference to the grounded structures.

This chapter is about failures of insulators of various materials such as ceramic, mica/glass, epoxy, or any other plastic or laminated construction. There are various shapes, ratings, types and makes of insulators in use. However there are only a few failure modes and mechanisms of failure. These are:

(a) Flashover in air across the surface of the insulator, between the live and earthed ends.
(b) Puncture between the steel fittings, through the material of the insulator.
(c) Mechanical fracture of the insulator itself or of the bond between the insulator and the steel fittings.

49.2 Failure Modes

Failures of type (a) cause the least damage to the insulator. It is the air around the insulator, which suffers a dielectric breakdown. If the protective equipment acts fast enough, there is

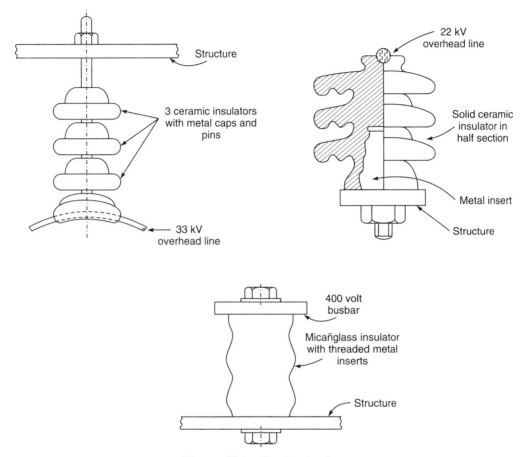

Figure 49.1 Bus-bar insulator

very little permanent damage to the insulators. The insulator can remain in service with a little cleaning after a flashover. Sometimes, in case of very high fault currents and delay in interrupting the power supply, there may be visible damage necessitating replacement of the insulator.

Type (b) failures viz. puncture through the insulators are rare because the flashover voltage of an insulator is less than the puncture voltage. In fact, to determine the puncture voltage of a ceramic insulator it is necessary during the test to immerse the insulator in transformer oil to prevent flashover occurring instead of puncture. However, puncture failures through mica-glass or epoxy insulators used indoors are not rare. They are generally due to degradation caused by partial discharges in voids or air bubbles in the body of the insulator.

In case of failures of type (b) and (c), the insulator suffers permanent damage. It has to be replaced. Type (c) failure, i.e. mechanical fracture may follow as a consequence of type (b) failure, i.e. puncture through the material of the insulator. Conversely, mechanical fracture of the insulator may result in contact between the live and grounded conductors and thereby cause a short circuit. It is necessary to examine failed insulators very carefully to determine the original mode of failure and the consequential damages before determining the root cause and its preventive measures.

Type (c) failures are of two types: Type (c_1) in which there is fracture through the material of the insulator, and type (c_2) in which the mechanical failure is in the bond material between the insulator and the steel fittings.

In some of the cases of failures type (a) and (c_2), it may be possible to repair a failed insulator, but it is generally advisable to replace even such insulators. If insulators are repaired they should undergo the same kinds of acceptance tests as the original or new insulators before they are re-installed on overhead lines or in machines.

49.3 Mechanisms of Failures

Corresponding to each type of failure there will be found in the specification a test to be carried out either on prototype or as routine tests. These tests are:

(a) Dry and wet flashover voltages.
(b) Puncture test.
(c) Mechanical tension, compression, bending and/or shear tests as may be specified depending on the normal loading conditions.
(d) Partial discharge tests on epoxy insulators.

Even when insulators are meant to be used indoors only, the dry and wet flashover voltages must be determined. There are possibilities of water film deposition (through condensation) on the surface of the insulator even inside buildings or cubicles.

If an insulator of inadequate strength, either electrical or mechanical, with reference to the electrical or mechanical stresses likely to be imposed on it, is selected it will fail sooner or later. The peak stress likely to be imposed must be calculated and an adequate factor or margin of safety must be included in the calculations to decide on the strength of the insulator to be specified. This factor or margin of safety must be allowed for two reasons. The first is to allow for some unknown or unpredictable factor that may be present in the service conditions. This is derived only from experience. The second is to allow for the possible degradation of strength that may occur in the course of service as a result of normal ageing, environment, etc.

If the strength of the insulator becomes less than the stress on it, the insulator will fail at once. Therefore insulator failures like that of any other component or material can occur mainly due to accelerated or excessive degradation that may occur in service due to any reason.

Prevention of failures of insulators, if designed or selected correctly at the outset, depends on the monitoring and prevention of the degradation of the properties of the insulator.

49.4 Degradation of Flashover Strength

The glazed surface of a ceramic insulator is generally immune to any permanent loss of strength against flashover, except as a result of damage due to earlier flashovers that may be caused by any temporary degradation of the surface. Insulators made of other materials like epoxy, rubber, plastic etc., are less resistant to the effects of ageing.

Insulators of any material including ceramic are prone to lose their strength against flashover as a result of deposits on the surface, of a variety of substances such as smoke, smog, acid rain, saline spray, dust, etc. This type of degradation is temporary. Removing the deposits and cleaning the surfaces can restore the original properties of the insulator.

Prevention of failures due to flashover is, therefore, based on only two lines of action. These are: (a) removing if possible the source of the pollution or (b) carrying out periodical cleaning of the insulators to remove the deposits before they reach levels where they can cause failures. Method (a) is a long-term solution and generally very difficult to implement. Method (b) is certainly practicable in principle but difficult in practice because it calls for shutting down the service and taking power blocks for some time. Since method (b) is the only practicable method, all the difficulties in getting power blocks must be solved by proper planning and preparation or developing hot-line maintenance practices.

The use of silicone based greases to coat the insulators often helps to increase the interval between cleaning of insulators. These compounds absorb the pollutants and offer a water-repellent surface on the outside. This helps to maintain and even increase both the dry flashover and the wet flashover voltages thereby preventing flashovers despite the deposition of some pollutants on the surface. However at some stage, the insulators have to be taken down and cleaned thoroughly, with detergents and plenty of water.

It is easier to replace insulators than to clean them in situ specially if the deposits are hard to remove. In such cases, the insulators should be taken down and replaced immediately by new or cleaned insulators. The dirty insulator taken down from line should then be thoroughly cleaned and inspected at ground level before being used for installation on the line again at some other location.

49.5 Insulator Flashovers due to Pollution

There are about 35 porcelain insulators per kilometre of electrified railway track. The operating voltage is 25000 volts AC and the dry flashover voltage is about 180,000 Volts. Despite this margin of safety on new insulators, flashovers in service are not uncommon, mainly because of atmospheric pollution by dust, smoke, moisture, salt spray, etc. As the pollution rate and type varies from place to place and season to season, it is necessary to study the conditions locally and determine the cleaning schedules.

CASE STUDY

The insulators in service in a railway traction division were divided, depending upon the rates of pollution, into three categories: (a) rapid pollution, (b) medium rate pollution and (c) little or no pollution.

Cleaning schedules were introduced as follows: every two months for category (a) every four months for category (b) and every year at the time of annual overhaul of the overhead equipment for category (c) insulators.

After implementing this type of planning the incidence of flashovers became negligible. It was necessary to keep the schedules under review and to investigate every case of failure.

49.6 Insulator Flashovers due to Partial Cleaning

CASE STUDY

Although the measures described in case study 49.5 controlled the problem, it was observed that in one section, flashovers continued to occur in the early hours of the day. When a vigil was maintained one night after midnight, it was observed that before the occurrence of flashovers, there was continuous corona discharge, but only at the grounded ends of all such insulators. On close examination of the insulators it was observed that the insulators had been cleaned only partially. Only those petticoats close to the grounded end had been cleaned fully and the others had been left without proper cleaning.

The insulators were cleaned fully using detergent first and then plain water. Cleaning of all the petticoats was ensured. Flashovers stopped completely thereafter. This experience emphasised two points. First, that keeping the insulators clean is not only necessary, but also sufficient to prevent flashovers. Secondly, the cleaning process must be thorough. If the insulator is only partially cleaned, i.e. if only a few petticoats are cleaned, it can be even more prone to flashover than an insulator which has not been cleaned at all.

This curious result is due to the fact that partial cleaning leads to increase in the voltage gradient across the petticoats, which have been cleaned thoroughly. If such insulators are tested in the laboratory, it is seen that the dry flashover voltage is much lower than that obtained on either new insulators or the old ones, which are cleaned fully. See Figure 49.2.

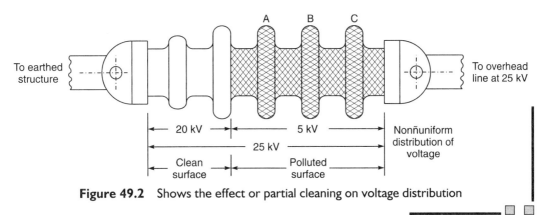

Figure 49.2   Shows the effect or partial cleaning on voltage distribution

49.7   Insulator Failures due to Defective Grouting

CASE STUDY

11 kV pedestal insulators were received as part of an imported consignment of electrical equipment. The steel pins of these insulators were grouted with lead in the ceramic insulator bodies. Lead is liable to creep and at the temperatures common in India, the insulator bodies soon became loose on the pins and even came off in some cases causing failures on the 11 kV lines.

Lead is perhaps a satisfactory material for this purpose in very cold climates but in India it is totally unsuitable. All the insulators were taken down, the lead melted out and the pins were grouted with the standard Portland cement–sand mixture.

49.8   Insulator Fractures due to Chip Inclusions

CASE STUDY

25 kV solid core insulators are used on the traction overhead lines. Some makes of these insulators have very high failure rates, the failure mode being brittle fracture across the insulator close to the end fitting. The fracture surface showed many inclusions of very small

chips near the surface. These were obviously the starting points of the brittle fractures. On deliberately fracturing (by applying excessive impact stress) insulators of some of the better makes which never failed in service, it was found that the fracture surfaces were totally free of any small chips.

The grinding and sieving process of the dry raw material powders must ensure the total absence of any chips, in the mix before water is added to form the thick 'dough' from which the insulator bodies are moulded. It must also be ensured that there are no such contaminants added to the mix during the process of mixing, kneading and moulding the 'dough'. Even air bubbles must not be allowed to form within the dough. The other precautions to be taken are:

- The coefficient of expansion of the glaze should be the same as that of the body material.
- A very tight finely ground high fired body is likely to be brittle and prone to failure under thermal shock. A coarse, low fired body may be good mechanically, but its electrical puncture strength may be poor. The best material for the electrical insulators has to be a compromise between mechanical strength, fracture toughness and puncture strength.
- The rates of drying of the insulators at their wet stage, and the rates of cooling and heating of the insulators during the firing stage are very important and must be kept low within limits.

As far as the purchaser is concerned, his safeguard is to carry out all the tests specified including heat shock test, flash-over proof test, tensile proof load test on every insulator, breakdown dielectric tests, tensile strength tests and impact breakdown tests on samples.

Statistics of fractures in service should be co-related with results obtained in acceptance tests and makes of the insulators. Porcelain insulator fractures due to poor fracture toughness cannot be verified by any non-destructive tests. It is necessary to maintain records of failure rates in service and during acceptance testing of different makes of insulators. Such data should be referred to while placing orders for new insulators.

49.9 Insulator Failures due to Corrosion of Pins

CASE STUDY

The suspension pins of suspension insulators are prone to corrode and fail. Despite galvanising of pins and use of stainless steel pins, failures of this type did occur mainly due to poor quality of galvanising and poor quality of the stainless steel.

It became necessary to monitor the condition of these pins periodically. Where signs of corrosion were seen, corrosion products were analysed to determine their causes. In order to prevent failures in service insulators with corroded pins were replaced. The complete insulator was changed to a better quality as it was not possible to change the make of the pin only. Corrosion tests were introduced on new purchases to prevent recurrence of this type of problem.

49.10   Insulator Failures due to Puncture through Epoxy

CASE STUDY

Epoxy insulators provided in a switchgear cubicle to support bus-bars failed due to puncture through the body of the insulator between the steel fittings. Partial discharge tests on other insulators of the same type and make showed varying levels of partial discharges. Insulators showing the highest partial discharge rates were subjected to continuous proof test voltages of about 50 per cent of the specified break down voltage. They failed in less than 100 hours.

All insulators were subjected to partial discharge tests and certain arbitrary limits were prescribed based on a statistical analysis of all test results. About 15 per cent of the insulators had to be replaced. The same limits were specified for new insulator purchases for the same application. The replaced insulators were used in other locations where the operating voltage was half the rated voltage.

49.11   Insulator Failures due to Birds

CASE STUDY

Some insulators flashed over and some even damaged beyond repair due to bridging of the insulators by stray wires carried by birds in their nesting seasons. This phenomenon is common in industrialised urban environments.

All kinds of measures were taken. None were totally effective, as the birds were very persistent. Some of the more successful measures were as follows:

(a) Removing the nests being built in the structures close to the insulators.
(b) Hanging small tubes just above the insulators. See Figure 49.3.
(c) *Not* removing the nests being built on Railway land in locations unlikely to affect the overhead lines or insulators, so that at least some birds stop carrying wires in their beaks.

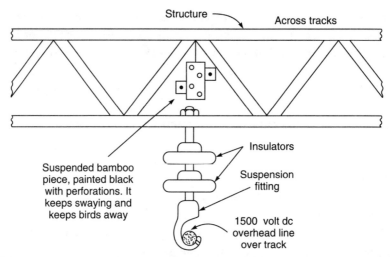

Structure — Across tracks

Insulators

Suspension fitting

Suspended bamboo piece, painted black with perforations. It keeps swaying and keeps birds away

1500 volt dc overhead line over track

Figure 49.3 Simple device to keep birds away from, 1500 volt dc insulators on railway track

49.12 Insulator Failures due to Carbon Dust

CASE STUDY

Mica glass insulators, which support brush-boxes in dc machines, are vulnerable to deposits of carbon dust from the carbon brushes. Dust in the cooling air also form deposits. These deposits reduce the flashover strength and cause failures in service.

These insulators are also subject to heavy transverse forces due to vibration of brush boxes, particularly in nose suspended, axle-hung traction motors.

In some cases, painting the insulator surfaces with anti-tracking varnishes was helpful. In some designs Poly-Tetra-Fluoro-Ethylene (PTFE) sleeves were shrunk on to the insulators to provide smooth, non-stick surfaces.

In all cases periodical and thorough cleaning of the tracking surface was essential.

The insulators must be subjected to proof load tests in the transverse direction before acceptance, in accordance with specified loads, which depend on g-levels in the motor, the mass of the brush-box and the margin of safety.

49.13 Do's and Don'ts for Preventing Insulator Failures

- Classify installed insulators based on pollution rates and adopt appropriate schedules for cleaning.

- Ensure thorough and full cleaning of insulators with detergents, since partial cleaning is worse than no cleaning.
- Carry out line patrolling in early pre-dawn hours of the morning to locate insulators with corona discharge at their end fittings. Clean or replace such insulators on priority.
- Replace lead seals by Portland cement/sand seals.
- Ensure during manufacture that there are no chips, air bubbles and other inclusions in the dough for ceramic insulators.
- Carry out tensile proof load testing on all insulators and fracture tests on samples to detect presence of chips and low fracture strengths.
- Monitor regularly, suspension insulator pins prone to corrosion and replace insulators that show signs of corrosion.
- Carry out partial discharge tests on epoxy insulators and determine limits on the basis of statistical analysis.
- Maintain detailed registers of all locations indicating: makes, dates of installations and dates of cleaning, for reference whenever there are any insulator failures.

Chapter 50

SOLENOID FAILURES

In this chapter we learn

- *Description and applications of solenoids.*
- *Failure modes and failure mechanisms.*
- *Preventive measures.*
- *Case studies illustrating five different failure mechanisms.*

50.1 Introduction

$\mathcal{A}$ solenoid is one of the simplest forms of electrical coils, the other forms being armature coils, stator coils, meter movement coils, transformer and reactor coils, and arc blow-out coils. Solenoids are generally used for electromagnets such as those in relays, contactors, circuit breakers, electro-valves, lifting magnets, etc.

Although so simple in regard to their construction, they are subject to a number of failure modes arising mainly out of non-compliance with a few simple rules of design and construction. In other words, it is possible to get zero failure performance from electrical solenoids if these rules are followed.

There are four main components of solenoids: the spool or the bobbin, the winding, the leads to the terminals, and the terminals. While 90 per cent of the initial cost is in the first two components viz. the bobbin and the winding, 90 per cent of the failures originate in the remaining two components viz. the leads and the terminals which account for only 10 per cent of the cost. Figure 50.1 shows the cross-section of a typical solenoid.

50.2 Failures due to High Voltage Switching Surges

CASE STUDY

There were many solenoid failures in an industry. The failure mode was the same in all cases. The solenoid was overheated and the top insulating layer charred. The resistance was very low and the coil drew excessive current.

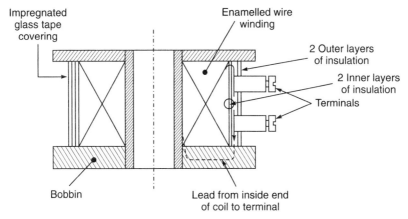

Figure 50.1 Cross-section of solenoid

On opening out a burnt solenoid very carefully layer by layer the following observations were made:

(a) The top layers of the winding were shorted and charred. On unwinding these damaged layers, the residual winding including the innermost layers was found to be intact and in good condition. This was significant because the normal operating temperatures of the bottom layers of the solenoid would have been higher than the temperatures of the layers near the surface.

(b) In a few coils with very little visible damage, it was observed that there was a short between the leads to the two terminals.

(c) Further examination showed that the lead to the inner end of the winding was brought out and run across the top surface of the winding to its terminal. The other or the outer end of the winding was connected to the other terminal. Both the terminals were placed side-by-side on the top insulating layer of the coil.

(d) There was an insulating layer between the top layer of the solenoid and the terminals but it appeared that this insulating layer was failing and short circuits were developing between the inner-end terminal and the top layer of the winding.

(e) It was also noted from the failure statistics that the solenoids that were switched on and off more frequently had higher failure rates.

(f) Measurement of the temperature rise by the resistance method, of the coil showed that it was well within the limits for the class of insulation used on the wire.

It was concluded, from an overall review of the above observations that these failures were not due to thermal degradation of the wire insulation. They were due to degradation caused by voltage stress. However the normal operating voltage between the terminals of the coil was very small (110 volts) compared with the breakdown voltage of the wire insulation and also of the thin layer of insulating cloth between the terminal and the top layer (about 7000 volts). The switching surge voltage at the terminals of the coil was then measured and found to be of the order of 5000 volts.

The mechanism of failure was as follows: every time the solenoid was switched off the inductive energy ($\frac{1}{2} LI^2$) of the coil charged up the interturn capacitance of the coil until the stored energy became equal to it ($\frac{1}{2} CV^2 = \frac{1}{2} LI^2$). A damped oscillation followed. The high voltage peaks created partial discharges in the top layer insulation between the inner-end terminal and the top layer of the coil. When the cumulative damage to this insulation reached the limit, failure occurred at normal operating voltage.

The action taken was to increase the BDV of the top layer insulation to about 11 kV. On some large coils, freewheeling diodes were fitted across the coil terminals. The strength of the insulation between the lead of the inner-end of the coil and the top layer of the coil was also increased to 11 kV.

On some makes of coils, it was observed that the coil was wound in two halves in opposite directions so that both the terminals were at the same potential as the top layer under it. See Figure 50.2. These coils had a zero failure rate.

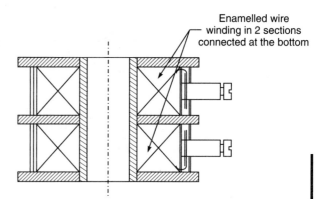

Enamelled wire winding in 2 sections connected at the bottom

Figure 50.2 Cross-section of 2-section solenoid

50.3 Damage to Copper Wire due to Acidic Flux

CASE STUDY

Some solenoid failures were due to open circuits. On opening out the solenoid carefully it was observed that the fine enamelled wire had parted at the point where it was soldered to the terminal plate. Flux residues were greenish in colour. Chemical analysis showed that the flux was unsuitable for this application. The action taken was to use rosin fluxes and to take great care in cleaning the enamel off the wire before soldering.

50.4 Failures due to Wire Damage During Winding

CASE STUDY

There were several cases of open circuits in solenoids of a particular make. Examination of failed coils revealed that the coils were wound with very fine wire in accordance with the design and some coils had developed breaks in the wire in the very first layer deep inside the coil. Inspection of the coil winding section in the manufacturer's works revealed that 5 to 10 per cent of solenoids had to be rejected during manufacture itself due to wire breakage while starting the winding. It was concluded that in some cases of solenoids, which went into service the wire, had been stretched but not to the breaking point. It was felt that the sudden rotary acceleration of the wire bobbin caused excessive tension in the wire during starting. Both the problems (viz. high breakage rate in manufacture and high open circuit rate in service) were solved by installing a soft starting device (a voltage variac) for the winding machine motor.

50.5 Damage to Copper Wire by High Operating Temperature due to Design Defect

CASE STUDY

In one particular application the solenoids in a motor starter had a high failure rate. Examination of failed coils showed that there was general overheating of the coils and the interturn shorts were deep inside the coils and not on the outer layers. Measurements, by resistance method showed that the temperature rise of the coil was nearly 80 °C at rated voltage. The hot spot temperature inside the coil would have been higher. It was also observed that the line voltages were often 5 to 10 per cent higher than the rated voltage. It was concluded that the coil design did not have adequate margin. It was seen that there was enough room in the contactor to house a bigger solenoid. The solenoid was, redesigned to operate at a lower temperature. The wire enamel was also up-graded.

50.6 Failure due to Incorrect Adjustment of Magnetic Core

CASE STUDY

In a certain factory, some of the asynchronous motors on the electric over-head travelling cranes were provided with electromagnetic brakes. The brakes were released when the motor started and applied when the motor was switched off. The brakes were clasp brakes on drums and a plunger, which was attracted into a large solenoid, effected the brake release. The solenoid was charged with an alternating voltage and the core was made of thin iron wires packed in a brass cylinder. The current drawn by the solenoid was initially high when the plunger was out; the current fell to a low value when the plunger was fully in.

The actual current in the solenoid during service was measured and found to be higher than the current indicated in the manufacturer's manual. The solenoids were operating with a temperature rise of over 100 °C. On investigation it was found that due to an error in the adjustment of the turn-buckles in the brake rigging, the plunger stroke was too small and the solenoid current was excessive. After correcting the adjustment procedure and training the maintenance staff, the solenoid currents dropped to the designed values and the solenoid failure rate became zero.

50.7 Do's and Don'ts for Preventing Solenoid Failure

- Provide adequate insulation level between inner end of terminals/leads and the top layer of the winding to withstand switching surges.
- If necessary, provide surge suppressors, capacitors or free wheeling diodes across the solenoids.
- Do not use acidic fluxes for soldering winding wires.
- Ensure, by proper design that the maximum winding wire temperature is at least 50 °C lower than the index temperature of the winding insulation.
- Ensure that the normal current in the solenoid or the voltage across it, are limited to the specified values in accordance with the design.
- Ensure that the winding wire is not stretched excessively even momentarily during the winding process. This needs to be verified particularly in the case of solenoids with wire thinner than 0.2 mm in diameter.

About Chapters 51 to 65

In chapters 51 to 65, we will discuss the failures of certain equipment or sub-assemblies of equipment where the failure mechanisms will be one or more of those described in the foregoing chapters relating to component failures; but the difference will be that the failure involves either combinations of two or more defects that would be acting in concert to cause the failure. The real or root causes of failures would be hidden by apparent or proximate causes.

For instance, in Chapter 61 we will discuss the explosion of an electrical circuit breaker due to its failure to clear a fault. One tends to think in such cases of issues such as short circuit MVA, rupturing capacity, etc. Actually the failure was due to a defect in machining the internal threads of a small component. The evidence of this defect was destroyed by the effects of the failure.

In Chapter 58, we will discuss traction motor failures of a certain type which always occurred on certain severely graded sections. This observation diverted attention to traction motor design, quality of manufacture etc., whereas the true cause was a defect in the ventilation system which could be very easily remedied.

The point sought to be made in the case studies which follow is that it is necessary always to think of all possible causes of failures. Apparent or immediate correlation should be considered with caution. Very often the same end result follows a number of possible defects.

FAILURES OF CARBON BRUSHES AND COMMUTATORS

In this chapter we learn

- *Applications and properties of carbon brushes.*
- *Failure modes in commutation and carbon brush problems.*
- *Causes of failures for different failure modes.*
- *Recommended sequence of checks when investigating carbon brush or commutation problems.*
- *Case studies.*   □ □

51.1 Introduction

$\mathcal{C}$arbon brushes are used in rotating machines to make connections between the stator and the rotor through the commutators or the slip rings. The carbon material is specially created to get the following desirable properties:

(a) High conductivity.
(b) Low wear while rubbing on the copper commutator or brass slip-rings.
(c) Low coefficient of friction.
(d) Low sparking while carrying high currents.
(e) High mechanical strength.

The chemical composition of the raw materials and the manufacturing processes are usually closely guarded secrets. The manufacturers offer a number of different grades to suit different operating and environmental conditions. The user has to select the required grade based initially on the recommendations of the manufacturer and later confirmed by tests and service trials over long periods under actual operating conditions.

Satisfactory performance depends not only on the quality of the brush itself but also on the design, manufacture, operation and maintenance of the machine. The electro-magnetic design of the machine is as important as the mechanical design of the commutator, the brush boxes and the carbon brushes. Chapters 52 and 53 about commutator failures may also be seen in this context.

This chapter will not cover the original design of the system. It will be presumed that the initial design and manufacture of the machine, the selection of the grade of the brush are all satisfactory and that brush failure is experienced on a system which has been working satisfactorily for some time. However, even when problems are experienced on new designs, the causes are likely to be one of those enumerated and discussed below, apart from several others applicable only to new designs.

If the defect lies in the electro-magnetic design of the machine the solution has to be found by the designer along with the maintenance engineer by checking out and eliminating the other possible causes which could lead to the same result. Before taking on a major review of the electromagnetic design of the machine, it is desirable to eliminate the other possible causes.

The final definitive test of the overall commutation design and manufacture of a dc motor is to carry out the black-band test, in which a small additional current is injected into the interpoles and increased until sparking is observed. The injected current is tested in both directions (i.e. aiding and opposing the motor current) in the complete operating range of the motor. The requirement is that the injected current in either direction should roughly be the same.

The most common cause of bad commutation is ovality of the surface of the commutator detected by the method shown in Figure 51.1.

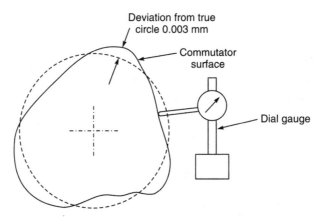

Figure 51.1 Uneven surface of commutator. In this figure the deviation from the true circle is magnified 1000 times. A diagram of this type can be prepared by measuring the deviation from the true circle with a dial gauge

51.2 Failure Modes

There are a number of common failure modes. Sometimes more than one of the following modes may be experienced simultaneously. The list of failure modes, generally in order of probability of incidence is given below:

(A) Excessive sparking, overheating of commutator, inter-turn shorts in winding behind commutator risers.
(B) Flashovers.
(C) Excessive wear on brushes.
(D) Scoring of the commutator or slip rings.
(E) Excessive wear of the commutator or slip rings.
(F) Chipping of brush edges.
(G) Disconnection of brush pigtails.

51.3 | Causes of Failures

The number of possible causes of brush failures is even greater than the number of failure modes. The same cause can produce different effects and different causes can produce the same effect. Every case may be different in regard to the combination of failure modes and causes. The following chart indicates some of the more common combinations:

Causes	Failure Mode as in Section 51.2
(a) High ovality of the commutator due to:	A, B, C, E
(i) Defective machining of the commutator and/or v-rings.	
(ii) Improper seasoning of the commutator.	
(iii) defective quality of mica segments and mica cones.	
(b) High or low bars due to:	A, B, C, F
(i) defective machining of the commutator.	
(ii) improper seasoning of the commutator.	
(iii) defective quality of mica segments and mica cones.	
(c) Mica undercut and segment chamfer defective.	A, B, C, F
(d) Incorrect quality or grade of the brush.	A, B, C, D, E, F
(e) Improper fit between the brush and brush holder.	A, B, C, D, E, F
(f) Improper angular/radial setting of brush holders.	A, B, C, D, E, F
(g) Brush pressure too low or too high.	A, B, C, D, E
(h) Shorts in stator coils.	A, B, C, D, E
(i) Shorts in rotor coils.	A, B, C, D, E
(j) Improper air gaps between pole faces and rotor.	A, B, C, D, E
(k) Improper tamping of pigtails.	G
(l) Improper drilling of brushes for pigtails.	D, G
(m) Defective electromagnetic design.	A, B, C, D, E

51.4 Recommended Sequence of Checks

Since several different causes can produce the same effect, it is possible to reach the correct solution by the process of elimination. Those defects which can be seen or measured easily should be checked out and eliminated first and the more difficult ones taken up later. The recommended sequence of checks based only on this principle is as follows: all checks should be made against standards, working drawings and maintenance manuals.

(a) Brush force.
(b) Brush dimensions.
(c) Quality, grade and source of the brush.
(d) Brush pigtails, their connections, pull out forces.
(e) Brush box dimensions and brush-box location with reference to commutator.
(f) Commutator ovality.
(g) Commutator surface.
(h) Commutator bar levels.
(i) Mica undercut and segment chamfer.
(j) Air gaps on main and commutating poles.
(k) Resistances of individual pole coils.
(l) Resistances between adjacent segments.
(m) Brush-box location with reference to magnetic neutral axis.
(n) Black-band test.

51.5 Failures due to Lacunae in Manufacturing Process for Commutators

CASE STUDY

In one particular make of traction motors there were many signs and symptoms of commutator defects. While the commutators were well machined and ran true while commissioning, they developed excessive ovality and high bars within a few months as detected by the method shown in Figure 51.1. The brush wear was excessive and there were flashovers. There was obviously some lacuna in the manufacturing process.

Investigation showed that the micanite segments used between the copper segments had a shrinkage (under pressure and heat) that was much higher than specified. The manufacturing process included only one cycle of heating, radial compression and cooling. It was found that the inner diameter of the commutator was reduced by about 1.5 mm on repeating the heating/ radial compression/cooling cycle of curing, twice. The commutator stabilised in diameter only after a total of three such cycles. The action taken was to reduce the bond content and the shrinkage of the micanite segments and to introduce three cycles of curing cycles in the manufacturing process. The performance of the traction motors became normal thereafter.

51.6 Failures due to Angular Shift of Brush-holder Ring

CASE STUDY

1000-volt dc compound motors drove the forced cooling fans of traction motors in a certain class of locomotives. While they gave good service for many years after commissioning, they started developing flashovers, excessive brush wear, excessive commutator wear and other signs of bad commutation. Checking of the commutators showed that their ovality and general surface condition was satisfactory. There was little sparking at the brushes during normal running but severe sparking was observed for a short time while starting the machines. The brush grade was the same as during the early years of satisfactory operation.

It was felt at this stage that there might be something wrong with the brush location as shown in Figure 51.2. The brush holders were mounted on a revolving ring, the angular position of which was locked by several nuts and bolts in circumferential slots. A 'kick' test was carried out as follows. A milli-voltmeter was connected across the brushes and a direct current injected into the series field. On switching this current on and off, deflections were observed in

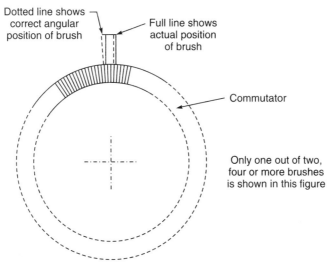

Figure 51.2 Angular shift of brush position. An error of a even a fraction of a degree is sufficient to cause bad commutation in modern machines

the milli-voltmeter. The brush holder rocking ring was shifted around a little until this deflection became either very small or disappeared altogether. The ring was then locked in this position. It was then observed that the sparking during starting had reduced significantly in intensity. There were no further cases of flashovers. See Figure 51.3.

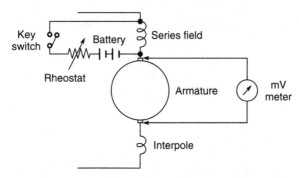

Figure 51.3   Kick test on dc motor to determine optimum angular position of brush holder ring

51.7 Various other Case Studies in Brief

51.7.1 Defect in Carbon Brush

CASE STUDY

Scoring of the commutator was observed on a few traction motors. The hole in the carbon brush for the pigtail was too deep. When the brush wore to its limit, the pigtail scored the commutator. The cause, was improper quality control of the carbon brushes by the manufacturer and by the user. See Figure 51.4.

All the brushes from the batch were X-rayed. Those with excessive depths of holes were discarded. Vendor's name was deleted from the list of approved vendors. Inspection procedure for brushes was modified to cover pigtail details.

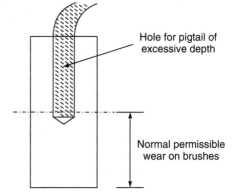

Figure 51.4   Shows how excessive depth of hole for pigtail, will lead to scoring of commutator

51.7.2   Defect in Carbon Brush

CASE STUDY

Carbon brush pigtails working out during service caused commutator flashovers on a few traction motors. In some cases, the carbon brush cracked across the hole. They touched adjacent steel parts and caused short circuits. The manual tamping of the pigtails in the carbon brushes was not sufficiently consistent. At times, the bond content in the tamping powder was excessive. All this was evidence of improper quality control during manufacture of carbon brushes.

Endurance tests including pull out tests on the pigtails of random samples were introduced in the specification and inspection. Vendor was changed.

51.7.3   Defects in Brush Boxes

CASE STUDY

There was excessive sparking and chipping of brush edges. Excessive clearance between the brushes and the brush boxes caused vibration of the brushes. It was clearly a case of improper quality control of the brush boxes.

Defective brush boxes were replaced, with new brush boxes double-checked for dimensions.

51.7.4   Defects in Interpole Coils

CASE STUDY

There was excessive sparking at carbon brushes though commutator condition, brush grade etc., were good. There were inter-turn shorts in some interpole coils due to presence of steel machining swarf between turns.

Poor layout of machine shop and coil shop, combined with bad house-keeping led to steel machining swarf finding their way into inter-turn insulation of some coils.

Coil shop layouts was changed. Entry to coil shop was restricted. Surge tests were introduced for testing of finished coils.

51.7.5   Defects in Brush Boxes

CASE STUDY

Excessive sparking and excessive wear of only two out of six carbon brushes was observed. The brush pressures were incorrect, they were too low.

The root cause was determined as poor quality of the spiral springs in brush boxes and inadequate inspection at the stage of purchase.

Vendor of brush boxes was changed and improvements were effected in the quality of inspection.

51.7.6 Commutator Defects

CASE STUDY

The soldered joints between armature coils and commutator segments were getting over-heated in some traction motors; and in some cases the solder melted and got thrown out. In other armatures there were shorts between coils behind commutator risers.

Measurement of commutator temperatures on several machines showed them to be much higher than in type tests on full load. Commutator profile was seen to be wavy. Carbon brush wear was excessive.

High temperature of the commutator and high wear rate of brushes were both due to commutator waviness caused by poor quality of mica cones and of seasoning process.

Both problems were solved when bond content of mica cones was regulated and seasoning process was improved by adding more cycles of heating and tightening. Commutator ovality and waviness were eliminated.

51.7.7 Defects in Commutator Maintenance

CASE STUDY

Excessive sparking and wear on commutator and chipping of brushes were observed on a group of traction motors. Careful examination of commutator surface showed mica undercutting was not satisfactory. Some segments were projecting. This was causing bad commutation and chipping of brushes. See Figure 53.3 and Figure 53.4.

Poor workmanship during overhauls due to inadequate training of workmen and poor supervision, was the root cause of the problem. Improved training and supervision solved these problems.

51.7.8 Defects in Brush Grade

CASE STUDY

The brush wear was excessive, in comparison with other traction motors in service, in one particular design of traction motor. There was no other problem regarding commutation. The grade and brush pressure were both in accordance with the recommendation of the manufacturers.

Comparison with other similar motors, showed that the specified brush force was about 50 per cent higher in the motors under consideration. There was also a difference in the recommended make and grade of the carbon brush. As there was no other commutation problem, it was decided to proceed cautiously with comparative trials with new grades and reduced brush pressures.

Three motors in a locomotive were fitted with brushes of the new grade and the other three were fitted with the grade already in use. The brush force was also reduced on trial brushes in consultation with the new brush manufacturer. After successful trials for about six months on three motors the trial was extended to thirty motors. A year later the new brushes with reduced brush force were converted to approved standard.

51.8 Do's and Don'ts for Preventing Carbon Brush Failures

- Consider very carefully, not only failures of carbon brushes themselves but also failures of commutation and armatures resulting from defects in carbon brushes.
- The performance of the commutator/carbon-brush system involves a very fine balance between electrical, thermal, magnetic and mechanical stresses. Most of these are at levels, which are close to their safe limits. Therefore design, manufacture and maintenance have to be precise and timely.
- Ensure brush grade, brush pressure, pigtail anchor strength and brush dimensions in accordance with design.
- Ensure clearances in brush boxes, clearance from brush box to commutator riser, and to curved face, brush box position relative to poles and cleanliness of brush box insulators in accordance with design.
- Ensure commutator surface finish, mica undercut, segment chamfer, commutator ovality when hot and when cold, and commutator seasoning process in manufacture, in accordance with design.
- Ensure evenness and accuracy of air gaps, absence of inter-turn shorts in field coils and correctness of inter coil connections are important.
- Verify sparkless commutation and adequate safety margin in the form of width of the black band throughout the operating range of currents.
- Measure the commutator surface temperature at full load and compare with type test results.
- Maintain records of brush replacements and monitor brush wear rate.
- Investigate every case of flashover and take corrective action.

Chapter 52

Commutator Connection Failures

In this chapter we learn

- *Two types of commutator connections viz. soldered and TIG welded.*
- *Failure modes for each type of connection.*
- *Failure mechanisms and preventive measures.*
- *Case studies.*
- *Condition monitoring of commutator connections.*

□ □

52.1 Introduction

$\mathcal{T}$his chapter deals with problems associated with only the connections between the armature coils and the commutator segments. Failures due to short-circuits and open-circuits in the armature coils and failures connected with current collection between the brushes and the commutators are dealt with separately in Chapters 54 and 53, respectively.

There are two types of commutator connections: soldered and TIG welded. Soldered joints were in use many years ago, but now they have been replaced by TIG welded joints in all new machines. The latter have a higher current carrying capacity and they are also more reliable.

52.2 Soldered Joints

Pure tin is generally used for soldering commutators because pure tin has a higher softening temperature than eutectic tin-lead solders.

Tin has a high creep rate at temperatures of the order of 100 °C, which is the usual operating temperature of commutators. Since thermal expansion and contraction of copper conductors produce stresses between the copper conductors and the copper risers, the combined effect of stress and temperature is to produce creep within the solder layer and this leads eventually to cracking and bad conductivity of the joints. Such joints overheat and finally the solder melts and is thrown out.

52.3 Failures of Soldered Joints through General Overheating

CASE STUDY

While soldered joints may fail after long service due to the inherent weakness of this system, premature failures occur within a few years of service if the operating temperatures increase due to any reason. The margin between normal operating temperatures and the dangerous temperatures is very small, i.e. about 40 °C. Some of these causes of overheating and failures of soldered joint on commutators are listed below:

(a) Reduced ventilation through the motor due to cracks in the bellows, blockage of air exit openings, operation of the motor at full load when the cooling equipment is defective, blockage of air passages within the motor, etc. are the common causes of overheating. (Falling out of brush box inspection covers may cause other types of failures but not commutator solder failures because ventilation of commutator is not affected.)

(b) Bad commutation due to any reason (such as those discussed in Chapter 51) increases the temperature rise of the commutator and this can precipitate failure of soldered joints on the commutators.

(c) Long and repeated starting efforts by locomotives to start heavy loads on gradients can lead to overheating of commutators and failures of soldered joints.

(d) Overloading of traction motors on long heavy gradients, while hauling full loads or overloads, particularly if combined with high ambient temperatures and even minor ventilation deficiencies lead to soldered joint failures on commutators.

52.4 Soldered Joint Failures due to Manufacturing Defects

CASE STUDY

In Section 52.3 we have discussed failures which occur despite proper manufacture of the motor. In this section we will discuss failures of commutator soldered joints that occur under normal operating conditions due to defects in the manufacturing process.

There are several types of defects in manufacture, which can either singly or in combination lead to solder throw from commutators. Each type of defect develops its own type of failure mechanism but in every case the final result is the same: overheating, melting of solder and solder throw. Commutation may become very bad, flashovers may occur and shorts may develop in the winding. These defects and failure mechanisms are described in the following table.

Table 52.1

CASE STUDY NUMBER	MANUFACTURING DEFECT	FAILURE MECHANISM AND REMEDY
52.4.1	Riser slot interior surface not clean enough and hence not tinned properly	The solder may fill the gap between the riser and the conductor palm but there is no proper bonding over some part of the area. Individual risers should be made free of any oily films by dipping in trichloroethylene, dried fully, then dipped in flux and finally in molten tin at 300 °C, before tapping to remove excess tin
52.4.2	Conductor palm surfaces not clean enough and hence not tinned properly	The conductor palms should be de-oiled by dipping in trichloroethylene, dried fully, then dipped in flux and finally in molten tin at 300 °C, before tapping and wiping to remove all excess tin
52.4.3	Low temperature of solder bath	The tin solder does not penetrate the space between risers and coil palms
52.4.4	Excessive clearance or inadequate clearance between risers and coil palms	In either case the tin solder penetration is not satisfactory
52.4.5	Flux quality unsatisfactory	Flux quality is very important. Only tested or proven fluxes should be used

52.5 Tig Welded Joint Failures

CASE STUDY

TIG welded joints are more reliable than tin soldered joints. TIG welding is usually done by special machines. While this helps to ensure consistency of quality, any error in the machine settings and parameters can lead to poor quality of a large number of joints. The points to be checked are as follows:

(a) The current cycle that is controlled electronically must be regulated correctly to ensure the prescribed magnitudes, durations and phases.

(b) The composition, the flow rate and the pressure of the inert gas must be correctly regulated.

(c) The angular alignment of the commutator should be adjusted to ensure correct positioning of the weld beads.

(d) The shape and size of the electrodes and the gas shield as also the operating clearance between the electrode and the commutator riser must be correctly adjusted.

52.6 Condition Monitoring of Commutator Joints

A very important test which monitors not only the quality of commutator connections but also, the internal connections of main coils and equalising connections, is the 'Commutator voltage drop test'.

In this test a small direct current is passed through the commutator and the connected winding as shown in Figure 52.1.

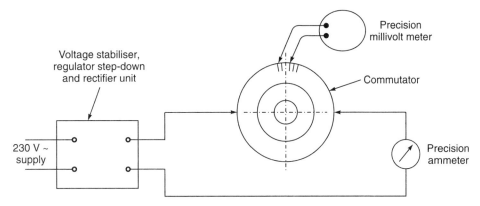

Figure 52.1 Arrangement for millivolt drop test on commutators of traction machines

The voltage drop (usually in millivolts) between consecutive segments is then measured with a sensitive and precision instrument. The circulating currents and the instrument range is adjusted to ensure that an accuracy of better than 1 per cent is obtained in the measurements.

The average and the standard deviations are worked out and all segment pairs, which give measurements outside the average +/– three times the standard deviation, are marked as suspect and investigated.

These tests should be carried out initially after manufacture, after every overhaul, and for investigating flashovers, armature coil shorts, etc.

52.7 Do's and Don'ts for Preventing Failures of Commutator Connections

For soldered joints:

- Ensure purity of solder alloy or tin as specified.
- Ensure coil-end and riser slot dimensions as specified.
- Clean, de-oil, apply flux, and dip-tin all coil-ends and segment slots before soldering in bath.
- Ensure purity of solder or tin and flux as specified in the solder bath.
- Regulate temperature of solder bath, and dip time as specified.
- After machining riser faces, verify, with the help of magnifying glass, quality of solder penetration and cleanliness of mica segments.

For TIG welded joints:

- Ensure TIG welding parameters, inert gas purity and current/time cycles as specified.
- Ensure electrode alignment, clearances and wire size as specified.
- Examine through magnifying glass and ensure correct bead size and overlap as specified.

For soldered and TIG welded joints:

- Verify good quality of soldering or welding by carrying out milli-volt drop test after completion of the work.
- Maintain records of milli-volt drop tests for comparison after overhauls.
- Ensure that temperature of commutator after full load run is within specified limits.

COMMUTATOR FAILURES

In this chapter we learn

- *A number of different failure modes in commutators of dc machines.*
- *Particularly difficult problem caused by defects in the manufacturing process.*
- *Common causes of commutator failures due to defects in commutators.*
- *Measures to prevent commutator failures.* ❏ ❏

53.1 Introduction

This chapter should be read with Chapters 51 and 52.

Commutator failures may appear in several different failure modes such as :

(a) Sparking in brushes and flash-overs;

(b) Carbon brush breakage;

(c) Excessive wear on carbon brush;

(d) Short-circuits in or over insulating cones;

(e) Short-circuits between segments.

Some of these failure modes can also be due to other causes such as defects in carbon brushes, defects in brush-gear, defects in the armature or stator windings but in this chapter we are concerned with failures in those modes caused by defects in the commutator. The commutator defects may be of four different types as follows:

(a) Commutator ovality or irregularity.

(b) Commutator surface finish unsatisfactory.

(c) Defects in insulating cones and segments.

(d) Defects in commutator seasoning during manufacture.

If the commutator is designed and built correctly there is hardly any need for maintenance, except periodical machining, undercutting and chamfering. These aspects are dealt with in Chapters 51 and 52.

53.2 Commutator Ovality or Irregularity

The commutator consists of a large number of hard drawn copper segments of tapered section and shapes as shown in Figure 53.1, and an equal number of insulating segments of the same shape but of uniform thickness.

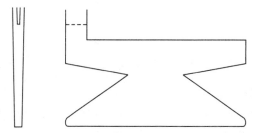

Figure 53.1   Copper segment for commutators

These are assembled alternately, around a circle and held together under very high compressive forces by steel vee-cones and insulating vee-cones as shown in Figure 53.2.

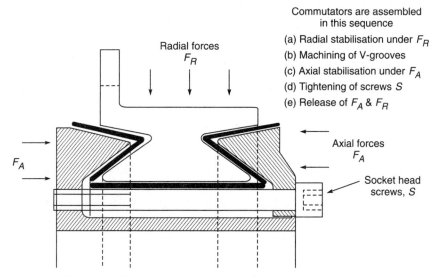

Radial forces
F_R

Commutators are assembled
in this sequence
(a) Radial stabilisation under F_R
(b) Machining of V-grooves
(c) Axial stabilisation under F_A
(d) Tightening of screws S
(e) Release of F_A & F_R

Axial forces
F_A

Socket head
screws, S

F_A

Figure 53.2   Schematic diagram of commutator assembly

The dimensions of the copper segments, the insulating segments and vee-cones as also the assembly processes are such that the final assembly is a solid and rigid component. The axial forces between the steel cones act on the tapers of the cones to develop very high, radially inward forces on the segments. These forces are strong enough to counter the centrifugal forces on the segments and the assembly remains firm and stable in dimensions even when rotating at high speeds.

If there are any errors or deviations from the designed magnitudes, in regard to the dimensions, or the mechanical properties of the insulating segments and cones, or in the forces applied during assembly the mechanical stability of the complete commutator is adversely affected. The commutator does not retain the perfectly cylindrical aspect of its active surface. The surface may develop irregularities, cause bad commutation, sparking, excessive wear, overheating etc.

One of the first few things to be checked when dealing with commutator problems is to check the ovality and the regularity of the cylindrical surface. The deviation from the perfect cylinder should be of the order of a few microns only as stipulated in the maintenance manuals or specifications. There must be no irregularities in the surface and no high bars and no low bars. See Figure 53.3.

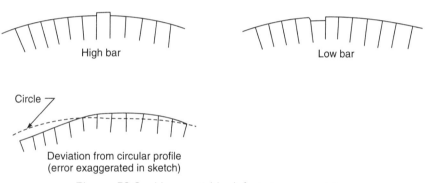

Figure 53.3 Unacceptable defects in commutator

If there are such defects, then the only thing to do is to check very carefully the raw materials, the dimensions of the components and the manufacturing process of the commutator against the stipulations in the relevant drawings and process sheets.

If a commutator which has been machined correctly to the truly cylindrical and smooth surface, develops ovality and high/low bars after some service it is definitely a case of defective design or defective manufacture.

The following are the more important requirements of good design and manufacture of commutators:

(a) The insulating segments and the insulating cones should have a bond content which is within the specified limits. The percentage reduction in thickness under repeated compression and heating/cooling cycles should also be within the specified limits.

(b) The dimensions of the tapered copper segments, the insulating segments, the insulating cones and the steel vee-rings should all be within the limits specified in the drawings. The hardness and surface finish of the copper segments is also important.

(c) The copper and insulating segments should be assembled in special fixtures to ensure symmetrical arrangement around the circumference and consolidated in pressing fixtures in which the segments are pressed inwards radially while being heated and cooled alternately. Several such cycles are completed until the bore of the commutator stabilises. The radial forces applied and the temperatures to which they are heated are all very important.

(d) After stabilisation the commutator, still held firmly in the pressing fixture, is machined to fit the vee-rings. Here again dimensional accuracy and surface finish is critical.

(e) The commutator is now assembled in the steel vee-rings with the insulating vee-rings and the whole assembly is again subjected to cycles of pressing/heating/pressing/cooling until the insulating cones are fully compressed and stabilised in thickness.

(f) At this stage the pressing fixture is removed and the active face of the commutator is machined, using first steel and then diamond tools, with reference to the bearing seat centre-line. The cylindricity, the ovality and the surface finish of the active face of the commutator must be ensured in this final operation.

(g) The insulating segments which would at this stage be flush with the copper segments are now undercut and the edges of the copper segments are chamfered as shown in Figure 53.4.

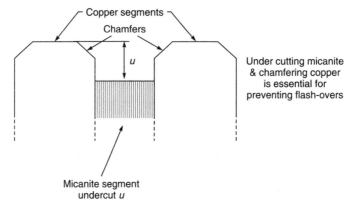

Figure 53.4 Undercutting of micanite and chamfering of copper

In short, for a commutator to give trouble free service, not only its final visible or measurable aspects must be satisfactory, the invisible features such as the stabilisation of the insulating cones and vee-rings must also be complete. If there is any development during service of ovality, high/low bars, protruding segments, etc., the design and manufacture become suspect.

53.3 Other Causes of Commutator Failures

There are many other possible causes of commutator failures. These are dealt with in greater detail in Chapter 51 and listed below generally in order of the frequency of their occurrence.

(a) Brush quality defective: Generally, there are no fool-proof specifications for carbon brushes. Brush quality has to be selected from amongst several branded brush grades available, after carrying out preliminary trials, service trials and service experience.

(b) Brush and brush-holder dimensions: These must be within specified limits so that the brushes can slide freely in the brush-boxes without being too loose.

(c) Brush holder position around the circumference must be correct with reference to the positions of the main poles and the inter-poles.

(d) The spring forces which cause the brushes to bear down on the commutator must be within the limits specified by the brush manufacturer and proven in service trials.

(e) The adjustment of the air-gaps and the number of turns in the inter-poles must be correct as designed and proven in service trials.

(f) The pig-tails in the carbon brushes must be well tamped and secured in the brushes to ensure that they do not come off during service.

(g) The size and design of the commutator should be such that its operating temperature during normal service does not exceed 85 °C.

(h) Certain types of defects such as inter-turn shorts in the armature windings or in the stator windings can lead to excessive sparking, wear and flashovers at the commutators.

53.4 Maintenance and Installation of dc Machines

Although Chapters 51, 52 and 53 are written with special reference to traction motors, it has to be emphasized that there is no difference whatsoever between dc traction machines and stationary dc motors and generators in-so-far as commutator design, manufacture and maintenance are concerned. Failure modes and failure mechanisms are also identical. The only difference is that traction machines are subject to vibration and the margins of safety in their design are smaller due to size and weight limitations.

53.5 Do's and Don'ts for Preventing Commutator Failures

- Refer to the do's and don'ts relating to brush-gear and commutator connections in Chapters 51 and 52, respectively.
- Ensure the quality of the micanite segments and the micanite cones as specified.
- Ensure that the commutator assembly is done generally as described in Section 53.2 and as specified in detail in the relevant manufacturing process sheets.
- Ensure that all dimensions of components and at various intermediate stages of assembly comply fully with the relevant process sheets.

Chapter 54

FAILURES OF ROTOR OR ARMATURE COILS IN DC MACHINES

In this chapter we learn

- *Similarities and differences between stator and rotor windings.*
- *Special factors applicable to rotor windings.*
- *The effects of thermal expansion on rotor windings.*
- *The effects of radial centrifugal forces on rotor windings.*
- *Rotor winding failures due to defects in commutator connections.*

54.1 Introduction

The insulating materials used on the conductors in the armatures or rotors of dc machines are very similar to those used in the stators discussed in Chapter 55. As the coils are of different shapes and sizes, and as there are severe space constraints on rotors, methods of insulating the coils and of assembling the coils in the rotors are quite different from those for the stator coils.

However, the failure modes and the factors that influence the failure modes, patterns and mechanisms are identical for both stator and rotor windings. For rotor windings, there are one or two additional factors such as the effects of centrifugal force on the coils and the effects of thermal expansion and contraction of the coils in the slots.

The general discussion in Chapter 55 regarding, effects of vibration, overheating and water penetration are fully applicable to rotor windings too. In fact whether a failure will occur in the rotor or in the stator due to these factors is simply a matter of chance if the other factors such as design margins and insulating material quality is the same for both. Thus, failures due to poor ventilation may occur either in the stator or in the rotor. The case studies in Chapter 55 may be extended to rotor failures too. Only those factors that are applicable specifically to rotors are discussed further in the following sections.

54.2   Effects of Thermal Expansion and Contraction

Short-circuit between the conductor and the steel core in its slot portion is a very common failure mode for dc machine armatures. The majority of such failures occur at the ends of the slots.

Figure 54.1 shows the manner in which the rotor winding is secured or held in the rotor slots. The temperature rise of the copper conductors is higher than the temperature rise of the rotor core. The temperature coefficient of copper is also higher than that of the core steel. The length of the copper conductor increases significantly in comparison to the length of the stator when the motor operates on load. This differential expansion produces shearing forces on the insulation held under compression between the copper and the steel. The direction of the forces is reversed when the motor having attained maximum temperature is switched off and begins to cool. These alternating movements are small but the forces are not. The effect is the highest at the ends of the slots and the failures of insulation are also the highest at the ends of the slots.

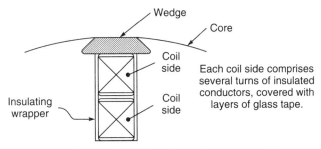

Figure 54.1   Armature conductors held firmly in slots, are subject to movement and forces due to differential expansion of copper and steel core

The adverse effect of reduced ventilation on the armature is therefore the sum of the effects of insulation degradation due to increase in its temperature and the effects of increased mechanical damage to the insulation caused by the differential expansion of the conductor in the slots. See Figure 54.2.

54.3   Effects of Radial Forces

The centrifugal forces acting on the armature conductors vary in proportion to the square of the speed of the motor. They are very high at the maximum operating speed and even higher if the motor over-speeds due to wheel slip. These forces would cause the conductors to fly away but for the counter forces exerted by the steel or glass banding around the portions of the conductors outside the core as shown in Figure 54.3. Inside the slots the

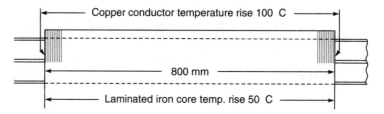

(a) For the typical dimensions and temperatures shown above, the difference in lengths of copper conductor and iron core is 0.9 mm.

(b) If this expansion is prevented totally by holding conductors in constraint, stresses of about 13.5 kg/mm^2 are developed in the copper.

Figure 54.2 Effects of thermal expansion

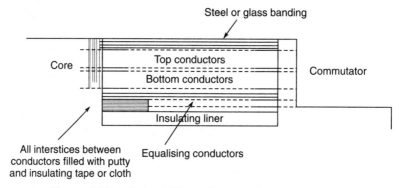

Figure 54.3 End windings of armature on commutator

centrifugal forces on the conductors are held in check by the reactions exerted by the slot wedges.

The forces exerted by the banding are usually higher than the possible centrifugal forces. There is little movement of the coils despite the magnitudes of these forces. If, however, this difference is not ensured due to inadequate banding tension, the conductors may actually move radially in and out every time the motor speeds up and slows down. The movements are very small but sufficient to cause damage. When such damage accumulates eventual failure of insulation is certain.

If the space below the conductors and between the conductors is not filled up fully by insulating putty, the conductors are likely to twist during the process of banding itself and this may damage the insulation during manufacture. If the damage is not severe enough to cause failures during stage inspection and testing, premature insulation failures are likely to occur during service.

The failures referred to in the two preceding paragraphs are generally of the inter-turn type. Unlike the stator coils, rotor coils have relatively higher inter-turn voltage due to the magnetic induction of electro-motive force. The voltages of the rotor coils to ground, i.e. the steel parts of the rotor are also much higher in the rotor than in the stator. Inter-turn short circuits in rotors will cause burning and arcing, which will produce a hole in the winding. Eventually a short circuit to earth may also occur and at that stage the protective system will act causing the supply to be disconnected.

Failures of the types described above occur regularly, when the defects mentioned above are allowed to develop in the manufacturing process.

54.4 Winding Failures due to Commutator Connection Defects

Armature conductors are connected to the commutator through soldered or TIG welded joints. The winding often comprises equalising or paralleling connections. If there are any defects in the quality of the joints, the current distribution between the various paralleled conductors in the rotor winding is uneven and this can lead to local over-heating of some coils. The insulation on such conductors may then fail due to excessive degradation by heating.

Failures of this type are relatively rare in the case of TIG welded commutators but not so in soldered commutators.

54.5 Practical Measures for Preventing Rotor Winding Failures

Due to the very large number of modes and mechanisms of rotor failures, actual cases of failures are often the result of a combination of circumstances and factors. It is difficult to conclude which defect was more crucial. In view of this and the relatively higher failure rates of rotor windings, it would be better to summarise the practical measures to be taken to reduce and to eliminate the failures of rotor windings.

Neglect of any one of the following measures may not lead to immediate failure but when a failure does occur it is certain that it is due to such neglect at some stage in the past. Lacunae in this regard can be considered to be seed defects. They are like seeds. Not all of them germinate and grow into trees or failures but some do. It is not possible to predict which seeds will germinate and which won't. So, the safest and also the only possible course is to avoid them all if you are determined to attain zero failure performance.

(a) Measures in routine inspection at intervals of the order of a few weeks, depending upon service conditions and observations during the inspections ensure that:

  (i) The cooling equipment is effective;

(ii) The rubber bellows are fitted correctly and not torn;

(iii) The inspection covers are tight in position;

(iv) The gaskets are good;

(v) The air exit openings from the motors are not blocked;

(vi) The commutator chamber pressures are OK;

(vii) The protective relays against air flow failure are OK;

(viii) There is no sign of water or oil entry into the motors;

(ix) There are no signs of pole screws being loose;

(x) The carbon brushes are not too short, not chipped;

(xi) The brush pigtails do not work out;

(xii) The commutator ovality is within limits;

(xiii) None of the commutator segments are either too high or too low compared to the adjacent segments;

(xiv) The commutator surface is smooth, even, of good colour;

(xv) The mica between segments is well undercut;

(xvi) The copper segments are well chamfered;

(xvii) There are no deposits of dust and dirt on the rotor or the stator;

(xviii) The vee cones and the insulator surfaces are clean and shiny.

(b) Measures to be taken during periodical overhauls at intervals to be based on the conditions of service and observations during routine inspections:

(i) Carry out all the checks included in the routine inspections. In addition, make the following checks;

(ii) Measure the mV drop between adjacent segments of the commutator, while carrying about 10 per cent of the rated current. Use precision instruments. If any of the measurements are outside the average plus/minus thrice standard deviation, investigate, determine and correct the causes;

(iii) Carry out diamond tool machining of active surface, mica undercutting and segment chamfering on commutators;

(iv) Carry out growler tests and surge comparison tests between adjacent segments of the commutator to detect any developing weaknesses in the inter-turn insulation;

(v) Check condition of banding and slot wedges and replace where necessary;

(vi) Work out carbon brush wear rates and look for excessive carbon dust. In case of excessive wear rates determine the causes and eliminate them;

(vii) Ensure specified clearances (radial and axial) between the brush boxes and the commutator;

(viii) Ensure specified clearances between the brushes and the brush boxes;

(ix) Ensure specified brush forces on all brushes.

54.6 Case Studies of dc Machine Rotor Failures

Investigations into hundreds of armature failures in dc machines always revealed lacunae in one or more of the requirements mentioned above in Section 54.5 to be the main cause of failure. Where the enumerated design and maintenance requirements were fully complied with, failures were very rare.

Problems were usually experienced on new designs of motors put into service for the first time, and these were usually of the following type:

(a) Brush grades and brush pressures as suggested by the manufacturer were not always the best. Changes had sometimes to be effected after carrying out trials.

(b) On some designs and makes of motors the insulation failure rates were high during the first few months but after these 'early' failures took place, the performance stabilised as shown by monthly monitoring of failure rates per 100 machines as related to the age since commissioning.

(c) Flash-overs, excessive brush wear, sparking and high commutator temperatures were due to improper curing and stabilisation of commutators during manufacture. After introduction of new procedures for commutator stabilisation, the failure rates dropped dramatically.

(d) In some cases the armature insulation seemed to absorb moisture if left unused for some time. These problems were overcome by the introduction of vacuum impregnation.

(e) In one case the armature conductors were made by paralleling thinner conductors in order to reduce the eddy current losses. There were many failures of such machines in the first few months of operation.

   ▪ These failures were traced to the effects of distortion of the coils under the pressure exerted by banding. Eventually the design was modified: larger solid conductors (instead of laminated conductors) had to be used. Failures due to coil distortion were thereby eliminated.

(f) One particular design of traction motors was prone to develop high failure rates after three or four years of service. Surge comparison tests on three-year-old machines showed higher incidence of weakened insulation as compared to two-year-old machines. These motors operated at voltages above 1000 and the windings were then not vacuum impregnated. It was surmised that the insulation was getting damaged through partial discharges in voids in the insulation. Vacuum pressure impregnation was introduced and the failure rate dropped significantly.

(g) Another make of traction motors had a high failure rate due to burning of armature coils just behind the commutator risers. In this failure mode there were inter-turn shorts between conductors in the upper and lower layers of the winding. These shorts usually occurred behind the commutator riser because, that was the location where the conductors in the upper and lower layers crossed each other and further where the conductors were twisted through 90° for insertion in the segment slots. Strengthening of the insulating wrapper between the layers and on individual conductor ends reduced the failure rate significantly.

(h) In one locomotive shed, an epidemic of traction motor failures was traced to defects in the inspection cover clamping mechanism which caused them to drop off on the run thereby allowing the cooling air from the blowers to escape without passing through the motors. Similar problems arose in other sheds due to defects in bellows between cooling air ducts and motor air inlets, accumulation of dirt in motor air passages, failures of cooling air equipment.

Bad maintenance of air cooling in forced cooled motors is probably the single largest cause of motor failures. Such causes are slow to make their presence felt because traction motors often continue to work even when their cooling is not adequate.

54.7 Do's and Don'ts for Preventing Armature Failures

- Investigate every case of armature failure and take corrective action on other traction motors in service if any maintenance deficiencies are detected. If design and manufacturing defects are suspected report the matter to the manufacturer and make suitable additions in the test program for future purchases.
- As a first step always look into the proper performance of the cooling system.
- As the second step always examine the commutation system fully.
- To the extent possible carry out the other checks mentioned in Section 54.5 on the armature that failed.
- If possible carry out all possible non-destructive tests on a few armatures of the same age and make before coming to a final conclusion about the cause of the failure.
- Remember that often more than one cause may contribute to the failure mechanism.

Chapter 55

FAILURES OF STATOR COILS IN ROTATING MACHINES

In this chapter we learn

- *Different types of coils in rotating machines and their general construction.*
- *Three failure modes of main pole and inter-pole coils.*
- *Ground insulation failures and their causes.*
- *Case studies relating to ground insulation failures.*
- *Inter-turn insulation failures and their causes.*
- *Case studies relating to inter-turn insulation failures.*

55.1 Introduction

There are three types of stator coils in rotating machines: interpole coils, main field coils and compensating windings. The first two of these three types viz. interpole and main field coils are constructed and fitted similarly and are dealt with in this chapter. Compensating windings are provided only in large machines and they are constructed in a manner similar to the rotor windings in dc machines. They are considered together in Chapter 54.

In general, the construction of interpole and main field coils is as shown in Figure 55.1.

The conductors are usually made of edge wound flat copper bars. The interturn insulation consists of impregnated paper made of either natural or synthetic fibre. The main or ground insulation consists of several layers of insulating tapes usually of glass fibre with mica flakes. The whole coil may be given a protective layer of impregnated glass fibre tape. In some machines the coil and the pole-piece may be moulded together in a solventless resin.

The coils are usually held firmly in the pole pieces or cores and these are screwed on to the stator frame by large steel screws.

There are three main types or modes of failure on these coils, generally in order of descending failure rate as follows:

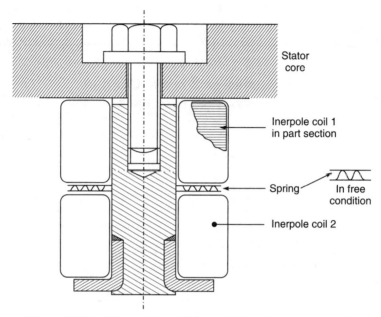

Figure 55.1 Construction of interpole with coils and springs

(a) Main or ground insulation failure.
(b) Inter-turn insulation failure.
(c) Fracture of the conductor at the brazed inter-coil connection or overheating of bolted or crimped joint.

Item (c) above has been discussed in Chapters 46 and 32 in detail and will not be referred to in this chapter.

55.2 Main or Ground Insulation Failure Rotating Machine Stators

Generally, these coils are insulated for the full motor voltage but the actual operating voltage between the main field coils and the frame is very low as they are often connected at the ground end of the circuit. The failure rate should normally be expected to be extremely low.

Interpole coils usually operate at the full motor voltage and they have to bear the brunt of surge voltages, which develop due to current interruptions. Despite these conditions of service even interpole coils can be designed to give very low failure rates. And if failures do occur in service the first conclusion should be that there is some lacuna somewhere in one or more of the following areas:

■ Design of the coil insulation or of the manner in which the coil is fixed in the stator.
■ Manufacture.

- Operation.
- Maintenance.

The case studies that follow illustrate the manner in which several degradation processes affect the insulation. These, in order of importance, are enumerated below:

- Vibration and relative movement between coils and stator frame.
- Overheating.
- Moisture or water penetration.
- Oil penetration.
- Normal ageing of insulation.

55.3 Case Study (Vibration)

Failure of insulation due to vibration and relative movement between coils and stator frame is common on traction motors on locomotives and motor coaches. Particularly vulnerable are those that are axle mounted and nose suspended. Similar failures may occur also on motors installed on machinery, which is subject to vibration.

Figure 55.1 shows flat springs below the coils or between sections of the coil. These springs must exert a force, which counteracts the effects of the maximum inertia forces on the coils due to the vibration.

If the acceleration of the motor in the direction of pole centre line is a and the mass of the (coils + springs) is m, the spring force in the assembled condition must be more than $m.\alpha$. Similarly if the acceleration due to vibration at right angles to the pole centre line is β, the spring force in the assembled condition must be more than $m.\beta/0.3$ where 0.3 represents the coefficient of friction. If these conditions are not satisfied, the coils will vibrate and develop relative motion between the stator and the pole piece. This causes the insulation to get abraded and to fail eventually due to electrical short circuit.

This is exactly what happened on many traction motors in some railways. There were several different causes for the absence of adequate spring force.

- In one case, the springs as originally designed were too weak for the conditions of service.
- In another case the design was satisfactory, but the material used for the springs was defective and, as a result, the springs took a permanent set.
- In yet another case the screws which hold the pole pieces down were not tightened fully.

The main reason to provide the springs is to compensate the expansion and contraction of the copper due to the cooling and heating of the coils with variations in the current. If the springs are not provided and the coils are given a tight fit between the pole piece and the stator, extremely large forces are developed on the copper. The forces are large enough to produce a permanent set in the copper thickness which then permits the coils to vibrate with relative movement between the coils and the stator frame or the pole piece. The insulation then gets abraded and eventually develops short circuits.

Finally it may be stated that if the coils are firmly held in the stator without relative motion, vibration does not lead to insulation failure. Direct compressive forces on the insulation are not harmful. However, it is necessary to allow for thermal expansion and contraction of the copper.

55.4 Failures due to Overheating

CASE STUDY

The copper conductors in all electrical machines get heated up due to the passage of current through them and also due to eddy currents produced by alternating magnetic fields. The insulating properties of the insulating materials around these conductors are adversely affected by the temperature rise on the copper conductors. These two basic natural phenomena are ever present and are absolutely unavoidable.

Higher the temperature attained; greater is the rate of degradation. On the other hand, lower the temperature limit adopted, the greater the size and cost of the equipment. In order to economise on the size and cost of the equipment, a fine balance has to be struck by the designer to limit the temperature rise to such levels as to get a reasonable life from the insulation at a reasonable initial cost. There is thus an optimum operating temperature for the conductors of all electrical machines.

If for any reason the temperature of the insulation rises beyond the designed limit, the life of the insulation gets reduced. For every 8 °C rise in temperature above the design limit, its life is reduced by half. A rise in temperature by 40 °C will reduce the life to about 3 per cent of the designed life. Failure of insulation within a few years of service is then inevitable. *And this rise in temperature does not have to be a general rise; it need only be a local rise at one point to cause an insulation failure.*

There were several cases of insulation failures of this type. These are summarised below:

CASE NUMBER	CAUSE OF OVERHEATING
55.4.1	The flexible bellows connecting the air duct on the locomotive under-frame to the air inlet opening on the traction motors were torn allowing the cooling air to escape without passing through the motor. (Defective design of bellows.)
55.4.2	The inspection covers on the traction motors below the air inlet openings had fallen off allowing the cooling air to escape without passing through the windings of the motor. Only the commutators were getting cooled. (Defective design of inspection cover latch.)

CASE NUMBER	CAUSE OF OVERHEATING
55.4.3	A cake of oil and dust blocked the air exit openings of a traction motor. (Bad maintenance.)
55.4.4	The spaces between the stator coils and the connections between the coils got blocked with a cake of oil and dust. (Bad maintenance.)
55.4.5	The cooling fan motor failed and the locomotive driver continued to operate the train at reduced speed to his nearby terminus, after isolating the protective relays. The relieving driver continued to operate the locomotive at full load and speed. Two traction motors also failed as a result. (Bad operation and bad communication.)
55.4.6	In one class of locomotives, two blowers cooled the six traction motors. There was a duct from each blower, which divided the air suitably between three motors. In one locomotive, one of the air distribution vanes in one duct got dislodged causing reduced air flow into one motor and increased airflow into the other two. The problem was exposed when it was noticed that motor failure rate in one position on the locomotive was significantly higher.
55.4.7	Due to a wiring defect in the field-diverter-contactor-control-circuit of a locomotive, one motor operated in weak field while the other five were operating in full field. This motor shared a higher load continuously and got overheated. The anomaly was detected only when two motors failed in quick succession in the same position of that locomotive.

55.5 Failures due to Degradation of Insulation Caused by Water or Oil

CASE STUDY

The adverse effect of water or moisture on insulation is almost instantaneous. Every possible precaution should be taken to ensure that water is not allowed to enter the windings of a traction motor. And if a motor is exposed to water immersion or condensation it must not be operated at full voltage unless it is dried out fully and normal insulation level *at operating temperature* is restored. If the water, which has entered the motor, is floodwater, the motors should be washed thoroughly with clean water before drying out. It is, possible to operate the motors at greatly reduced voltage but at normal current.

The effect of lubricating oil on insulation is relatively slower and uncertain. The insulation gets softened, collects dust and it may also lose dielectric strength. It is best to avoid exposure of motor insulating materials to stray oil. Here are some case studies in brief.

CASE	CAUSE OF WATER OR OIL PENETRATION
55.5.1	In one incident all 12 traction motors of a 3-unit EMU set got submerged in floodwater. The train was hauled dead to the car shed and the motors were dried out by passing hot air through them. However, the insulation resistance was measured only after the motors had cooled and found to be adequate. After re-commissioning, two out of the twelve motors failed. On measuring the insulation resistance of hot motors it was found to be much lower than when they were cold. They had to be dried out further until the insulation resistance was satisfactory while the motors were at operating temperature.
55.5.2	Two locomotives which had been under repair in a repair shop where the temperature was 8 °C, and R.H.50 per cent, were moved, one operating and the other dead, to a station 300 km away, temperature 35 °C, and R.H. 85 per cent. Both were put into service for hauling a train. The traction motors on the dead locomotive developed low insulation resistance. Investigation and experiments led to the conclusion that condensation on the cold motor insulation of the dead locomotive was the cause of failure. The machines in the other locomotive in use were warm, and they were not affected.
55.5.3	A large dc motor had oil lubricated sleeve bearings. Leakage and overflow of oil led to the accumulation of oil in the lowest part of the motor. The stator field coils at this level became partially immersed in the lubricating oil. The oil softened the insulation and perhaps the oil was not free from moisture. The field coils at the bottom level of the motor developed shorts to the frame.

55.6　Normal Ageing

CASE STUDY

Even when the insulation temperature is within the permissible limits, there is a slow degradation of the insulating properties of the material. The design life of the machine may be taken as 20000 hours with a multiplying factor. The multiplying factor depends on the margin between the maximum, hot spot, temperature of insulation, t_1, during operation and the index temperature, t_2, of the insulating material.

$$\text{Multiplying factor} = 2^m \quad \text{where} \quad m = (t_2 - t_1)/8$$

Thus if the temperature margin is 16 °C, the life expectancy is 80000 hours. It must be emphasised here that this is at best only an estimate. The operating temperatures vary from time to time and the insulation properties are also subject to much variance. If all other factors such as excessive vibration, overheating, and water/oil penetration are eliminated, and if the failures have occurred after more than 75 per cent of the estimated life, it may be possible to conclude that the failures are due to normal ageing.

The same formula applies to the degradation of insulation due to insulation temperatures being higher than the designed temperature. The margin m becomes negative and the factor 2^m becomes less than 1. If, for instance the coil hot spot temperature is 40 °C higher than the index temperature of the insulation, the life expectancy will be $20000 \times 1/32 = 625$ hours.

There are some methods for monitoring the condition of the insulation. These are (a) measurement of partial discharges, (b) surge voltage tests and (c) tan-delta measurement. These can be used to determine the priorities for taking out working machines for rewinding, in order to prevent failures of over-aged machines in service.

55.7 Inter-turn Insulation Failures

The interturn insulation between turns of the stator coils is usually in the form of impregnated paper or similar synthetic material. The voltage between turns is very small—usually a fraction of a volt—and the breakdown voltage of the inter-turn insulation is very high in comparison. Failures of inter-turn insulation due to thermal degradation or ageing are extremely rare.

However, inter-turn insulation failures do take place sometimes. They are then usually due to mechanical damage caused by foreign metallic particles or sharp edges/corners on the copper bars. Often there is no direct sign of overheating at the point of the short-circuit but there may be indirect effects on the commutation as indicated in case study 51.7.5 in Chapter 51.

Machine tool swarf, copper chips and similar material are usually strewn around in machine shops where motor components are machined. They can find their way into coil insulating shops in unpredictable ways. It is best, to maintain high standards of house keeping in coil insulation shops where the inter-turn insulation is placed between turns and the ground insulation is applied.

Bad maintenance of air cooling in forced cooled motors is probably the single largest cause of motor failures. Such causes are slow to make their presence felt because traction motors often continue to work even when their cooling is not adequate.

55.8 Do's and Don'ts for Preventing Stator Winding Failures

- Investigate every case of stator winding failure and take corrective action on other traction motors in service, if any maintenance deficiencies are detected. If design and manufacturing defects are suspected report the matter to the manufacturer and make suitable additions in the test programme for future purchases.
- As a first step always look into the proper performance of the cooling system.

- Regarding failures of connecting bars and crimped sockets refer to the Do's and Don'ts at the end of Chapters 46 and 32, respectively.
- Look for weakness of springs, abrasion of insulation and looseness of the coils on the cores in case of coil insulation failures.
- Look for signs of moisture entry, softening of insulation by oil and accumulation of dust etc., around the coils.
- If possible carry out all possible non-destructive tests on a few stator of the same age and make, before coming to a final conclusion about the cause of the failure.
- Remember that often more than one cause may contribute to the failure mechanism.

Chapter 56

STARTING RESISTOR FAILURES

In this chapter we learn

- *A description of the starting resistor system on dc EMUs.*
- *A description of a problem of resistor failures in service.*
- *The manner in which the problem was analysed.*
- *The remedial measures which solved the problem.*

□ □

56.1 Introduction

$f$igure 56.1 shows a schematic circuit illustrating the principle of a typical arrangement of starting resistors for dc electric multiple unit motor coaches. Initially, the contactors r_1, r_2, etc., are open and all the sections of the resistor are in circuit, in series with the traction motor. As the train picks up speed, the contactors r_1, r_2, etc., are closed, one-by-one, by the control circuits, thereby cutting off sections of the starting resistor until finally the motors are connected directly to the overhead lines.

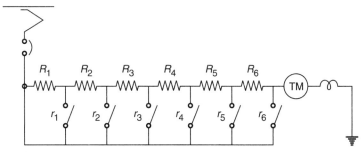

Figure 56.1 Schematic circuit to show principle of operation of starting resistors and conductor in an EMU

These starting resistors and this arrangement of gradual reduction of the resistance in the motor circuit, is aimed at preventing excessive current flowing through the motors at any stage. The current is maintained within certain safe limits by the control circuits. See Figure 56.2.

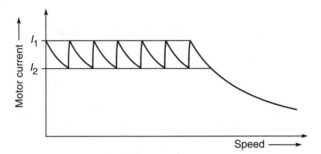

Figure 56.2 Shows how motor current is kept within limits by sequential shorting of starting resistors

While maintaining the current through the motors as described above, the resistors themselves get heated up by the current passing through them. The size and the mass of the resistors are so selected as to ensure that the temperature does not exceed, at any stage the safe limits for the material of the resistors and the insulating materials around them. As the resistors are in circuit for short intervals, in the range of 10 to 20 seconds only, during starting, they are usually designed for such short time loading.

56.2 Failures of Starting Resistors

In one particular design of starting resistors, failures of certain sections of the resistors were reported and it was seen that these were increasing in frequency year after year. Initially it was thought that these failures were due to the increase in the loading on the coaches as a result of which the accelerating time of the units had been increasing which meant also that the starting resistors were in circuit for longer periods.

The sections of the resistors that were actually failing were not the ones, which were in circuit the longest.

Examination of the failed resistor sections and measurement of the time in circuit for different sections, showed that there were actually two designs of the resistor sections one of which had a rating nearly double that of the other. The higher rated sections were used for the resistor sections 5 and 6, which remained in circuit longer and the lower rated ones were used for Sections 1 to 4. It was also seen that all failures were in Section 4 only. Refer to Figure 56.1.

56.3 Analysis of the Problem

The time for which each section of the resistor remained in circuit was determined with ordinary stopwatches during acceleration of the fully loaded train. The values of an arbitrary index $W = (I^2 R \, t/m)$ were calculated for each of the resistor sections where;

R = resistance of section (ohms)

I = average current through the resistor = $(I_{max} + I_{min})/2$ (amps)

t = time for which the resistor section remained in circuit (seconds)

m = mass of the resistor section (kg)

It was found that $W_4 > W_6 > W_5 > W_3 > W_2 > W_1$, while

$$t_6 > t_5 > t_4 > t_3 > t_2 > t_1$$

The anomalous position of R_4 was confirmed by actual tests. Measurements of temperature after the coach was in service for an hour of operation, i.e. making about 10 starts and stops, showed that the temperature of Section 4 was significantly higher than that of any other section. It was higher than the temperatures of even Sections 5 and 6, which were in circuit for a longer time.

56.4 Remedial Measures

Modification of all motor coaches by the replacement of the lower rated resistor Section 4 by the higher rated ones solved the problem and there were no further failures of resistors.

Measurement of the temperature rise of the different sections after the 1 hour run similar to the earlier test showed that the temperatures of the various sections were now all within limits (200 °C). In particular, the temperature of Section 4, which was excessive before modification, was now well within limits and less than the temperatures of Sections 5 and 6.

From the above analysis and results of action taken it became clear that while the increased loading of the trains was the proximate cause of the failures the root cause was the error in the initial design of the resistor sections. As there was adequate margin of safety, this deficiency was not exposed when the train loading was below normal; but when the loading started increasing with the growth of traffic, the resistor with the highest specific loading started to fail.

This episode shows how observation of the components that fail, the determination of failure mechanisms and scientific analysis of failure data and test results, point the way to effective and economical solutions to reliability problems.

56.5 Do's and Don'ts for Preventing Starting Resistor Failures

- When investigating resistor failures, (or for that matter, when investigating failures of any equipment or component), examine the equipment and its surroundings very carefully not only on the shop-floor, but also immediately after full load operation for several hours.
- If the component gets heated during operation check the temperatures at different parts and particularly at the usual location of failure.
- If possible, watch live operation (while taking adequate precautions against electric shock) of the equipment that fails.
- Compare the failure rates of different components of the same type and see whether higher failure rates can be correlated to any external factors.

Chapter 57

TAPCHANGER CONNECTION FAILURES

In this chapter we learn

- *A description of the on-load tapchanger and its associated equipment on 25 kV ac electric locomotives.*
- *A description of conductor fractures experienced on these locomotives.*
- *The manner in which the problem was analysed.*
- *The remedial measures applied.*

❑ ❑

57.1 Introduction

Figure 57.1 shows an outline sketch of a transformer fitted with a on-load tapchanger used on a 25000 Volt electric locomotive. It consists of three units fitted on the main transformer. These three units are (a) the tapchanger itself, (b) the transition resistor unit and (c) the on-load transition circuit breakers.

It will be noticed from the above figure that the main tapchanger unit is mounted directly on the transformer while the other two units (b) and (c) are supported on insulators. Copper conductors that are at a potential of approximately 19000 Volts are used to provide connections between the transformer and the three tapchanger units.

This was the arrangement originally provided by the manufacturers of the tapchanger.

57.2 Failure Modes and Failure Mechanisms

There were many failures due to fractures of the high voltage copper conductors shown in Figure 57.1 above. These failures usually occurred while the locomotives were in service. They resulted in short-circuits and even a few fires.

Examination of the fractured conductors showed that they were due to fatigue. The occurrence of fatigue fractures suggested the presence of vibration. The equipment being inside the locked high-tension compartment, it was not possible initially to verify whether

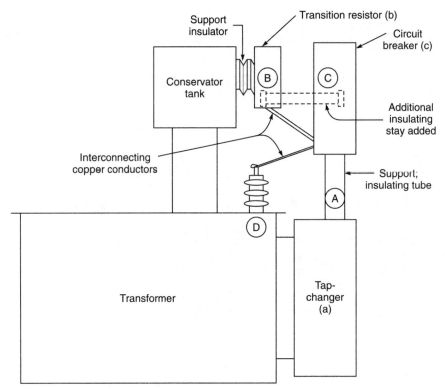

Figure 57.1 Block diagram of locomotive transformer, tapchanger and transition resistor. Additional insulating stay added after investigation is shown in dotted lines

there was such vibration. It was decided to run the locomotive dead by hauling it with another locomotive. The vibration levels were then measured with the help of a sensitive vibration meter at the points A, B, C and D as in Figure 57.1.

It was observed that the horizontal vibration amplitudes at A, B, and D were similar while the amplitude at C was 4 to 5 times the amplitude at the base viz. A. There was obviously flexing or bending of the insulating tubes T which supported the heavy mass of the circuit breakers C.

It was clear from these measurements that there was significant horizontal, relative vibration between the units C and (A, B, and D) due to the flexing or bending of the insulating tubes T. There was little relative vibration between the units A, B, and D. The fractures of the conductors followed the same pattern.

57.3 Remedial Measures

The problem was completely overcome by providing insulating stays in the direction of the vibration between the units B and C depicted by dotted lines in Figure 57.1. The circuit breaker unit C acquired the necessary rigidity in the required direction and its relative vibration with the other units ceased. Failures of the conductors due to fatigue fractures also ceased completely.

57.4 Conclusion

This investigation highlighted the usefulness of making direct observations and measurements under actual working conditions. In this particular case as the equipment was normally inaccessible due to its being at high voltage, it became necessary to haul the locomotive in the 'dead' condition by another locomotive in order to simulate the normal operating conditions of vibration.

Vibration is the first condition to be checked when investigating fractures of any components on electric locomotives and motor coaches.

Chapter 58

ELECTRICAL MACHINE FAILURES
DUE TO VENTILATION PROBLEMS

In this chapter we learn

- *The economic considerations underlying the use of forced cooling on transformers and motors.*
- *The need for protective devices to detect the failures of forced cooling systems.*
- *Case studies of electrical machine failures due to defects or deficiencies in the ventilation protective system.*
- *Case studies of electrical machine failures despite the provision of protective equipment.*
- *Prevention of protective system failures.* ❑ ❑

58.1 | Introduction

*M*odern electrical machines such as motors, transformers and generators are usually cooled by forcing air through the machine. Forced ventilation helps to increase the output of the machine. The additional size, weight and cost of the cooling system is comparatively much less than the savings in the size, weight and cost of the main machine itself when it is cooled by forced air. It makes sound economic sense to provide forced cooling, particularly for large machines.

Since the reliability and the durability of insulating materials depend on the temperature of the windings, failure of the cooling system can and often does lead to failures of insulation on the electrical machines. It is usual, to provide airflow or air pressure detectors and relays to switch off the machines in case of any failure in the cooling system.

It is difficult to provide for all types of failures of the cooling system. Two case studies which follow demonstrate how, despite the provision of an automatic cooling system failure detection system, electrical machine failures did occur in electric locomotives of a certain class, on account of poor ventilation through the motors.

58.2 | Failures of Bellows

CASE STUDY

One particular class of locomotives has six motors per locomotive, mounted in two sets of three each for two bogies. One blower cools each set of three motors which forces air through ducts and individual bellows over each motor as shown in Figure 58.1. The output duct of the blower is fitted with airflow relays, which cut off the supply to the locomotive in case of any reduction in airflow below a certain limit. This system can guard against only total failure of the blower as would be evident from the following case:

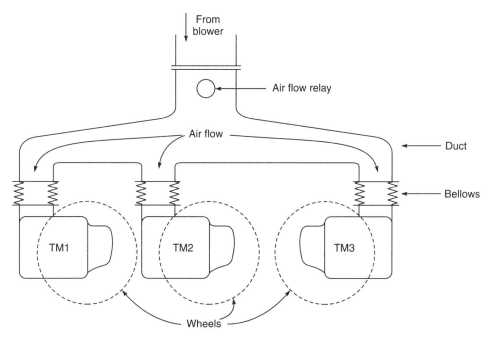

Figure 58.1 Cooling system for traction motors

In one case, the defect was bursting open (due to poor quality of the material) of one bellow connecting one of the three motors cooled by one blower. The result was that there was little air passing through that motor while the air flow output of the blower actually increased due to the reduced resistance to air flow. Naturally, there was no tripping of the airflow relay and the locomotive continued to operate in the defective condition. The traction motor got overheated and eventually failed due to an electrical insulation failure. The most vulnerable part of the cooling system was the bellow and there was in effect no protection against failure of this component.

The preventive action in this case consisted of the following steps:

(a) Development of an improved design of bellow, which could withstand the rigours of service without failure for at least three years—the interval between intermediate overhauls.

(b) Systematic replacement of all bellows every three years during the intermediate overhauls.

(c) Visual inspection of bellows during all inspection schedules.

(d) Rigorous inspection of supplies of bellows received from vendors.

58.3   Failures of Motors due to Bad Maintenance

CASE STUDY

In another case, the defect in the cooling system was of a different type. Due to negligence on the part of the maintenance staff, the growth of caked dust and oil blocked the air outlets from one of the three motors. The airflow through the motor was greatly reduced and it got overheated. Eventually the insulation deteriorated, developed short-circuits and the motor failed.

The air flow relay common to three motors did not operate as the normal air flow through the other two motors and the reduced air flow through the defective motor, was together more than the limiting flow adjusted on the air flow relay.

Some locomotives were fitted with air pressure relays, which sensed the air pressure developed by the blowers. This pressure develops as a result of the resistance to airflow through the motors. If the pressure drops below a certain limit, it is an indication of reduced air flow through the motor. The relay operates and switches off the supply to the locomotive. This type of relay operates in the event of a defect of the type described in case study 58.2 but it would be ineffective in the case of blocked ventilation even if all the three motors had the same defect.

The action taken in this case was (a) to improve the skills of the maintenance staff through job related training in this regard, (b) to increase the supervision on this aspect and (c) to increase the size of the mesh openings on the air outlets of the motors.

58.4   Motor Failures due to Defects in Inspection Cover Latches

CASE STUDY

In yet another case the defect in the ventilation system was that the inspection covers on the underside of the commutator chamber of one traction motor had fallen off due to a defect in the cover latch mechanism. This allowed all the air entering the traction motor just above the commutator to escape to the outside. Very little air flowed through the motor windings, which got overheated and eventually failed by developing short circuits.

The action taken in this case was:

- to replace all the inspection covers by new ones with an improved latch mechanism.
- to train the maintenance staff in regard of the dangers of inspection covers falling off.
- to introduce a ban on the operation of locomotives with defects in the ventilation system.

58.5 System Design Improvements

The three case studies in Sections 58.2 to 58.4 highlight the need for improvements in the protection system. In very large, force ventilated electrical machines, the sensing device provided is not on the air system but on the temperature of one of the windings. By monitoring its resistance continuously through a relay operated by the current through and the voltage across the winding, it is possible to detect any kind of defect in the ventilation system.

58.6 Motor Failures due to Defect in Ventilation Duct

CASE STUDY

In one case, after commissioning in a thermal power station, it was observed that during the initial trials, the exciter attached to a large turbo-generating set was running much hotter than the main generator. This was initially explained away by the manufacturer as a feature of modern design. The maintenance engineers, however, decided to investigate it further. On opening out the machine and inspecting the ventilation system it was found that:

(a) The exciter was self-ventilated by an internal fan on the main shaft.

(b) The exciter had two chambers on either side of the fan, separated by a partition cast into the exciter casing. This partition had a large rectangular opening with machined edges below the fan.

(c) The rectangular opening below the fan was obviously meant to be closed by a steel sheet. In the absence of such a closure the air blown by the fan was short-circuited through the opening between the suction and the delivery chambers of the fan. Very little air flowed through the machine and overheating was a consequence of this defect. See Figure 58.2.

The action taken in this case was to provide a sheet steel closure over the opening. The temperature of the exciter immediately dropped by more than 50 °C. This was not a case of

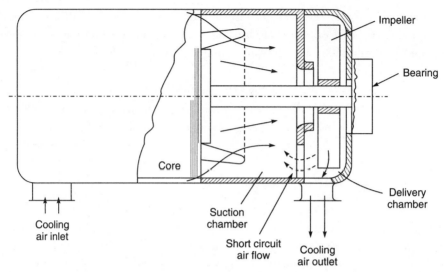

Figure 58.2 Cooling system of a self ventilated exciter

failure, because the defect was detected and corrected in time. Otherwise, the winding insulation would have failed prematurely.

58.7 Power Transformer Failures due to Defects in the Cooling System

CASE STUDY

Large power transformers are invariably provided with forced air or water cooling and also forced oil flow over the windings. Both these systems are very important for ensuring the reliability and the durability of the transformer coil insulation.

The most effective protection for large transformers is through the oil or winding temperature relays. These will operate eventually in the case of any failure in the cooling system; but it is desirable to provide relays to detect reduction in the airflow and in the oil flow. These will act faster and give more time for effecting repairs.

In one particular case an oil pump was used to circulate the oil through the transformer and the radiator. The oil pressure at the delivery end of the oil pump was used to monitor the proper functioning of the oil pump. There was no temperature-sensing device on the transformer. The valve on the suction side of the oil pump was inadvertently kept nearly closed. The oil flow through the radiator was greatly reduced and the transformer became overheated.

Fortunately the defective setting of the valve was discovered within a few days. It was also observed that the transformer oil temperature was well above the permissible limit. Opening of the valve fully led to an immediate reduction in oil temperature. A failure was averted but it was clear that if the defect had not been detected the transformer might have failed sooner or later.

58.8 Prevention of Failures of Protective Devices in Ventilation Systems

Since failures of protective devices in combination with defects in the ventilation systems can lead to expensive damages to large electrical machines like motors, transformers and rectifiers, it is necessary to ensure the installation of the appropriate protective systems and then to prevent failures of such systems.

Protective device failures are usually due to failures of items like springs, wires, sockets, solenoids, contacts, etc. and these subjects are dealt with in Chapters 18, 34, 32, 50, and 43.

58.9 Do's and Don'ts for Preventing Cooling System Failures

- Ensure that maintenance of cooling systems is given the same kind of importance as the maintenance of the main machines.
- Ensure that maintenance check lists include all possible ways in which cooling air flow can be affected. Provide job related training to all the operating and maintenance staff specially in this regard.
- In the case of large machines it is desirable to provide additional protection in the form of temperature sensing devices on the main machines.
- Ensure that protective gear fitted on the cooling system can detect all possible types of defects and arrange to test it periodically.
- Ensure that radiator and pump valves that control oil flow through the cooling systems are fully open, before re-commissioning any large transformer after maintenance shut-downs.
- Monitor the full load oil temperatures of the transformer and if there are any changes in comparison with the normal pattern as recorded in the log books, investigate them fully to determine the causes.

POWER TRANSFORMER EXPLOSIONS

In this chapter we learn

- *Transformers have the inherent potential for very high reliability, as they are static devices.*
- *The main cause of transformer explosions is inter-turn shorts due to shrinkage of insulating coil separators.*
- *Use of pre-compressed separators, curing of coil assemblies, provision of pressure plates at coil-stack ends and the periodic tightening of the coil pressure screws will prevent inter-turn shorts.*
- *12 other miscellaneous causes of transformer failures.*
- *Case studies.* ❏ ❏

59.1 Introduction

$\mathcal{P}$ower transformers are static devices with no wearing parts. Yet their failures are not rare and are due to only one simple deficiency. The cost of eliminating the cause of failure is negligible in comparison with the total cost of repairing the transformer.

This main cause will be discussed in detail here in the Sections 59.2 to 59.6. The other causes and the failure mechanisms are, however, briefly enumerated in Section 59.7.

59.2 Design Consideration Relating to Wedges or Separators in Coil Stacks

The coil stacks in power transformers are usually in the form of pancake or disc shaped coils, assembled one over another. Wedges or separators separate the pancakes. See Figure 59.1.

The main purpose of these wedges or separators is two-fold. Firstly they provide increased insulation between the turns of adjacent coils, secondly they provide passages for the flow of the insulating oil, which helps to reduce the operating temperature of the coils. If any of these separators become loose and fall out, turns of adjacent pancakes may not only

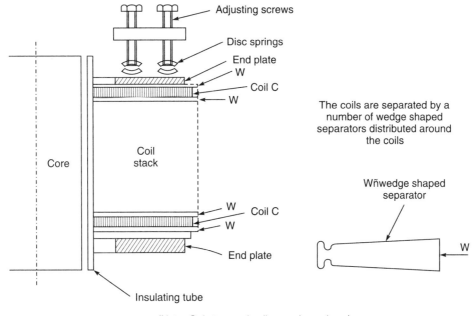

(Note: Only two end coils are shown here)

Figure 59.1 Schematic arrangement for keeping transformer coils under pressure

touch each other but also rub against each other thereby causing wear and tear on the conductor insulation.

The maximum inter-turn voltage between the conductors of adjacent coils is $2 \times n \times v$, where v is the voltage per turn and n is the number of turns per pancake. Thus, in the event of coil separators falling out, the maximum voltage between turns that may touch each other may increase 20 to 100 times as compared to the normal level.

The insulation covering the copper conductors is usually a spirally wound paper strip. This strip is very weak in tension specially when soaked in transformer oil. However it is extremely strong and durable when under compression. Hence the pancakes of insulated copper conductors are held under compression which is heavy enough to immobilise the conductors and prevent even the smallest relative movement between conductors. If this compression between pancakes gets reduced due to shrinkage or falling out of the separators, the coil assembly begins to develop relative movement between conductors.

The combination of increased voltage between turns that touch and reduced insulation strength due to wear and tear is certain sooner or later to cause a short circuit, thereby leading to heavy circulating currents, overheating/arcing and possibly an explosion.

The separators or wedges are made of pre-compressed paperboard. Although, these separators are pre-compressed, they continue to shrink in thickness, slowly but surely, under the influence of heat and pressure. Hence spring loaded pressure plates are provided at the tops of the coil stacks to compensate the shrinkage and to maintain the coil pancakes and their separators under a heavy and continuous pressure.

59.3 Failures due to Defective Material of Separators

CASE STUDY

In one case of a transformer, the coil separators used by the manufacturer were not pre-compressed. As a result they shrank very rapidly and greatly. Within a few months of service after commissioning, the coil stack assembly became loose and some separators fell out due to vibrations. The coil conductors of adjacent pancakes came in contact with each other. Due to vibrations and abrasion, the insulation was damaged and there were inter-turn shorts, overheating and arcing. As this transformer was not provided with Buchholz relays, the failure resulted in an explosion, spilling of oil and fire.

A Buchholz relay is, basically, a float switch that operates and causes the tripping of the circuit breaker in the main power supply line to the transformer as soon as a small quantity of gas enters it. This relay is mounted at the highest point in the transformer. Any little gas that may be produced anywhere in the transformer due to overheating or arcing would bubble to the top and enter the Buchholz relay. Thus, this relay will operate when the fault is still in its nascent stage thereby preventing a major explosion as a result of an inter-turn short in a power transformer.

It may be recalled that over-current, earth leakage and differential relays will not operate even when the local inter-turn currents are extremely high. It should also be noted that Buchholz relays do not prevent inter-turn shorts. They only act fast causing the breaker to trip and prevent explosions. When this relay operates, the transformer has to be opened out for repairs.

59.4 Failures due to Lack of Pre-curing of the Coils

CASE STUDY

Even if the coil separators are of the required pre-compressed quality, they do shrink a little over years of operation, in situ between the coils after assembly. To allow for this further shrinkage it is normal practice to subject the coil stacks to three or four cycles of alternate heating, pressing and cooling. The coil stacks shrink by a few per cent during this process but stabilise thereafter. These 'cured' coils are assembled in the transformer. The shrinkage thereafter is very slow and is looked after by the spring loaded pressure plates on the tops of the coil stacks.

In this particular case, the manufacturer did not carry out this curing process satisfactorily before assembly of the coil stacks in the transformer. The shrinkage that accumulated over the next four years of service was sufficiently high to cause the separators to become loose in their settings and to fall out. The mechanism of failure thereafter was the same as discussed in case study 59.3.

59.5 Failures due to Neglect of Shrinkage Compensating Devices

CASE STUDY

In yet another similar case, the coil separators were of the required quality, and the manufacturer had carried out the curing process described above but even these 'cured' coils were subject to some very slow shrinkage over a long period. As the coil pressure screws were not re-tightened during the general overhaul after six years of service, the total shrinkage exceeded the limit allowed by the deflection of pressure plate springs. The failure mechanism thereafter was the same as discussed in case study 59.3, the only difference being that the failure occurred after nearly ten years of service instead of a few months as in case study 59.3.

The lessons from the three case studies discussed above are that:

(a) Pre-compressed dovetailed separators should be used.

(b) The coil stacks should be cured in three to four cycles—each comprising heating, pressing and cooling—before final assembly in the transformer.

(c) The pressure plate screws should be re-tightened during each general overhaul.

If these three precautions are followed accurately, transformer failures due to inter-turn shorts will be totally eliminated.

59.6 Other Causes of Transformer Failures

We have discussed in the foregoing sections the causes and the failure mechanism of one common type or mode of failure viz., explosion of the transformer due to shrinkage of the coil separators. There are many other causes and failure mechanisms but these are individually not frequent. All of them put together may contribute about ten per cent of transformer failures and explosions. If the kind of defect discussed in the foregoing sections is not present, the following possibilities may be considered when investigating transformer failures. They should be considered also when designing transformers or when carrying out reliability reviews of transformer designs.

All the probable causes enumerated below with their failure mechanisms are based on actual cases. The nature and magnitude of the consequences or effects of the failures varied from case to case. In some cases there were explosions. In some cases there was only a little overflow of oil. In some cases protective devices operated fast and there were no external damages. It depends partly on chance and partly on the efficacy of the protective gear; but the fact remains that every case of an electrical short-circuit inside a transformer tank is a potential cause for an explosion, especially when the insulating fluid is of a petroleum base oil. Even when there is no explosion, the transformer has to be opened out for repairs.

(a) Accumulation of sludge (a product of the normal degradation of the transformer oil) at the bottom of the tank due to poor maintenance; interference with the convection of oil; overheating of conductor insulation; inter-turn short; an explosion.

(b) Design or manufacturing defects in the arrangement of insulating cylinders, baffles, wedges, etc. around or inside the coil stacks, leading to pockets or zones where oil flow or convection is restricted; local overheating of conductor insulation; inter-turn short-circuit; hot spots; and finally an explosion.

(c) Failure of crimped joint in internal connection between coils, to terminals or to tap-changer (See Chapter 32 crimped joint failures); local overheating in coil; possibly insulation failure; short-circuit; an explosion.

(d) Failure of bolted joint in internal connections between coils, to terminals, or to tap-changer (See Chapter 43 regarding contact failures); local overheating; possibly an insulation failure; a short-circuit; an explosion.

(e) Fracture of internal connection bus bar at bend or of brazed/bolted joint (See Chapters 46 regarding bus-bar failures); arcing; short-circuit and explosion.

(f) Tracking over or between the laminae of laminated wood cleats. (See Chapter 39 regarding laminated board failures due to tracking); short-circuit and explosion.

(g) Inadequate contact force between moving and fixed contacts of tap-changer due to design or manufacturing defect; overheating; charring of insulating materials; short-circuit and explosion. (See Chapter 43 regarding contact failures).

(h) Sustained overloads exceeding the overload capacity of the transformer; overheating of conductors and insulation; degradation of insulation; inter-turn short-circuits; arcing; explosion.

(i) Mechanical damage to conductor insulation during manufacture; inter-turn short; over-heating; short-circuit; arcing; explosion.

(j) Local overheating of conductors due to proximity with steel components and eddy currents; degradation of insulation; interturn shorts; short-circuits; arcing; explosion.

(k) Sharp edges on conductors (defect in material), damage to conductor insulation, inter-turn short-circuits, arcing, explosion.

(l) Moisture penetration due to leakage in transformer tank conservator and defective silica-gel breather; moisture entry into inter-turn insulation; degradation of insulation; inter-turn short-circuit; arcing; explosion.

The remedies in the above cases are obvious. In fact, in any failure of any type in any equipment, truly effective remedies become obvious when the true cause and failure mechanism is established. This is crucial because the failure mode and the later stages of the failure mechanism are generally the same or very similar. It is very important, therefore, to establish clearly the early stages of failure mechanisms.

59.7 Do's and Don'ts for Preventing Transformer Failures

- Ensure that coil separators or wedges are made of pre-compressed hardboard of reputed make.
- Ensure that coil stacks are cured in three or more heating/pressing/cooling cycles until stack length stabilises.
- Ensure that the coil stacks are held under very high axial pressure and that the end plate is held in place through disc springs and screws to compensate for residual shrinkage.
- Ensure that the end plate screws are re-tightened ever six years.
- Ensure that all the materials used in the manufacture of the transformer and the finished transformer are subjected to acceptance tests in accordance with the appropriate specifications.
- Minimise the use of bolted joints inside transformers. Brazed joints are preferred.
- In investigating transformer failures and in reviewing designs and maintenance practices look for and eliminate the defects enumerated in Section 59.6.

Chapter 60

FAILURES OF OHE
REGULATING EQUIPMENT

In this chapter we learn

- *Description and applications of OHE tension regulating equipment on electrified railway track.*
- *Failure modes in tension regulating equipment.*
- *Failure mechanisms.*
- *Preventive or remedial measures.*

❏ ❏

60.1 Introduction

Overhead lines for carrying electric current usually consist of copper or steel cored aluminium wires strung between poles or towers. The wires are stretched tight in order to minimise the sag. The magnitude of the sag is adjusted so that the tension in the wires is kept within the limits determined by the tensile strength of the material of the wires. Due to the expansion and contraction of the wires caused by temperature changes, the sag has to be correlated to the temperature of the wire.

In short therefore, the sag as also the tension in the wires keeps changing with changes in temperature. The designer has to determine the spans between towers and the heights of the towers for minimum initial cost while ensuring that:

- The required minimum clearances to ground are ensured at the highest operating temperatures of the lines.
- The tension in the wires does not exceed the safe limits for the material at the lowest operating temperatures of the lines.

In the case of track overhead lines on electrified railways, there are certain additional requirements that must be ensured. The level and the stagger of the line from the centre line must remain within certain specified limits for satisfactory current collection. This is possible where the temperature variations are not very high as in the coastal regions. However, in many parts of India where summer temperatures are very high and winter temperatures are very low, it is not possible to balance the requirements. It then becomes necessary to provide devices to maintain a constant tension in the overhead lines despite the changes in their lengths with changing temperatures. This is done by the regulating equipment, which consist of a system of rope and pulleys, and counterweights. See Figure 60.1 for a schematic view of this device.

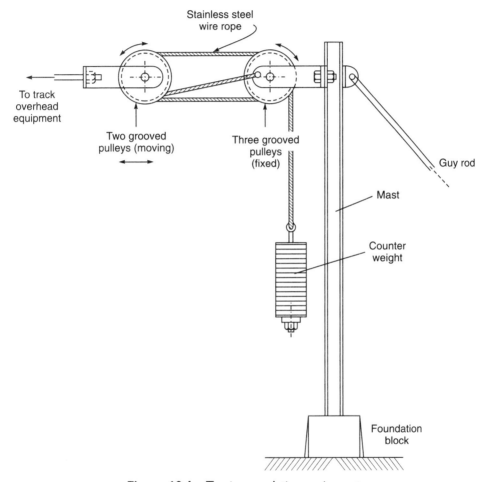

Figure 60.1 Tension regulating equipment

60.2 Failures of Regulating Equipment

The function of the regulating equipment is to maintain a constant tension in the wire. It does this by moving out or in as required. Failure in this regard can occur due to any one of the following reasons:

- Bearing failures.
- Wire rope failures.

These two failure modes have been discussed in Chapters 17 and 10, respectively. There may occasionally be other component fractures due to corrosion or flaws in the metal. Defective manufacture or defective installation can also lead to failures of components.

Special care has to be taken when regulating equipment is installed, because any failure in it can make matters worse than conditions where there is no regulating equipment at all. For instance, if the bearings seize when the temperatures are high, the tension can become excessive when temperatures fall. Similarly if the bearings seize when the temperatures are low, the sag can become excessive when the temperatures rise to high levels.

60.3 Do's and Don'ts for Preventing Failures of Regulating Equipment

- Regulating equipment defects can lead to pantograph/OHE entanglements. Examine the concerned equipment carefully whenever there is any such accident. If any defects are seen, look for similar defects in other regulating equipment installed in the same Division. Eliminate all such defects.
- Refer to the Do's and Don'ts in Chapters 10, 11, 17 as relevant if there are defects in the wire rope or the bearings. If there are signs of frayed ropes, replace the rope. If the bearings have seized replace the bearings and check the seals that are supposed to stop water entry into the bearings.
- Test the regulating equipment periodically by raising and lowering the counter-weight slightly by hand and assessing the required force.
- Check the position of the counterweight with reference to the ambient temperature. Do this in the evenings or early in the mornings when there is no direct heating of the wire by the sun.

Chapter 61

MINIMUM OIL
CIRCUIT-BREAKER FAILURES

In this chapter we learn

- *A description of the protection system on track over-head lines of electrified railways.*
- *A description of one particular type of failure of circuit breakers.*
- *The manner in which the failure was investigated and the root cause of the failure as surmised after investigation.*
- *The corrective measures taken to prevent further cases of this type.*

61.1 Introduction

Although this chapter deals with minimum oil circuit breakers used in electric sub-stations, that supply power to track over-head equipment on electrified railways, the observations and conclusions there-in are equally relevant to circuit breakers provided in any electrical distribution system. These circuit breakers automatically cut off power supply to the systems, if and when they develop faults or short circuits. The circuit breakers detect and isolate such faults within a fraction of a second thereby minimising the damage at the point where the fault has occurred. The circuit breakers are specially designed to interrupt the very high fault currents, which may be ten or more times the normal operating currents. There is some wear and tear inside the circuit breaker but it can do such operations many times before they have to be opened out for maintenance, when worn parts can be replaced.

There are many types of circuit breakers, e.g. Oil-, Minimum Oil-, Air Blast-, Vacuum-, SF6-, — etc. This list is generally in order of their development and increasing fault rupturing capacity, reliability and maintainability. We will discuss here certain failures of the second type listed here viz. Minimum Oil Circuit Breaker (MOCB).

61.2 Failures of MOCBs due to Use of Defective Spares

CASE STUDY

When a short-circuit occurred on a track section controlled by one such MOCB, it failed to clear the fault. In its attempt to do so, the MOCB itself exploded and created another short-circuit. The situation was prevented from further aggravation by the correct operation of other back-up protection circuit breakers in the substation.

Initially, it was thought that the circuit breaker did not have the required rupturing capacity since the substation was close to a large thermal power station and the fault it was trying to clear was also close to the substation. The fault MVA was, therefore, very high.

On collecting the various components of the exploded circuit breaker and re-assembling them as far as possible it was observed that the replaceable tip of the moving contact was stuck in the fixed contact and there was extensive arcing and melting of the moving contact rod as shown in Figure 61.1.

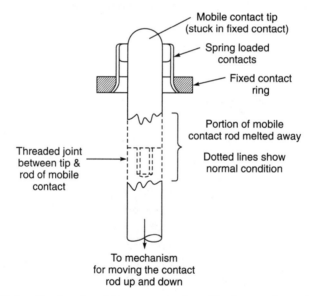

Figure 61.1 Fixed and mobile contact as found in damaged circuit breaker

It was obvious that the threaded joint between the replaceable tip and the moving contact rod had failed by stripping of the threads. On checking similar tips obtained from the stores with the rod of another circuit breaker, it was seen that some of the tips were very loose on the threaded ends of the rods until it was finally tightened after full penetration. On checking the internal thread dimensions, it was seen that the internal diameter of the thread on the contact tips was excessive. Clearly, the machining of the internal thread of the replaceable tip was defective. Further investigation revealed that these tips were manufactured locally and the internal threads had not been inspected correctly.

All the moving contact tips in stock and in the circuit breakers were checked and most of them were discarded. New tips were obtained with special emphasis on inspection of all dimensions and materials. The tips were tightened carefully and fully. There were no further failures on these circuit breakers.

61.3 Conclusion

Careful investigation showed that the failure to clear the fault was due to a manufacturing defect in the circuit breaker and not due to any deficiency in its rupturing capacity. The failures were due to a defect in the manufacture of the spares used during maintenance.

This case illustrates the need for careful investigation before arriving at any conclusion regarding the root cause of a failure. Many different defects lead to the same mode of failure and it is only after examination of all the physical evidence and results of inspection and tests on other unaffected equipment in the same zone that it is possible to reach the root cause of the failure.

This case also highlights the need for care in the procurement of spare parts used in the maintenance of sophisticated equipment. Spares should be either purchased from the original equipment manufacturers or obtained against drawings and specifications giving full details such as material specifications, manufacturing tolerances on all dimensions and other quality assurance details.

Chapter 62

LINE BREAKER FAILURES

In this chapter we learn

- *Description of line breakers provided on one class of DC EMUs (Electric Multiple Units) for the protection of the high voltage equipment.*
- *Modes of failures which had been occurring on these circuit breakers from the time they were installed many years earlier.*
- *Revelation of certain unexplainable patterns in the dates of failures.*
- *Eventual discovery of a rational explanation for the observed patterns led to the provision of surge absorbing capacitors and subsequent satisfactory performance.* ❑ ❑

62.1 Introduction

*T*he term line breaker is applied to high voltage (1500 volts) circuit breakers provided in DC EMUs for the protection of all the high voltage circuits and the traction motors. They are opened out automatically when any fault is detected by over-load or earth fault relays provided in the high voltage circuit. The relevant part of the schematic circuit diagram is shown in Figure 62.1.

There were, from time to time, incidents of short-circuits on the line side of line breakers provided in one particular class of 1500 volt dc EMUs.

As the shorts were on the line side, the faults could get isolated only by the operation of the roof fuses. These fuses sometimes did not operate soon enough, and the circuit breakers in the substations opened to clear the faults. There were cases where even these circuit breakers failed to clear the faults due to some weaknesses in the protection system and in such cases there were fires in the motor coaches.

There were two to four such incidents every year and these continued for many years almost from the time these coaches were commissioned. The line side components of the circuit breakers were given special attention with regard to cleaning of tracking surfaces, tightness of fasteners, replacements of old insulation etc. but all efforts to determine the causes and to prevent their recurrence proved unsuccessful.

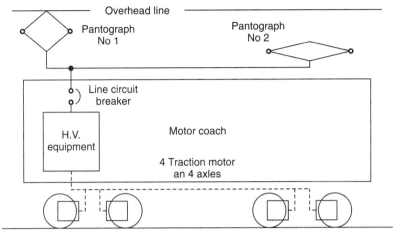

Figure 62.1 General arrangement of high voltage equipment in motor coach

62.2 Failure Investigation through Statistical Analysis of Failure Data

CASE STUDY

As meticulous records of all failures on the EMU stock were maintained since their commissioning, failure reports were available for nearly 23 years when this subject was taken up for investigation. All reports of similar failures were collected. There were more than 50 such reports. These reports included such details about each incident as the name of the motorman, date, time, place of occurrence, last date of inspection, last date of overhaul, serial number of the equipment, observations by motorman, observations by maintenance foreman, etc.

All the available data were tabulated to see whether there was any common factor between these 50 odd failures. At first nothing special was noticed but a very peculiar and unexplainable observation was that all these failures except one or two had occurred on even dates such as 2, 4, 6, 8 etc. No rational explanation could be given for this bizarre fact until after some considerable time, when it was noted that there was indeed a rational explanation, which led to the correct remedial measure.

These motor coaches were all fitted with two pantographs each. In order to equalise the wear on these two pantographs, there was a traditional operating procedure in force over the previous twenty three years, according to which the pantographs on the north side of the coach were utilised on even dates of the month (such as 2, 4, 6 etc.) and those on the south side were utilised on odd dates. The pantographs of the north side were close to the high tension compartments which housed the line breakers. A very short lead from the north side pantograph took the supply to the line breakers while the south side pantographs were connected in parallel through a 15 metre long cable running the length of the coach.

It was concluded tentatively that the long cables from the south side pantographs helped to absorb or quench high voltage switching surges from the traction lines, and that the failures on the line breakers were due to these high voltage surges.

62.3 Preventive Measures

It was therefore decided to install 4 µF, 2000 volt, dc capacitors on the roofs and to connect them to the north side pantographs. On completing these modifications, the failures of line breakers were completely eliminated thus confirming the earlier tentative conclusion. In addition a few other types of failures of equipment in the high-tension compartments were reduced in frequency. The choice of the capacitor ratings was based on reports of such equipment fitted on some other foreign railway.

62.4 Conclusion

This episode highlighted again the need for maintaining detailed failure data, and the usefulness of the method of tabulation of failure data in determining the true failure mechanism and the appropriate remedial measures.

Chapter 63

Squirrel Cage Failures in Asynchronous Motors

In this chapter we learn

- *Failures of squirrel cage induction motors due to fractures of rotor bars.*
- *Failure mechanisms involved in these fractures.*
- *Condition monitoring of rotors to detect bar fractures inside the slots.*
- *Design and manufacturing features, which can help to minimise rotor bar fractures.*
- *Failures of die cast rotors.*

63.1 Introduction

$\mathcal{T}$he most common and popular electric motor is the squirrel cage induction or asynchronous motor. This type of motor has no commutator, no brush-gear and no wearing parts except the bearings. Its advantages include: nearly constant speed (a speed reduction of less than 4 per cent from no load to full load); good reliability and good maintainability.

There are five failure modes that account for more than 95 per cent of all failures on squirrel cage induction motors. These are: Stator winding failure (30%), rotor squirrel cage failure (30%), bearing failures (30%), stator lead failure (3%), terminal board failure (2%).

In this chapter we will discuss only one failure mode viz. fracture of squirrel cage. This type of failure was observed on motors used for driving auxiliary machines like compressors, blowers, oil pumps and exhausters. Bearing failures have been discussed in Chapter 10, terminal failures in Chapter 39 and stator winding failures in Chapter 64.

63.2 Mechanical Stresses in a Squirrel Cage

The squirrel cage consists of a cage like structure built with straight copper bars brazed into slots in two copper end rings. The straight bars of the cage are embedded in the magnetic core of the rotor and the rings are outside the core. See Figure 63.1.

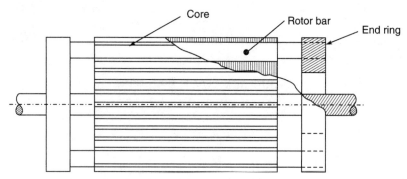

Figure 63.1 Squirrel cage rotor of induction motor

The cage constitutes a shorted winding. Currents in the straight bars are induced by and interact with the rotating magnetic field produced by the stator winding. This interaction produces the driving force of the motor. The voltages between the ends of the conductors are extremely low because the induced voltages in the conductors are used up almost fully against the conductor impedance. Hence the conductors do not need any insulation.

While insulation failures on the rotor are thus eliminated, it is observed that sometimes the rotor bars develop cracks and this leads eventually to failure of the motor to develop the required torque. The failure mechanism is as follows:

63.3 Failure Mechanism in Squirrel Cages

The copper bars of the cage are usually designed to operate at temperatures which are some-what lower than 150 °C, the temperature at which copper begins to succumb to metal creep. However there is another physical phenomenon which is ever present: thermal expansion and contraction. If a copper bar is allowed to expand or contract freely, there is no mechanical stress in the copper; but if the bar is physically restrained, a compressive stress is produced on heating and a tensile stress is produced on cooling. The magnitude of the stress is sufficient to cause fatigue fracture when the temperature cycle amplitude is of the order of 50 °C or more.

Motors, which start and stop frequently or those subject to constant variations in load, develop constant changes in the temperature of the rotor bars. As the bars are often held firmly in the slots of the magnetic core and as the expansion/contraction of the copper is more than that of the core, the copper bars are subject to alternating tensile and compressive stresses. These lead to fatigue fractures of the copper bars within the slot. If these are not detected in time and the motor is allowed to continue in service, the electrical and thermal loading on the remaining bars increases. Eventually there will be so many breaks in the cage that the motor will not develop enough torque to start and run as desired.

There is another similar but more severe failure mechanism for squirrel cage bars. The rotor bar currents pass through the end rings and they also get heated up and cooled down depending upon the load variations. Since this is much more than the heating/cooling of the core, the cantilevered ends of the copper bars are subjected to alternate bending cycles as shown in Figure 63.2. This produces bending stresses in the copper bars, which are highest at the point of entry in the core and into the ring. Fatigue cracks and fractures of the copper bars at these points are the result of this failure mechanism.

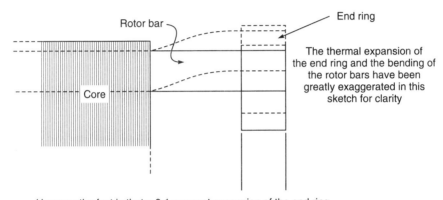

Rotor bar

End ring

Core

The thermal expansion of the end ring and the bending of the rotor bars have been greatly exaggerated in this sketch for clarity

However the fact is that a 0.1 per cent expansion of the end ring diameter is enough to produce a breaking stress in the copper bar

Figure 63.2 Effects of end ring expansion

63.4 Detection of Cracked Bars in Squirrel Cages

As the copper bars are inside the core slots it is often not possible to detect bar cracks inside the slots. Even the cracks in the cantilevered ends of the bars outside the slot are not easy to detect by visual examination.

It is possible to detect cracks by measuring the impedance of one phase of the winding while turning the rotor slowly by hand. There will be some variation at slot pitch intervals but if a significantly higher variation is seen at some locations around the full circle, it can be presumed that there is an open circuit in one of the slots. It is then necessary to remove the rotor from the motor. The defective slot can be determined with the help of a 'growler' or by measuring the impedance of a specially wound stator with a coil around only two slots.

When cracks in rotor bars are detected, the only course open is to replace either the complete rotor or the complete cage. No effort should be made to replace individual bars since all the other bars would be in various stages of fatigue fracture and will fail sooner or later.

63.5 Prevention of Cracks in Rotor Bars

It is possible to minimise or even to prevent cracking of rotor bars by selecting the dimensions of the slots, rings and bars so that:

(a) The rotor bars have a tight fit in the slots in the centre of the core at AA as shown in Figure 63.3.

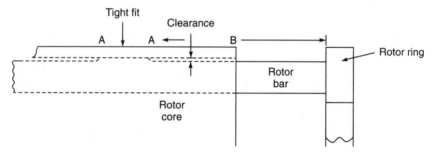

Figure 63.3   Provision of radial clearance in slot to minimise bending stress

(b) The rotor bars have a transition fit in the slots circumferentially in the rest of the slot on either side of AA.

(c) The rotor bars have an adequate clearance radially on the outer side of the slots as shown in Figure 63.3 to permit the outward bending of the bars over part of the slot length AB, thereby reducing the magnitude of the bending stress.

(d) The rotor bars are pressed down in the slots while brazing to the end ring, thereby ensuring that the radial clearance is fully on the outer side.

(e) The temperature rise of the end ring should be less than the temperature rise of the core, by a magnitude such that the bending stresses in the rotor bars are kept very low.

The above measures, which can be taken only by the manufacturer or at the time of cage replacement, will enable the cage bars to expand and contract freely and to bend over a longer length thereby reducing the stress level. It may be emphasised that thermal expansion or contraction cannot be prevented. Any effort to do so produces very high stresses. The best course is always to allow the expansion and contraction to take place with least constraint.

63.6 Die Cast Aluminium Alloy Squirrel Cages

In many designs of induction motors, aluminium alloy die-castings are provided around the core to form the squirrel cage. Due to the very close contact between the core and the cage, the temperature differential between them is smaller. This design is generally used for small motors and the temperature rise is also limited to a lower value. Cracking of the cage is noticed very rarely. There have been some cases, of voids in the die-castings due to defects in the die-casting process. These can be detected by the same method as described in Section 63.4.

63.7 Conclusion

Failures of rotor bars usually occur amongst motors which start and stop frequently and on high power motors designed to operate at high temperatures by the use of class H and Kaptan insulating materials. The root cause is the fluctuating temperature of the rotor bars and the end rings in comparison with the core temperature. Thermal expansion and contraction of the squirrel cage creates high stresses in the cages, which are often prevented from expanding and contracting. The only way to prevent such failures is to provide room for free expansion and contraction of the cages while at the same time preventing the bars and the cage from vibrating in the slots.

Chapter 64

FAILURES OF MUSH
WOUND STATOR WINDINGS

In this chapter we learn

- *Important constructional features of stator windings.*
- *The possibilities of high voltages appearing between adjacent turns of mush windings.*
- *The generation of very high transient voltages when induction motors are switched off.*
- *Mechanism of winding failures due to the above factors.*
- *Measures that can be taken to prevent such failures.*
- *Other causes of stator winding failures and the appropriate remedial measures.*

64.1 Introduction

$\mathcal{S}$tators of small and medium sized induction or asynchronous motors are usually provided with mush windings. These are windings in which the conductors in a slot are not arranged to conform to any specific pattern. Such mush windings are usually prepared by winding the wire (which may be enamelled, silk or cotton covered) either directly or manually into the appropriate slots in the stator or onto a diamond shaped coil on a wooden mandrel. In the latter case, the entire bunch of coils is later placed manually in the slot. Figure 64.1 shows the coils and how they are placed in the stator. Figure 64.2 shows, for comparison, the manner in which coils made from rectangular conductors are placed in the slots of larger machines.

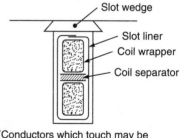

(Conductors which touch may be more than one turn away)

Figure 64.1 Mush-wound coil in slot

The slots are usually lined with insulating liners, which provide the main or ground insulation between the winding and the grounded core. The enamel or the silk/cotton cover on the wires is mainly for inter-turn insulation. The voltage between turns is generally much smaller than the voltage between the winding and the core.

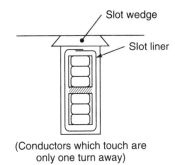

(Conductors which touch are only one turn away)

Figure 64.2 Rectangular conductors in slot

64.2 Failure Modes and Mechanisms in Mush Windings

In general the ground insulation (the slot liner) fails very rarely. The most common failure mode of mush wound stators of asynchronous motors, which contribute about 30 per cent of all failures on these machines, is inter-turn short across the conductor insulation (enamel, silk or cotton covering).

The maximum voltage between consecutive turns of the coil is usually less than 1 per cent of the inter-turn break-down voltage; but due to the random nature of the winding, the voltage between turns which touch each other inside the slot could be ten times this voltage per turn. Even then there is more than adequate margin of safety. However, it is possible and actually very probable that in the end portion or involute portion of the winding where the coil bunch turns round and returns to a slot one pitch away, the first turn in a coil may fall adjacent to the last turn of the next coil. The inter-turn voltage at such points may now reach a value, which is about 30 per cent of the B.D.V. of the insulation. This is still not enough to cause immediate breakdown. If now there is a contributory factor such as local damage to the insulation in the critical zone, or generation of high voltage standing waves in the coil, winding failure can occur after some service. This latter factor viz. high voltage standing waves calls for some explanation given in the next section.

64.3 High Voltage Standing Waves

The winding of an asynchronous motor has its own internal inductance and also its own inter-turn capacitance. When the motor is running there is a constant interchange of energy between the magnetic core and the supply system. When the motor is switched off, depending upon the magnitude of the instantaneous current at that instant there is a certain amount of energy in the magnetic field. When the motor is disconnected from the supply system all this magnetic energy is unable to return to the supply system. It begins to charge the capacitance of the system to convert it into electrostatic energy. The inductance/capacitance system then forms a high frequency oscillatory circuit. The peak voltages generated are about ten times the normal voltage but of very short duration.

Degradation of the insulation due to these high-voltage, short-time, surges now begins at the points where even the normal inter-turn voltage is very high as explained in Section 64.3. Every time the motor is switched off, a certain amount of damage to the insulation takes place due to partial discharges. When the accumulated damage reaches a point where the residual strength of the insulation is inadequate to withstand even the normal operating voltage, a short-circuit is developed. Heavy current, overheating, possibly arcing and charring of the insulation follows.

64.4 Measures to Minimise or Prevent Mush Winding Failures

The following steps should be taken to minimise or prevent failures of mush windings of asynchronous motors:

(a) The intrinsic quality of the insulated wires should be checked by carrying out cut-through tests and twisted wire B.D.V. tests on baked samples in accordance with the relevant standard test methods.

(b) Care should be taken during coil winding to ensure that the conductor insulation is not damaged by scraping, sharp bending, hammering etc.

(c) Insulating sheet liners should be placed between the coils of different phases.

(d) The end or involute portions of the coils should be taped to prevent direct contact between the turns of one coil with turns of the next coil.

(e) After completing the winding, surge comparison tests at a voltage roughly ten times the normal voltage should be carried out on the winding. Motors, which display irregularities in the waveforms, should be rewound.

(f) The windings should be vacuum impregnated with solvent-less varnishes.

(g) In the case of motors, which start and stop frequently, capacitors of voltage rating at least three times the normal operating voltage should be connected in parallel to the windings. See Figure 64.3. These capacitors should be placed close to the motor terminals.

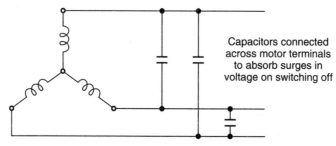

Figure 64.3 Switching surge absorption capacitors

64.5 Other Causes of Asynchronous Motor Failures

There have been cases of asynchronous motor failures, in which the stator windings were charred or burned out in a manner similar to those resulting from the failure mode described in Sections 64.2 to 64.5; but the root cause was different. In these cases the initial short had occurred between the leads and the winding. Often the leads are tied on to the end windings and taken round to a location in line with the terminal box. The leads are insulated but the level of insulation is not enough. Since the voltages developed by the standing waves described in Section 64.4 affect the leads, it is necessary to ensure that the level of insulation on the leads is adequate to withstand such voltages.

Another cause of failure of stator windings is single phasing. This could be due to an open circuit anywhere in the long line up to the main supply point. This could be due to a defect or failure in any cable, terminal board, contactor, fuse, cable termination, overhead line, bus-bar, transformer, driven machine, etc. Such cases should not be considered to be motor failures because the motor failure is only a consequential damage due to some other failure. They should be investigated with reference to the component or equipment, which failed first. As far as preventive action is concerned, consequential damages on the motor due to failures of other components are irrelevant.

64.6 Motor Failures due to Poor Ventilation, Overloading, Excessive Voltage, Low Voltage, Low Frequency

The five types of causes, mentioned above in the title of this section, are considered together as they have similar failure mechanisms. In all the five cases there is a general overheating of the insulation, followed by degradation and eventually its failure at normal operating voltage.

Poor ventilation can be the result of accumulation of dust and grime on the windings, and in the air passages. Input of heat from external sources of heat such as ovens, furnaces, and resistors is also another possible cause which has the same effect as poor ventilation.

It must be remembered, that short time deviations in loading, input voltage and supply frequency are not necessarily harmful. They may reduce the durability somewhat but immediate failures are unlikely.

The crucial points in all the above cases are (a) the maximum, sustained operating temperature of the winding and (b) the maximum or hot spot temperature.

64.7 Do's and Don'ts for Preventing Stator Failures

- Ensure that the insulated wires comply fully with the appropriate specification. In particular, cut through tests and twisted wire heat shock tests must be carried out on samples.
- Ensure that the coils are not damaged during the process of coil winding and while placing the coils in the slots.
- Ensure that the stator windings are vacuum-pressure impregnated with a solvent-less varnish.
- Ensure that the coil ends, which are formed into the end connections between the slots are taped to separate the conductors with high potential difference, at the crossing points.
- Ensure that the end connections are well secured and not allowed to vibrate, by tying them up with glass cord.
- Ensure that leads to the terminals are well insulated from the conductors of the coils.
- Ensure that the terminals are designed for reliability. Refer to Chapter 39 in this regard.

Chapter 65

PANTOGRAPH OHE ENTANGLEMENTS

In this chapter we learn

- *The general classification of pantograph/OHE entanglements.*
- *Entanglements due to defects in pantographs.*
- *Entanglements due to defects in OHE.*
- *Preventive measures that can be taken.*

❏ ❏

65.1 Introduction

The track Over Head Equipment (OHE) in an electrified railway consists basically of a contact wire suspended at a height of approximately five metres above the centre line of the track. The current collection surfaces of the pantographs of electric locomotives and motor coaches press on to the underside of the contact wire with a force of approximately 7 kgf for collecting a current of about 150 amps, at about 25000 Volts from the OHE.

While the basic function as indicated above is simple enough, there are many complex requirements that must be satisfied by the design of the OHE and the pantograph.

- The level of the contact wire with reference to the rail level must remain constant except that at certain locations as in tunnels and under road bridges it may go down. Similarly, at level crossings the level may go up.
- The changes in level on either side of tunnels, bridges and level crossings must be very gradual. In other sections, the level should remain constant with permissible variations, which may be less than a millimetre per metre along the track.
- The lateral deviation or distance of the contact wire from a centre line exactly perpendicular to the plane of the two rails should swing about 150 millimetre on either side, between the supporting structures that may be located every 60 to 70 metres.
- There must be no fitting or projection on the OHE below the level of the contact wire within the profile of moving dimensions. Similarly, there must be no projection above the surface of the pantograph pan.

- The contact wire should be suspended freely and should be capable of getting lifted straight up by the force exerted by the pantograph until the force is balanced by the weight of the contact wire and a component of its tension.
- The upward force exerted by the pantograph should be fairly constant throughout the range of the contact wire height above rail level and the range of operating speeds, which may reach 160 km/h.

If any of the requirements mentioned above are not met at any point, there would be problems such as sparking, arcing and excessive wear. In extreme cases, there would be entanglement of the OHE and the pantograph. There would be extensive damage to the OHE over a distance of several kilometres, i.e. until the train is brought to a stop. The pantograph would also be badly damaged.

This chapter is about such cases of severe damage to the OHE and the pantographs. This type of failure is the worst that can happen on the OHE. Trains are held up for many hours not only at the point of failure but also on both sides for hundreds of kilometres. While there are generally no injuries to staff or passengers, the effects on traffic are similar to those caused by derailments and collisions of trains.

65.2   General Classification of Causes of Pantograph/OHE Entanglements

There are many possible causes that can and do lead to pantograph/OHE entanglements. These can be classified into a few groups as follows:

- Catastrophic component failures, i.e. fractures of contact or catenary wires and other suspension or anchoring fittings like insulators, tubes, splices, suspension and anchor fittings, dropper wires, etc.

  It must be noted here that these kinds of damages also occur as the effects of pantograph/OHE entanglements. These are then not relevant for determining the causes of such incidents. It is important to see whether there was any such failure before the incident due to other reasons such as those discussed in Chapters 7, 8, 17, 18, 22, 32, 33, 38, 43, 49. It is often not easy to distinguish between cause and effect. Nevertheless every effort must be made, during investigation to examine all the fractures very carefully. Those amongst the 'effects' are likely to show characteristics of ductile fracture or brittle fracture whereas those amongst possible 'causes' are likely to show fatigue fractures. While this distinction is not conclusive it is one of the points to be considered.

  In general it can be assumed, that the defects to be encountered in the first few spans along the track in the direction of motion of the train is more likely to include the cause of the failure at the point at which the entanglement started. Beyond this point all damages are likely to be part of the consequential damages, which have no relevance in the investigation.

■ Levelling and alignment defects in the OHE that could be due to one or more of the following:

- Defective installation of the OHE.
- Shifting of the track (i.e. rails) relative to the masts.
- Shifting of the masts relative to the track.
- Defects in the regulating equipment.

■ Defects in the pantograph such as:

- Pantograph force too low or too high.
- Pantograph strips or their screws loose or projecting.
- Fracture of pantograph springs, horns or other parts.
- Lateral deflection of pantograph excessive.

65.3 Prevention of Pantograph/OHE Entanglements

The only way to prevent pantograph/OHE entanglements is to prevent the above mentioned and similar causes from occurring. And this calls for every possible precaution being exercised at every stage:

■ The ordering of the materials from the suppliers.
■ The selection of the suppliers.
■ Inspection of the materials at purchase.
■ Inspection before installation or erection of the OHE.
■ Inspection and adjustment of the OHE periodically during service.
■ Inspection and adjustment/repair of the pantographs of all the rolling stock in service.
■ Investigation of every case of entanglement and corrective action.

The last three details are of particular importance to the engineers in charge of maintenance. While the lower levels in the organisation carry out the bulk of the work, it is necessary that random checks are made, by all the supervisory officers merely to ensure that the lower levels are actually doing their jobs correctly.

Records should be maintained of all such inspections. These records should be reviewed periodically by the senior engineers in charge of maintenance to verify that the entire section is in fact being checked by the line staff. Further, random inspections at higher levels should always be carried out soon after the checking has been completed on any sub-section by the front line staff and by the junior supervisory staff to verify the quality of inspections done by them.

The purpose of investigations is mainly to discover if any material quality deficiencies or design deficiencies are present so that corrective action can be taken in other locations where similar materials are in use. Investigations should not be aimed at fixing staff responsibility for workmanship defects. The quality of workmanship whether during installation or in maintenance is to be ensured only through random inspections by supervisors and engineers.

If it is not possible, as is often the case, to determine the true cause of the entanglement, the investigating officer should indicate a few probable causes, one of which may be the root cause. The reasons for inclusion of such items should be indicated. Action should then be taken to review all specification, inspection, maintenance and training procedures in respect of each of the probable causes.

Chapter 66

FAILURES AND ACCIDENTS
IN CHEMICAL PLANT

In this chapter we learn

- *Failures and resulting accidents in chemical plant.*
- *Chemical plant failures, due to failures of mechanical and electrical components and equipment discussed in detail in other chapters.*
- *Chemical plant failures, which are due to unexpected or uncontrolled reactions.*
- *Lessons learnt from a review of chemical plant failures and accidents.*
- *Basic principles of maintenance for obtaining plant reliability.*
- *Basic principles of operation for obtaining plant reliability.*

❑ ❑

66.1 Chemical Plant Failures due to Mechanical/Electrical Failures

This chapter is about failures and accidents in the chemical industry. Failures and accidents are discussed together because the root causes of both are often the same. Whether a particular incident remains a mere failure or it escalates into a major accident is often a matter of chance, depending on fortuitous circumstances that are not connected with the root cause.

Many chemical equipment failures (and accidents) are due to component failures, which have been discussed in previous chapters such as failures of fasteners, gaskets, solenoids, gears, valves, etc. Chemical plant also include many items, which are normally classified under mechanical or electrical equipment such as compressors, switch-gear, motors, boilers, ball mills, etc. In many cases there may be no adverse effect other than the cost of repairing the failed item. However, in other cases there may be interruption in production, the cost of which may be many times the repair cost. Chemical plant involving continuous processing are particularly vulnerable in this regard.

In some cases, such as those described below, the consequences of component or equipment failure may result in accidents involving loss of property and life. The only way to prevent such cases is to monitor all types of defects and failures, investigate them fully and take effective measures to prevent their recurrance. In all such cases the failures were due to causes that had nothing to do with chemical reactions but in some cases the effects were chemical in nature.

CASE STUDY

- A mixing vessel of 1 cubic metre capacity was fitted with a hinged and counter-weighted lid. The vessel was swivelled on trunnions for unloading. Residues sticking to the sides were manually shovelled out. One day, the weld on the hinge between the lid and the vessel fractured, the lid fell down and injured the worker. Investigation showed that the ten-year-old vessel had been subjected to some repairs on the hinge welds and the quality of welding was poor.

- A reactor vessel, which was one of a number of similar vessels connected in series, developed a crack and was taken out for repairs. In order to keep the plant in operation a temporary pipe was manufactured locally and fitted to replace the reactor vessel. Due to deficiencies in the design of the pipe and the flexible bellows, it broke under the effect of the high pressure of the gas within. The highly inflammable gas (cyclohexane) poured out and produced a vapour cloud. It ignited when it reached an operating furnace in the plant and 28 workers were killed. A major part of the plant was destroyed.

- A pilot operated relief valve in the cooling circuit got stuck in the open position due to an internal mechanical defect. The indicator light on the control panel was operated by the voltage at the solenoid of the relief valve and not by the actual position of the valve. Thus, the control panel light indicated the valve to be closed, whereas it was actually open. This caused confusion amongst the operating staff in the control room, and the steps they took to deal with a minor problem were incorrect. This led to a number of malfunctions and eventually it led to the release of radioactive material into the atmosphere. This is the notorious Three Mile Island incident, which did not result in any casualties but caused a great deal of consternation amongst the public.

It will be seen from the above three examples that mechanical or electrical failures of components, machinery and plant can produce failures and even accidents in chemical plant. Chemical plant comprises essentially mechanical and electrical equipment such as switchgear, motors, pressure vessels, valves, pipe-work etc., etc. Chemistry and chemical engineering is involved in what happens subsequently to and between the materials, which flow through the plant, but not in the initial stages of the failure mechanism.

66.2 Learning from Accidents

The three cases mentioned above have been taken from the book *Learning from Accidents* referred to in Section 66.8. Some of the accidents listed in Section 66.4 are also from the same book which should be referred to for details. Trevor Kletz, the author of that book, does not favour the use of the term "cause of failure or accident". In fact he discusses in some detail the reasons for his aversion to this term. Briefly, they are:

- Many of the so-called causes are beyond control.
- Determination of 'causes' may inhibit further investigation because it leads immediately to 'fixing responsibility,' which leads further to people becoming defensive and reticent.
- Causes are often given in abstract terms such as System Failure, Acts of God etc.

Kletz suggests that instead of trying to determine causes and responsibility, it would be more fruitful and effective to aim at the determination of preventive measures.

Fixing of responsibility and apportionment of blame is often the main objective of investigations. Disciplinary action against somebody high up in the hierarchy, is what the media and the public find interesting and satisfying.

It is usually difficult to fix responsibility on one or two persons. Most accidents are the result of several deficiencies or actions none of which by themselves or individually would have caused an accident. Responsibility is very diffused both vertically and horizontally in the organisation and also over a long period of time.

Preventive measures, which constitute a secondary objective in investigations, are actually far more important because they show the only effective way to accident free performance.

66.3 Chemical Plant Failures due to Chemical Reactions

An attempt is made in the remaining part of this chapter to consider those types of failures which are due to unexpected chemical reactions caused by defects in the system design or to operational or maintenance failures. These could be classified into different groups:

(a) Accelerated reactions due to increase in temperature resulting from failures in cooling equipment, temperature controls, etc.

(b) Unexpected reactions due to incorrect mixing of chemicals.

(c) Corrosion and chemical reactions between the plant components and the chemicals within.

66.4 Case Studies Involving Chemical Reactions

(a) A pump handling an inflammable liquid was to be replaced. The fixing bolts/nuts had seized due to rusting. It was decided by the maintenance staff to cut the bolts using an oxyacetylene cutting torch after the inflammable liquids were drained out. The drain, which contained some residues of the inflammable liquid due to level defects, was merely covered by a polyethylene sheet (instead of a steel sheet). The hot metal thrown out during the cutting operation burnt through the polyethylene sheet and ignited the inflammable liquid residues in the drain. The fire was extinguished and there were no casualties in this incident but it was a narrow escape from a major fire.

(b) An organic solvent, which was inflammable, was being manufactured in a plant. As the electrical switchgear in its control panel was not flameproof, its cubicle was pressurised with nitrogen, which was available in the plant. The idea was to prevent entry of inflammable vapours from entering the cubicle. There was no problem until one day the nitrogen pressure became too low due to leakage, and inflammable vapour did accumulate in the cubicle. An explosion occurred when the switch was operated. One operator was injured. The maintenance staff had isolated a low-pressure alarm switch provided by the original designer of the system because it was operating frequently and this was affecting production!

(c) A crude oil distillation plant had a re-flux drum, which often got filled with water that remained below the oil. This was usually pumped out through a drain at the bottom of the vessel and sent to a scrubber unit. During a major overhaul schedule, when the plant was about to be re-started it was discovered that a slip-plate, which had been inserted above the drain valve for isolating the drum during overhaul, had been left in place inadvertently. The foreman in charge thought he could remove the slip plate and retighten the flange bolts in less time than it took for the large quantity of accumulated water to leak out while this was being done. He might have succeeded in doing so but the gasket tore during the work and a new one had to be brought from the store and fitted in its place. Not only all the water but also a great deal of oil sprayed out. Some of it reached the oil burners of a nearby boiler and ignited. A major fire ensued. Efforts to close the burners failed because the maintenance staff had isolated the remote control devices provided for just such emergencies.

(d) In an ethylene plant, a ½" ID pipe carrying fluids at very high pressures was provided with joints of a very special and precise design. These called for special skills and the staff had to be given special training. As a result of staff re-shuffles included in cost reduction schemes, the work got re-assigned to staff without adequate training and experience. One badly fitted joint developed heavy leaks, leaked inflammable gases and these eventually ignited causing an explosion, four workers were killed and several more injured. There were no devices for detecting inflammable gases that may have leaked, although the plant was dealing with highly inflammable gases.

(e) In a plant using flammable hydrocarbons, a very heavy leak was the result of failure, to provide slip plates in the valves leading to a pressure vessel under overhaul and an inadvertent opening of a valve which should have remained closed. The flammable liquid and vapour spread around and eventually reached a diesel engine. The engine began to race when it sucked in the vapour air mixture. The operator could not shut down the engine. Finally the engine backfired and the massive explosion that followed killed four persons and caused plant damage exceeding 40 million pounds.

(f) A tank contained a volatile hydrocarbon. Incoming fluid was being splashed down from an inlet pipe high in the tank. Normally the free space in the tank above the liquid was filled with nitrogen but on that day the nitrogen had been disconnected. The staff had become complacent having done the same thing many times earlier without any adverse effect. On the fateful day a pulley supporting a steel rope inside the tank for a swivelling inlet pipe had seized and the friction between the rope and the pulley generated enough heat to ignite the vapour air mixture inside the tank above the liquid level. The resulting explosion killed several staff and caused expensive damages.

(g) A tank used to store a material with melting point 100 °C was heated by a steam coil to liquefy the material. The tank had a six-inch vent and it was designed for a bursting pressure of 1.3 bar. At some stage a three-inch vent replaced the 6-inch vent. Over the years, solidified material blocked the three-inch vent. The radiated heat from the liquid and the hot sides of the tank could not melt the blockage. A very slightly leaky compressed air device led to the build up of pressure of several bar level. The tank exploded spilling out hot molten hydrocarbon. Two workers were killed. Losses of material and plant was valued at 20000 Pounds.

(h) The Chernobyl Nuclear Power Plant accident occurred during experiments being made to assess the intrinsic safety of the system. But the root cause of the accident was partly the design of the experiment itself. In addition, the automatic trip system which would have inserted all the graphite rods into the reactors when the temperature started rising in the reactor, had been isolated previously and this was not checked before starting the experiment.

(i) A large tank containing a highly inflammable and toxic gas was to be cleaned periodically. Before men could enter the tank, all the vapour had to be purged. This was being done by pumping large volumes of air through the tank. The pure vapour in the absence of air is not explosive. Obviously the pure air at the end of the purging process is also not explosive. At some intermediate stage of the purging process there is a highly explosive mixture in the tank. One day, unfortunately, at this precise stage, the fan impeller broke and the sparks caused by the pieces impacting the tank sides ignited the mixture and caused an explosion.

(j) Open ladles were being used for the transport of molten metal from one part of the plant to another part. A large, closed, and torpedo shaped ladle was being tried out with a view to reduce costs. 2 or 3 tonnes of water entered the ladle from leakages in the plant pipe work but it continued to boil over the insulating layer of slag and dross. This was a normal feature of operations but as it did not seem to cause any problems when open ladles were in use, the water leakages were neglected. However when the large, closed ladle was moved, the splashing around of the metal inside, caused a large quantity of water to vapourise very quickly. The pressure inside the large closed ladle with a relatively small vent rose to such levels as to cause the ladle to explode and throw out large quantities of molten metal. 11 workers were killed and many more injured.

(k) The factors that contributed to the terrible accident at Bhopal were first, entry of water into a storage tank for MIC, a chemical known to react violently with water to produce a toxic gas. Secondly, excessive storage of this hazardous chemical in an intermediate stage of the process. Thirdly, disconnection of protective gear and equipment such as scrubbing plant, flare, high temperature alarm.

66.5 Lessons to be Learnt from Review of Chemical Plant Failures

(a) Whenever new chemical plant is being designed and installed hazard analysis should be undertaken in all seriousness. In this analysis the likely results of failure, in every known mode, of every component and every piece of equipment should be considered in 'thought experiments' and measures to prevent serious accidents decided upon. The results should be documented in the forms suggested in the F-MECA (Failure Mode Effect Criticality Analyses) forms given in the US MILITARY specification of that name, for review if failures and accidents do take place later at any time. These measures could be of the following forms:

 (i) Electrical or mechanical interlocks to prevent mal-operation
 (ii) Protective devices to detect rising (in some cases falling) parameters such as pressures, temperatures, voltages, currents, flow rates, vibration, rpm, inflammable or toxic gas content, etc.
 (iii) Redundant components or equipment.
 (iv) Increased margins or factors of safety in vulnerable or critical areas.
 (v) Improved or increased maintenance.
 (vi) Improved materials.

(b) The F-MECA analysis mentioned above should be made for all operating conditions and also for conditions obtaining during scheduled maintenance.
The cost of the effort involved in such analyses is negligible in relation to the total cost of the plant.

(c) A maintenance and operation culture should be developed and nurtured by top management, in which very great importance and sanctity is attached to documented operating procedures and protective devices. If any of these are considered to be irksome, counter-productive or redundant, it should be possible for anyone to recommend changes but no one should have the authority to implement such changes without thorough consideration at high levels in the hicrarchy. And even then, such changes and the reasons therefore should be recorded and new operating procedures issued.

(d) Every case of failure or accident, howsoever minor in effects, must be thoroughly investigated by a department not under the control of either design, or operation or maintenance. The F-MECA records and operating manuals should be up-dated to incorporate the lessons learnt from the accident. This is more important than action against errant staff.

(e) Systematic and positive training should be provided for all operating and maintenance staff. It should not be left to the staff to learn for themselves on the job from their predecessors or others doing similar jobs. Senior and experienced supervisors should train them. Course materials should be documented. Course material should be prepared by them and got approved at higher levels. The course material should be an important part of the basic technical documentation for the plant and it should be updated from time to time to cover new equipment, changes in procedures, results of failure investigations, etc. Training should cover (a) chemistry of reactions involved, (b) safety hazards, (c) operating procedures, (d) maintenance instructions, (e) control systems, (f) safety devices, (g) first aid, (h) case studies of accidents and failures world wide.

(f) In plants where toxic, inflammable or hazardous chemicals are produced in intermediate stages of the processes, the plant should be designed to minimise the quantities of such materials in process or storage. Where some storage is unavoidable the storage should be isolated and provided with double protection. Where either the raw material or the final product is itself toxic, inflammable or hazardous, the inventory of such items should be carefully regulated to the minimum possible levels. Their transportation, purchase and sale should be controlled for 'just-in-time' receipts and despatches.

(g) Safety engineers with no direct responsibility for either quality or quantity of production should be appointed and they should report directly to top management, where their recommendations are not accepted by the production departments. The safety engineers should be charged with the responsibility for making the F-MECA analyses, vetting documentation, course material for training, operating instructions, maintenance instructions, and investigations into all failures and accidents. They should also be responsible for making regular inspections throughout the plant to check that approved procedures are being followed and that protective devices are in operation.
Small plants, which may find it too expensive to maintain an independent safety department, may have safety and reliability audits carried out periodically by consultants.

(h) Top management should ensure that the role of the safety engineers, is properly appreciated by the production departments. On the other hand the safety engineers must see that they take a firm stand only on really dangerous conditions.

(i) Care must be taken in repairs on the plant to ensure that the standards of design and workmanship are the same as in the original installation. Special care must be taken while repairing stressed components as in lifting equipment and pressure vessels.

(j) Plant dealing with hazardous or inflammable gases should be either in open air or in structures without walls. Leak detectors should be installed near vulnerable points.

(k) Maintenance work should be organised on the basis of permits to work being issued by the operations department to the maintenance departments. All valves that have to remain closed must be provided with danger boards and locks/collars to prevent inadvertent operation while permits to work are with the maintenance department. Slip plates should be inserted where valve leakages are possible.

(l) No modifications in the plant should be made without careful examination of the designs and the effects of the modifications on reliability and safety. Qualified mechanical and electrical engineers should design, install and maintain the plant.

(m) Special steps must be taken to prevent entry of extraneous materials, which could start or accelerate dangerous reactions.

(n) At the design stage heat, temperature, pressure and such other parameters should be checked to see whether runaway reactions are likely to take place in any part of the plant due to the limits for these parameters being exceeded for any reason. Suitable safety devices should be provided where necessary.

66.6 Basic Principles of Maintenance to be Followed in the Design of Maintenance Schedules

- All parts or components which wear, corrode or degrade during service or even mere passage of time must be listed.
- All such items should be replaced at appropriate intervals.
- Items, which may suffer invisible damage during maintenance or repairs, should be replaced as a rule (e.g. gaskets, seals, split pins, cotters).
- Protective devices should be subjected to periodical tests.
- Every case of unscheduled repair, defect and failure should be investigated and steps taken to prevent recurrence.
- All critical components (e.g. highly stressed parts, linings in contact with corrosive fluids, parts with a history of failures, active surfaces of bearings, etc.) which are likely to develop incipient cracks must be thoroughly examined periodically, either visually or by non-destructive test methods.

66.7 Basic Principles of Operation to be Followed

- Control panel instruments which indicate operating parameters such as voltage, pressure, temperature etc., should be accurate, of appropriate range for ease of reading, with red lines to indicate upper and lower operating limits.

- Indicating lights which are meant to indicate status of valves, pumps, and other active components should be activated by the actual status and not by command status. (Thus, solenoid valve position should be indicated by a contact operated by the valve mechanism and not by the supply voltage to the solenoid).

- Indicating lights should be dim in the off condition and bright in the on condition of the device being monitored. A dark indicating light will then show failure of the light itself.

- Operation of protective devices should trigger red lights. If audible alarms are provided, alarm switch off should trigger red lights on the panel.

- All critical operating temperatures, pressures and flow rates should be indicated continuously on the control panel.

- Infra red temperature monitoring devices should be used regularly by staff monitoring the plant operation to check critical parameters.

- All pipes, valves, pumps, reactor vessels, etc., should be provided with nameplates and serial numbers which should be accurately reflected in the mimic diagrams of the control panels.

- Telephone sockets should be provided at close intervals throughout the plant and all operating and maintenance staff who walk through the plant should carry telephone sets which can be plugged anywhere for instant communication with the control room. Alternatively, they may all be given wireless mobile sets.

- Control room logbooks should record all abnormal and unusual conditions observed by the control room and by the operating/maintenance staff walking through the plant. It should also record separately all repairs and operations carried out during each shift.

References

1. Kletz, Trevor, *Learning from Accidents,* Butterworths Heinemann, 1994.
   This book should be considered as required reading for all engineers in charge of chemical plant. The case-histories referred to in this chapter are discussed in considerable detail in this book. This book also gives a comprehensive list of references.

About the Author

A. A. Hattangadi is an electrical reliability consultant, specializing in prevention and investigation of electrical fires and failures, with a diverse international clientele. Until recently, he was general manager at Chittaranjan Locomotive Works in India.